UMPOLED SYNTHONS

UMPOLED SYNTHONS

A Survey of Sources and Uses in Synthesis

Edited by

TAPIO A. HASE

University of Helsinki
Helsinki, Finland

A Wiley-Interscience Publication

JOHN WILEY & SONS

New York Chichester Brisbane Toronto Singapore

Library of Congress Cataloging-in-Publication Data:

Umpoled synthons.
"A Wiley-Interscience publication."

Bibliography: p.
1. Chemistry, Organic—Synthesis. I. Hase, Tapio A., 1937–

QD262.U45 1987 547′.2 86-26702
ISBN 0-471-80667-6

Printed in the United States of America

10 9 8 7 6 5 4 3 2 1

CONTRIBUTORS

DAVID J. AGER, Department of Chemistry, University of Toledo, Toledo, Ohio

TAPIO A. HASE, Department of Chemistry, University of Helsinki, Helsinki, Finland

MICHAEL KOLB, Strasbourg Research Center, Merrell Dow Research Institute, Strasbourg, France

JORMA K. KOSKIMIES, Department of Chemistry, University of Helsinki, Helsinki, Finland

JOSEPH E. SAAVEDRA, Frederick Cancer Research Facility, National Cancer Institute, Frederick, Maryland

NICK H. WERSTIUK, Department of Chemistry, McMaster University, Hamilton, Ontario

FOREWORD

In 1981 Professor Tapio A. Hase and his associate Jorma K. Koskimies published an excellent compilation of acyl anion equivalents in *Aldrichimica Acta.* During a vist to Helsinki in June of 1982, I encouraged Dr. Hase to undertake the writing of an extensive and more general publication on synthetic equivalents, because the earlier brief review was so useful. The present volume represents the completion of the first phase of this major effort. It places at the disposal of synthetic chemists a large and well-organized body of information drawn mainly from developments in synthesis over the past 20 years.

Contemporary activity in organic synthesis is impressive by any standard. Each year scores of new syntheses are described which are remarkable for the ingenuity underlying their design, the elegance and inventiveness with which known and new chemistry are applied, and the effectiveness of laboratory execution. Reviews such as this text play a significant role in the achievement of this level of accomplishment.

Most contemporary syntheses are planned with the guidance of retrosynthetic analysis and appropriate higher level strategies, e.g., topological, stereochemical, or those centered on the identification of key-reactions or key-intermediates. The chemist must perceive critical information extracted from the target structure and each precursor structure generated in the analysis, and couple this with the vast body of chemical knowledge, to derive potentially valid synthetic sequences. The number of possibilities to be analyzed is awesome. Yet, before proceeding with the execution of a particular plan, the chemist is well advised to carry out a very extensive analysis. This is one of the areas in which critical and comprehensive compendia of synthetic information are of great importance.

I sincerely hope that practicing synthetic chemists will utilize fully this valuable source of information on synthetic equivalents.

E. J. COREY

PREFACE

Following E. J. Corey's introduction in 1967 of the concept of synthons, it has become common among practitioners of organic synthesis to treat numerous synthetic reactions as simple combinations of the appropriate synthons instead of writing out the actual reagents in full. Thus, for a Grignard reaction, it is convenient to write

$$^{+}CHR{-}OH + {}^{-}R' \longrightarrow R'{-}CHR{-}OH$$

and to dispense with the Mg and the halide anion that do not appear in the final isolated product. Similarly, an alkyl halide RX (an electrophilic reagent) may be simplified to R^+ which, barring side-reactions, clearly will combine with another synthon bearing a negative charge. For synthetic purposes this can be done irrespective of S_N1/S_N2 mechanistic considerations, although in practice the various limitations of a particular synthon, reaction conditions, and reactivities must of course be kept in mind.

The emergence of the concept of "umpolung" (Seebach) or "symmetrization of reactivity" (Corey) has made the synthon approach even more important. At the same time, the use of synthons has become more demanding because the new umpoled species no longer necessarily look very much like the actual reagents. A case in point is the acyl cation/acyl anion synthon pair. Although it is obvious to treat an acyl halide as an acyl cation synthon, it may be less clear that, say, 1-methyl-3,5-dithiazolidine is in fact a source for the formyl anion synthon, $^{-}CH{=}O$. Also, unlike the classical synthons, which are usually obtained by a simple disconnection from a reagent, many of the new umpoled synthons are more or less obviously incapable of actual existence ($^{-}COOH$, for example) and are seldom available by a simple disconnection.

Synthons originally were introduced as a vehicle for performing the analysis of organic syntheses. If one were to use synthons in planning an actual synthesis, it would be convenient to have access to a catalog of synthons, showing the various sources (i.e., reagents) for any given synthon, along with key information on reaction conditions, reactivities, and possible limitations.

However, professional synthetic workers do not need to be reminded that

aralkyl ketones, for example, can be made in a Friedel-Crafts reaction using an acyl cation synthon derived from the acyl halide. Indeed, most synthetists could probably run a simple Friedel-Crafts acylation without consulting the literature. In any case, most of the reliable classical reactions have been discussed thoroughly and reviewed over the years, and ample digested information is available on these reactions.

The purpose of this book is to provide chemists who plan their synthetic routes manually with information on some of the new umpoled synthons. The introductory chapter examines, in general, the use of synthons in planning organic syntheses. Concluding the introduction is a fairly detailed list of generally convenient synthons that are, however, extraneous to the main contents of the book. A major portion of the book is devoted to acyl anions, hydroxycarbonyl and related anions, carbonyl α-cations, and carbonyl β-anions. A further group comprises carbanionic synthons in which the carbon atom carries a singly bonded heteroatom (O, S, N, or halogen). In appropriate chapters, the above synthons are surveyed with emphasis on the aspects relating to synthetic use. The book concludes with a tabular presentation of known applications of these synthons, with relevant data on the sources, auxiliary reagents, yields, and restrictions. Each table has been prepared by the author of the corresponding chapter, with the exception of Tables 6.1.–6.9., which were prepared by Dr. Koskimies.

Although well over 2000 references have been cited, it is quite possible that some important references have been overlooked, and sincere apologies are offered to anyone feeling neglected in this respect. However, in addition to plain ignorance, the omissions may also be due to the Editor's feeling that the most recent work should be cited preferentially, and references to previous work then can be easily located. For the same reason, references are usually omitted to preliminary communications that have been followed up by full papers.

In closing, it is a great pleasure for me to thank Professor E. J. Corey, without whose initiative and encouragement this book would not have been started. The members of the Organic Chemistry Laboratory at the University of Helsinki who have been involved with the editing and writing of the book are most sincerely thanked. Professor Gösta Brunow and Dr. Jorma Koskimies read through all chapters and provided criticism and valuable suggestions. Kristiina Wähälä, M.Sc., was exceedingly helpful and resourceful in checking and sorting out the literature references and taking care of numerous other editing details. Lasse Koskinen and later Hannu Hirvonen gave invaluable assistance in the operation and maintenance of our laboratory text processing system for the needs of compiling this book.

TAPIO A. HASE

Helsinki, Finland
January 1987

CONTENTS

TABULAR SURVEY OF UMPOLED SYNTHONS

ABBREVIATIONS AND CONVENTIONS

Ac	Acetyl
AcF_3	Trifluoroacetyl
AcOH	Acetic acid
tAm, Am-t	1,1-Dimethylpropyl
Ar	Aryl
Bu	Butyl
iBu, Bu-i	Isobutyl
sBu, Bu-s	*sec*-Butyl
tBu, Bu-t	*tert*-Butyl
Bzl	Benzyl
cat.	Catalysis, catalyst
Cb	Carboxybenzyl
Cp	Cyclopentadienyl
Cx	Cyclohexyl
d	Day
DABCO	1,4-Diazabicyclo[2.2.2]octane
DBU	1,8-Diazabicyclo[5.4.0]undec-7-ene
DDQ	2,3-Dichloro-5,6-dicyano-1,4-benzoquinone
DME	1,2-Dimethoxyethane
DMF	*N,N*-Dimethylformamide
DMSO	Dimethylsulfoxide
DNPH	2,4-Dinitrophenylhydrazone
ee	Enantiomeric excess
El	Electrophile
Et	Ethyl
h	Hour
(H)	Reduction
$h\nu$	Photolysis, photochemical reaction
het	Heterocycle
Hex	Hexyl
HMPA	Hexamethyl phosphorictriamide

KDA	Potassium diisopropylamide
LAH	Lithium aluminium hydride
LDA	Lithium diisopropylamide
LDCA	Lithium dicyclohexylamide
LDMAN	Lithium 1-dimethylaminonaphthalenide
LN	Lithium naphthalenide
LTMP	Lithium 2,2,6,6-tetramethylpiperidide
MCPBA	*m*-Chloroperbenzoic acid
Me	Methyl
MEM	2-Methoxyethoxymethyl
Met	Metal
Mes	Mesityl (2,4,6-trimethylphenyl)
Ms	Methanesulfonyl
NBS	*N*-Bromosuccinimide
NCS	*N*-Chlorosuccinimide
NMP	1-Methyl-2-pyrrolidone
Np	Naphthyl
Nu	Nucleophile
(O)	Oxidation
PbTA	Lead tetraacetate
Pe	Pentyl
Ph	Phenyl
Phth	Phthalyl
Pr	Propyl
iPr, Pr-i	Isopropyl
PTC	Phase-transfer catalysis or catalyst
py	Pyridine
R	Alkyl
R—Ni	Raney nickel
rt	Room temperature, ambient temperature
Tf	Trifluoroacetyl
TFA	Trifluoroacetic acid
THF	Tetrahydrofuran
THP	Tetrahydropyran
TMEDA	Tetramethylethylenediamine
TMSCN	Trimethylsilyl cyanide
Tol	*p*-Tolyl
TosMIC	Toluene 4-sulfonylmethyl isocyanide
Ts	*p*-Toluenesulfonyl
X	Halogen
xs	Excess
$\rightarrow$	Reaction (synthetic direction)
$\Rightarrow$	Disconnection (retrosynthetic direction)
Δ	Heat

In some structural formulas, unessential substitutents are stripped for clarity. Such formulas apply to cases where the structure given is general in terms of degree of alkyl substitution. Thus, C═C—COOH is short for R—CR′═CR″—COOH where R, R′, and R″ can be any combination of H and alkyl groups. Cases where generality is lacking or was not studied, or where just a single structure is possible, are given in full.

In indicating ring sizes, the "n" accounts for all ring atoms:

(n ring with CR═CR) cyclopentene and larger if $n \geq 5$

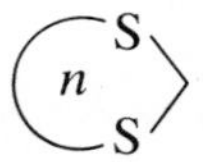

(n ring with S–C–S) 1,3-dithiane if $n = 6$

UMPOLED SYNTHONS

1

INTRODUCTION: CLASSICAL AND UMPOLED SYNTHONS

Tapio A. Hase

In 1967, E.J. Corey[1] introduced the new concept of *synthons,* defining them as "structural units within a molecule which are related to possible synthetic operations." He went on to point out how the synthetic chemist has learned by experience to recognize within a target molecule these units that can be synthesized, modified, or joined by known or conceivable synthetic operations. Certainly, any reasonably proficient pre–1967 synthetic chemist requiring an alkyl aryl ketone would immediately have thought of joining an aryl fragment to an acyl group by using the Friedel-Crafts reaction of an arene with an acyl halide. Whether having thought or not in terms of the aryl and acyl fragments, that is, synthons, he or she would have written out the synthesis as

$$\mathrm{ArH} + \mathrm{RCOCl} \xrightarrow{\mathrm{AlCl_3}} \mathrm{ArCOR}\ (+\mathrm{HCl})$$

Synthons are formally obtainable from a molecule by the disconnection of a single bond. Although any such bond may be disconnected, as far as actual synthesis is concerned the most useful disconnections will involve carbon–carbon bonds. Thus, an alkyl aryl ketone can be seen to furnish the synthons Ar, RCO, R, and ArCO, with Ar and RCO being used in the above example. In this context, it should be pointed out that disconnections of double bonds, although feasible, actually correspond to two-step synthetic processes and do not give conveniently handled synthons.[2]

Synthons were first introduced as a means for performing the analysis of organic syntheses, that is, for retrosynthetic operations. As originally defined, synthons were not visualized as carrying a charge or an unpaired electron. Later, as the use of synthons in planning actual syntheses became widespread and many previously unexpected synthons were being developed deliberately and were in routine use, it became useful to view disconnection as a heterolytic

(or homolytic) process that gives two cation/anion pairs (or two radicals), as shown in Scheme (1).

$$\mathrm{ArCOR} \begin{cases} \nearrow \ \mathrm{Ar^+ + {}^-COR\ (or\ Ar^- + {}^+COR)} \\ \searrow \ \mathrm{ArCO^+ + R^-\ (or\ ArCO^- + R^+)} \end{cases} \qquad (1)$$

Now, the above Friedel-Crafts acylation can be represented as

$$\mathrm{Ar^- + RCO^+ \longrightarrow ArCOR} \qquad (2)$$

Unquestionably, this is a very efficient and concise way of indicating a Friedel-Crafts reaction, and our pre–1967 chemist would presumably have understood what is meant by this expression. However, he or she might also have made two pertinent remarks in this connection. First, it would have been noted that the alkyl aryl ketone can be made from the same acyl halide under at least one other set of conditions, that is, by using the diarylcadmium reagent Ar_2Cd. This then must be another source for the Ar^- synthon. Possibly, yet other routes from the acyl halide to the ketone could have been found. Obviously, there can be several sources for any given synthon; thus it is clear that one must have a good grasp of the vast number of organic reactions to be able to use synthons effectively in planning syntheses. Alternatively, a "menu" of the available synthons would be highly useful. A further difficulty arises when one realizes that there are many synthons that clearly cannot exist, or in any case are unlikely to exist as such, for example, ^-COOH or $^-CH_2CH_2Cl$. This however is not a problem as long as the menu provides the synthetist with suitable multistep sequences that correspond overall to the change required, for example, R—Br → R—COOH.

Second, reaction (2), even if taken to represent a Friedel-Crafts acylation, tells one nothing about the compatibility of existing functionality in the aromatic ring, for example. To take this and similar aspects into account, one must either be very familiar with the reaction in question, or consult the literature. Most conveniently, the menu referred to previously should include information about the applicability and restrictions of the various synthons.

The bulk of this book is intended to serve as precisely such a menu, at least for some of the less accessible but important synthons of recent origin. Partial listings of various types of synthons, based on one unifying theme or another, have appeared in the literature.[3–9] Additional references to previous reviews are given in the introductory remarks in the chapters that follow.

DEFINING SYNTHONS

We visualize synthons as purely conceptual entities that may or may not actually exist, however fleetingly, in a reaction. For our purposes, synthons

are defined as cationic or anionic fragments of a neutral molecule, formally obtained from the latter by heterolysis of one single bond. Two pairs of synthons can always be written for the heterolysis, or disconnection, of a single bond (Scheme 1). Therefore, the combination of a cationic and an anionic synthon gives a neutral molecule. In this book, a full positive or negative charge will always be shown irrespective of whether in the actual reagents a full or partial charge would be more appropriate.

It is unfortunate that in current usage the word "synthon" sometimes has quite another meaning. Instead of indicating fragments of a molecule, as pointed out above and set out below in more detail, the word is used as a synonym for "reagent" or "synthetic equivalent," or "convenient starting material," such as in the expression "mannitol is a synthon for (R)-glycer-aldehyde." This practice is redundant, does not agree with the original definition, and should be abandoned.

There is also another matter of definitions we would like to bring up at this point. Homologations of carbonyl compounds, such as those based on the Wittig and related reactions,

$$\text{R—CHO} \longrightarrow \text{R—CH}_2\text{CHO } (\textit{via}\ \text{RCH=CH—OR}')$$

are sometimes referred to as corresponding to operation with an acyl anion synthon (here $^{-}$CHO). This is not entirely correct because the addition of the formyl anion synthon to another aldehyde will obviously give, after proton quench, a homologated α-hydroxyaldehyde (Scheme 3):

$$\text{OHC}^- + \underset{\text{H}}{\overset{\text{R}}{\diagdown}}\!\text{C=O} \longrightarrow \xrightarrow{\text{H}^+} \text{OCH—}\underset{\text{H}}{\overset{\text{R}}{\overset{|}{\underset{|}{\text{C}}}}}\text{—OH} \qquad (3)$$

It is as if the starting aldehyde RCHO in the Wittig homologation reaction acts as a source for the RCH_2^+ synthon. However, in Grignard reactions an aldehyde's synthon equivalent normally is RCH^+OH, as is easily seen by writing down a Grignard reaction of this sort, and as is in fact apparent from Scheme (3). It is true that any aldehyde RCH_2CHO can be disconnected to give the $^{-}$CHO synthon (which is to say it can be thought to be formed using the formyl anion synthon). However, the cationic counterpart to $^{-}$CHO is then necessarily RCH_2^+, available from RCH_2Br, for example. The discrepancy is now apparent: The nonphosphorus component in the Wittig reaction is RCHO, not RCH_2Br. The problem only arises because the Wittig reaction is a multistep process involving an elimination step, the latter being a reaction that cannot be handled using synthons. This example shows that careless use of terminology will very easily obscure the admirably simple yet useful basic concept of synthons.

It follows from our definition that only single-step substitution and addition

reactions can be treated using synthons. The above Wittig-type aldehyde homologation reaction involves the breaking of two bonds and forming of two bonds and therefore cannot be handled in terms of synthons. It may be possible to apply the synthon treatment to other homologation cases, for example,

$$RCH_2Cl \longrightarrow RCH_2CH_2Cl$$

$$\text{i.e., } RCH_2^+ + {}^-CH_2Cl \longrightarrow RCH_2CH_2Cl$$

using lithiated chloromethyl phenyl sulfoxide (see Chapt. 4). As already mentioned, multiple bond-forming reactions such as eliminations, condensations, and dehydrogenations also involve the breaking and forming of several bonds and usually proceed in more steps than one, thus they are not directly amenable to synthon presentation. The same applies to numerous multicenter reactions, cycloadditions, rearrangements, and so forth. Still, a good deal of mainstream synthetic organic chemistry remains, and within its area of applicability, the synthon approach is a powerful and natural way of discussing organic synthesis as demonstrated by Warren.[4,10] It is perhaps appropriate to note here that disconnections at multiple bonds are best handled by first modifying the C═C-containing moiety. Thus, anyone wishing to disconnect between α and β of an enone, to see if useful synthons emerge, should probably do so by first adding a molecule of water across the double bond. The β-hydroxyketone is then readily disconnected to give routine synthons:

$$RCO{-}CH{=}CH{-}R' \Rightarrow RCO{-}CH_2{-}CHOH{-}R' \Rightarrow$$

$$RCO{-}CH_2^- + {}^+CHR'{-}OH \simeq RCOCH_3/\text{base} + R'{-}CH{=}O$$

Reagent/synthon relationships are often very easily identified just by looking at the reaction partner. An alkyl halide, polarized as $R^{\delta+}$ $X^{\delta-}$, is a source of the R^+ synthon. Therefore any reagent that can be used with this alkyl halide to give the aldehyde R—CHO is a source of the $^-$CHO synthon; this relationship is not dependent on the number of steps required or on the reaction mechanism.

Above, we concluded that only single-step reactions can be handled in the synthon mode. It should be understood that this restriction really applies to the simplified synthon presentation of a reaction. In actual practice, a procedure that corresponds to the use of a given synthon may require several synthetic steps before the entire reagent/synthon relationship is fully unmasked. Multistep synthetic sequences are common with the umpoled synthon species in particular. For example, the various dithioacetal-based sequences that correspond to the use of acyl anion synthons (see Chapt. 2) must include

as the final step a dithioacetal → carbonyl group conversion. Furthermore, to effect an overall umpolung, the starting aldehyde must be converted to its corresponding dithioacetal in the first place. Thus, for the synthesis

$$RCO^- + {}^+R' \longrightarrow RCOR'$$

four operations will be required (Scheme 4).

$$\begin{array}{ccccc} R\text{—}CHO & \longrightarrow & RCH(SR'')_2 & \xrightarrow{\text{Base}} & {}^-CR(SR'')_2 \\ & & & & \downarrow \\ & & R'\text{—}CO\text{—}R & \longleftarrow & R'\text{—}CR(SR'')_2 \end{array} \qquad (4)$$

All this brings us to the question of just how many synthetic steps are reasonable, or permissible, in synthon expression. In principle of course, practically any functionality can be made from any other type of compound, given enough steps. We think it is pointless to stretch sequences in excess of two or three synthetic steps, and we rely on our readers to understand that standard reactions, oxidations and reductions in particular, should be applied as required to obtain a suitable intermediate or to modify further a synthon-derived product. In this context, it is also pointed out that additions of nucleophiles to carbonyl compounds have been treated in a liberal manner, to furnish not only the alcohols (A) but ethers (B), amines (C), halides (D), and so on.

$$R_2CO \longrightarrow \begin{cases} Nu\text{—}CR_2\text{—}OH & (A) \\ Nu\text{—}CR_2\text{—}OR' & (B) \\ Nu\text{—}CR_2\text{—}NR'_2 & (C) \\ Nu\text{—}CR_2\text{—}X & (D) \end{cases}$$

Finally, it may be asked is there any justification in restricting ourselves to disconnecting just one single bond at a time—why not two or more? Certainly this can be done, but it will become increasingly difficult to decide just where to write the emerging plus and minus signs, because there are two ways to do this for each single bond disconnected. Also, in many of these cases it will not be very clear what actual reagents might serve as the synthon sources. There are exceptions, however. Cyclohexenes can be opened up by disconnecting two single bonds in such a way that one of the resulting pairs of bidentate synthons is identical with the very familiar Diels-Alder diene and

dienophile reagents (Scheme 5). Nevertheless, it is doubtful if this is a practical way of analyzing a cyclohexenoid target molecule.

MeO COOMe ⇒ MeO CH ⊕ CH₂ ⊖ + ⊖ COOMe ⊕CH₂ ↕ ↕ MeO COOMe (5)

CLASSIFYING SYNTHONS

When first introduced, the principle of umpolung[11] or symmetrization of reactivity[1] as applied to synthons merely meant that every cationic synthon has its anionic counterpart, and vice versa, whether any of these species is known or not. Later, organic chemists began to think of certain synthons as classical, or normal, and of their less commonplace counterparts as umpoled. Although there does not seem to be much difference in this sense between R^+ and R^-, it is clear on the other hand that RCO^+ would be a classical synthon and RCO^- an umpoled one. There is certainly some confusion about the use of the words "umpolung" and "to umpole." The reason is perhaps that one can umpole a synthon, that is, change its polarity to the opposite, simply by changing the sign of the electric charge, as when going from $^+CH_2OH$ to $^-CH_2OH$. On the other hand, umpolung operations with the actual reagents usually require deep-seated (although reversible) changes at the heteroatom.

In 1979, in connection with a sophisticated review on reactivity umpolung, Seebach[6] introduced a definitive system of nomenclature for reagents and synthons, based on the location of an acceptor (a) or donor (d) center in regard to a heteroatom at C—1. Thus RCO^+ is an a^1 synthon whereas RCO^- is a d^1 synthon; a^0 and d^0 refer to the electrophilic or nucleophilic properties of the heteroatoms themselves. In general, synthons belonging to the series $a^{1,3,5..}$ and $d^{0,2,4..}$ show normal reactivity, and interchanging the a and d symbols gives the umpoled synthons. This rule stems directly from the electronegative nature of the heteroatoms commonly encountered in organic molecules (N,

O, halogens). The classical, umpoled nature of acyl cations and anions follows from this scheme in an unambiguous manner, and the same is true for most synthons. However, some of the well-known electrophilic aromatic substitution reactions, for example, would seem to require special treatment under the a^n/d^n system. A more obvious gap in the scheme is the fact that it cannot accommodate simple nonfunctional alkyl cations or anions; although the alkyl bromide R—Br is an a^1 reagent, an alkyl cation R^+ is not an a^1 synthon. This is a result of the requirement for the presence of a heteroatom for operation with the a^n/d^n definitions. Yet from the purely practical point of view, the basic poled/umpoled pair R^+/R^-, whichever way one looks at them, must be a part of any system of synthon presentation because of their ubiquitous nature and close relationship. It is very easy to see that the polarity of R in RBr can be readily reversed, that is umpoled, just by converting the alkyl halide to the Grignard reagent.

A further inconvenience in using the a/d labels is that they do not distinguish between the varying oxidation states of heteroatoms: $^-CH_2OH$, ^-CHO, and ^-COOH are all d^1 synthons, and although in the end products these functionalities may be accessible from each other, it is by no means futile to consider which one would be optimal for a particular synthesis. Planning the chain extension of a sugar aldehyde R—CHO to the polyhydroxyacid R—CHOH—COOH, one would probably not start by using $^-CH_2OH$ to make R—CHOH—CH_2OH, because it would invite difficulties in view of the selective CH_2OH to COOH oxidation required.

For these reasons, instead of applying the a^n/d^n notation, we will show in our formulas the appropriate positive or negative charge, or use their verbal equivalents. These full charges are attached to synthons, and not necessarily the actual reagents. As long as one does paper chemistry with, say anionic synthons, it is immaterial to what extent the corresponding reagents possess true carbanionic character. In fact, an alarmingly increasing number of standard organic reactions, such as those involving Grignard and organolithium reagents, dialkylamides, alkoxides, thiolates, Wittig reagents, complex metal hydrides, and so on, have been proposed in recent times not to be ionic at all but to involve single electron transfer (SET) mechanisms with subsequent radicaloid steps.[12,13] These claims in no way interfere with the ionic synthon presentation of the usual single bond-forming reactions; radicals or not, carbonyl compounds still give alcohols in Grignard reactions just as before.

The synthons that we discuss in this book are mostly of the umpoled variety. It is perhaps illuminating to draw up a list of the synthons treated, contrasting them with their nonumpoled and more familiar counterparts:

Chapter 2:	Acyl anions HCO^-, RCO^-	Acyl cations HCO^+, RCO^+ (typical source: acyl halides RCOCl)
Chapter 3:	Hydroxycarbonyl anion ^-COOH, etc.	Hydroxycarbonyl cation ^+COOH, etc. (typical source: CO_2)

Chapter 4: Heteroatom substituted C-anionic synthons $^{-}CH_2OH$, etc.	Heteroatom substituted C-cationic synthons $^{+}CH_2OH$, etc. (typical source: $CH_2{=}O$)
Chapter 5: Carbonyl α-cations $^{+}CH_2COCH_3$, etc.	Enolate anions $^{-}CH_2COCH_3$, etc. (typical source: CH_3COCH_3)
Chapter 6: Homoenolate anions $^{-}CH_2CH_2COOR$, etc.	Carbonyl β-cations $^{+}CH_2CH_2COOR$, etc. (typical source: $CH_2{=}CHCOOR$ in Michael additions)

This introduction concludes with a selection of recent literature references to a number of useful synthons that fall outside the scope of this book. However, the selection will not deal with some of the simplest and widely used species, such as alkyl cations and anions, formyl and acyl cations, and enolate anions and related systems. These synthons, or at least the well-known reactions where they would be implied, have been thoroughly treated elsewhere. Also, extensive reviews are available for the following species of synthons: $C{=}C^+$ (ref. 14), $C{=}C^-$ (refs. 15–21), $C{=}C{-}C^+$ (ref. 22), and $C{=}C{-}C^-$ (refs. 23–25). In any case, many synthons of this type are available in a quite trivial manner—for example, synthons $C{=}C^+$ are involved when the moiety CH—CO in a carbonyl compound reacts with Grignard reagents and the product alcohols are dehydrated.

The principal organization is by functional groups with subdivisions into cationic and anionic systems in that order, and, last, bidentate synthons. On the next level, the synthons are listed in order of increasing chain length, that is, increasing separation between the charge and the functional group (in some of the larger synthon classes, the number of C atoms separating the charge and the functional group is given to facilitate locating the subclasses). Finally, synthons containing unsaturation or some other additional functionality appear after the simpler cases. For the actual sources to these synthons, consult the original literature.

The following notation for structural formulas is used throughout the book: "missing" H or R groups at carbon atoms indicate generality. Thus, the expression $CH_2{=}C{-}C{=}CH_2$ refers to the series $CH_2{=}CR{-}CR'{=}CH_2$ where either R may be a hydrogen or an alkyl group.

The following symbols have been used to identify reaction partners (i.e., nucleophiles or electrophiles), and they apply to this table only.

A Aldehydes

C Acyl halides, anhydrides, amides, nitriles

D CO_2

E Enolate anions (including malonates, etc.) or enol silyl ether

G Organometallics including Grignards, lithium and cuprate reagents, and acetylides

K Ketones

L Michael substrates

M Enamines
N Amines
O Epoxides
P Alcohols and phenols including the anions
Q Thiols and thiophenols including the anions
R Alkyl halides
S Disulfides
Si Silyl halides
T Acetals, ketals, orthoformates
V Aryl or vinyl halides
W Water (and D_2O)
X Halogen molecule (i.e., source of Br^+, etc.)
Y Initial product from the first addition cyclizes, as for example in the reaction of a $^-$C—C—C$^+$ synthon with a Michael substrate, giving a cyclopentanoid
Z Other (electrophiles or nucleophiles)

For bidentate synthons, reported nucleophiles or electrophiles are given in the order they appear in the synthon as written, and are separated by a semicolon. Thus, the entry

$^-$CR=CH$^-$ A, R; C, R

indicates that the CR anionic site was reported to react with aldehydes or alkyl halides, and the CH anionic site with acyl or alkyl halides. With some bidentate synthons, discrimination between the two sites is not possible in practice. Just one electrophile can then be used for reaction with both sites, and this situation is indicated as 2R, for example.

For synthons that contain C=C unsaturation and that react stereospecifically, the double-bond geometry is indicated by the usual *E* or *Z* notation. Substituent priority is sequenced for the synthon before reaction with an electrophile or a nucleophile: a − charge at a doubly bonded C is deemed lower than H, and a + charge lower than a − charge. This convention prevents irrelevant changes in the *E* or *Z* labels due to fortuitous variations in existing or incoming R groups. For the same reason, any counterions or more covalently held metals that the actual reagents may contain are ignored.

LIST OF CONVENIENT SYNTHONS

ALKENES, BI- AND POLYDENTATE

C=CH—CH^{2+}	G;G	26
$^+$CH=CH$^+$ (*E* or *Z*)	G;G′	27
	E,G	28

$^{+}CH_2CH{=}CHCH_2^{+}$ (*E*)	E,N;N	29
$^{+}C{-}C{=}C{-}C^{+}$	Review	30
$CHR{=}C^{\pm}$ (*Z*)	G;R	31,32
$C{=}C^{+}C^{-}$	Review	33
$^{+}CR{=}CH^{-}$	G;C,L,O,R	34
$^{+}CH_2CH{=}CH^{-}$	R;A,K	35
$^{+}CH_2C({=}CH_2)CH_2^{-}$	L(Y)	36
	A,K(Y)	37
$^{+}CH_2C({=}CH_2)CH^{-}R$	L(Y)	38
$^{+}CH_2CH_2C^{-}{=}CH_2$	A,K,L(Y)	39,40
	L(Y)	41,42
$^{+}CH{=}CHCH_2^{-}$ (*Z*)	G;A	43
	E;A	44
$^{-}CH{=}CHCH_2^{+}$ (*Z*)	A,C,K;E,G	300
$^{+}C{-}C{=}C^{-}$	E;L,R	45
$^{+}(CH_2)_3C^{-}{=}CH_2$	L(Y)	46,47
$^{+}(CH_2)_nC^{-}{=}CH_2$ $n = 1–6$	Not given	48
$CH_2{=}C^{2-}$	R;R′	49
	2R(Y)	50
$RCH{=}C^{2-}$	R;R′	51
$^{-}C{=}C^{-}$	R;R′	52
$^{-}CR{=}CH^{-}$ (*Z*)	A,R;C,R	53
$^{-}CH_2{-}C^{-}{=}CH_2$	A,K;Z	51
	K;L(Y)	54
$^{-}C{-}C^{-}{=}C$	R;C,K,R	55
$^{2-}C{=}C^{2-}$	R	56
$^{-}CH_2CH{=}CH^{-}$	L;C,X	57
$CH_2{=}C(CH_2^{-})_2$	A	58
	2R	59
	2(A,K,O,R)	60
$^{-}CH_2CH{=}CHCH_2^{-}$ (*Z*)	2R	59
$^{-}CH_2CH^{+}CH{=}CH_2$	O(Y)	61

ALKYNES, BI- AND POLYDENTATE

$^{+}C{\equiv}C^{+}$	G;G′	62
$^{+}C{\equiv}C^{-}$	G;A,R,Si,W	63
	G;R	64
$^{-}C{\equiv}C^{-}$	R;R′	64

Synthon		Ref.
$^{-}C\equiv C-CH_2^{-}$	Review	65
	A,C;R	66
$^{-}CH_2-CH^{+}C\equiv CH$	C;E	67
$^{-}CH_2-CMe^{+}-C\equiv CH$	C,R;P,W	68
$^{-}CHR-C\equiv C^{-}$	2Si	69

ALLENES

Synthon		Ref.
$^{+}CH=C=CHMe$	G	70
$^{+}CH=C=CR_2$	G	71
$^{+}CH_2CH=C=CH_2$	G	72
$^{+}CR=C=CHR$	G	73
$^{+}CHR-CH=C=C$	E	74
Organometallic allenes	Review	75
$^{-}CH=C=CH_2$	R	76
$^{-}C=C=CH_2$	T	77
$^{-}CH=C=CHR$	O,R	78
$^{-}CH=C=C-R$	V	79
$^{-}CMe=C=CH_2$	A	80
$^{-}CR=C=CH_2$	A	81
	O,R	78
	A,D,K,R	82
	A,K	83
$^{-}CH=C=CMe_2$	D,K,O,S,Si	84
$RCH_2-CR'=C=CH^{-}$	D,R,Si	85
$^{-}C=C=C$	C,D,K,O,S,Si	86
	C	87
	R	88,89
	Review	90
$^{+}CH_2-CR'=C=CH^{-}$	G;D,R,Si	85

1,3-DIENES

Synthon		Ref.
$CH_2=CH-CH=CH^{+}$ (Z)	E,G	91
$CH_2=CH-C^{+}=CH_2$	G	92,93
$CH_2=C^{+}-C^{+}=CH_2$	2G	94
$^{+}CH=CH-CMe=CH_2$ (Z)	E,P,Q	95
$CH_2=CH-C^{+}=CHR$ (Z)	G	96
$CH_2=C^{+}-CH=CHPh$ (E)	G	93
$C=C^{+}-CH=CH$	G	74
$^{+}C=C-C=C^{+}$	2G	97

$^{+}CH_2CH{=}CH{-}CH{=}CHR$ (2*Z*,4*E*)	G	98
$^{-}CH{=}CH{-}CH{=}CH_2$ (*Z*)	R	99,100
$^{-}CH{=}CH{-}CH{=}CH$ (1*Z*,3*E*)	A,K	101
$^{-}CH{=}CH{-}CMe{=}CH_2$ (*Z*)	A,R	102
$^{-}CH{=}C{-}C{=}CH_2$	R	103
$CH_2{=}C^{-}{-}CH{=}CH_2$	A,K,O	104–106
	A	107
	R	108
	L	109
	A,K	110
$^{-}CH_2C({=}CH_2){-}CH{=}CH_2$	A,R	111
	R	112
	A,K	113
	O,R	114
$^{-}CH_2CH{=}CHCH{=}CH_2$ (*E*)	A,L	115
	A,K	116
	A	117
	A,C,K,T	118
	A,K,T	119
	Review	120
$^{-}CHMe{-}CH{=}CHCH{=}CH_2$ (*E*)	A,L	115
$^{-}CMe{=}CH{-}CH{=}CH_2$ (*E* or *Z*)	A,K,R	121
$(^{-}C{=}CH_2)_2$	D,R,S,Si	122
$^{-}CH{=}CH{-}CH{=}CHR$ (*Z*)	D,L,R	123
$Me_2C{=}C^{-}{-}CMe{=}CH_2$	C,R,Si	1829
$^{-}CH{=}CH{-}CH{=}CHCH_2^{-}$ (1*Z*,3*E*)	R;R	124
$CH_2{=}CH({-}CH{=}CH)_2^{-}$ (1*Z*,3*E*)	R	99

OTHER DIENES

$^{-}CH_2C({=}CH_2)CH^{-}C{=}CH_2$	2R	59
$^{-}CH_2CH{=}CHCH^{-}CH{=}CH_2$ (*Z*)	2R	59
$^{-}CH(CH{=}CH_2)_2$	Review	120
	R	125
	A,K	126

ACIDS, ESTERS, AMIDES

$^{+}(CH_2)_nCOOH$ $n > 1$	G	127
$CH_2{=}C^{+}CH_2COOH$	G	128,129
$^{+}CH_2CH_2CH^{-}COOMe$	E;R	130

$^{+}CH_2CH{=}CHCH_2COOH$ (*E*)	G	131,132
	E	133
$^{+}CH{=}C{=}CHCH_2COOH$	G	134
3		
$^{-}CH_2C({=}CH_2)COOH$	K	135
	A,K,O,R	136
	A,K	54,1830
assymmetric addition to	A	137
$^{-}CHMe{-}C({=}CH_2)COOH$	A	138
$^{-}C{-}C(CONR_2){=}C$	C,K,R,S,Si	139
$^{-}C{\equiv}C{-}COOR$	A,K	140
4		
$^{-}CH_2CH{=}CHCOOH$ (*Z*)	A	141
$^{-}CH_2{-}C{=}CHCOOMe$	L	142
	A,C,K,R	143
$^{-}C{-}C{=}C{-}COOR$	A,C,K,T	144
	C,R,T	145
$^{-}CH_2C({=}CH_2)CH_2COOH$	A,K,R	146
$^{-}CH_2CMe{=}CHCONHMe$ (*Z*)	R	147
5		
$^{-}(CH_2)_nCOOH \quad n > 3$	A,O,R	148
$^{-}CH_2{-}C{=}CH{-}CH_2COOH$	E	149
$^{-}CH_2{-}C{=}C{-}CH_2COOH$	R	150
$^{-}CH_2CH{=}CHCH_2{-}CZ{-}SR \quad Z = O,S$	A,K,O,R,Si	151
$^{-}CH{=}CH(CH_2)_2COOH$	R	152
$^{-}CH{=}C{-}CH_2CH_2COOEt$	R	153
$^{-}CH{=}CR(CH_2)_2CSSMe$ (*E*)	R	154
≥6		
$^{-}CH{=}CH(CH_2)_nCOOH \quad n > 2$	R	155
$^{-}(CH_2)_7COOMe$	A	156

ALCOHOLS, ETHERS

$C{=}C{-}C^{+}{-}OAc$	E	157
$^{+}C{-}C{-}OH$	Review	158
$CH_2{=}CH{-}CH^{+}CH_2OAc$	G	159
$^{+}CH_2{-}CHOH{-}CMe{=}CH_2$	E,N,P,Q	160
$HOCH_2CH{=}C^{+}CH_2OH$ (*E*)	G	161
$^{+}CH_2CH_2CH_2OH$	Review	65

$^{+}CH_2CH{=}CHCH_2OAc$ (E or Z)	G	159
$^{+}CH{-}C{=}C{-}C{-}OH$	E	162,163
$^{+}CH_2CH{=}CHCH_2OH$ (E)	G	164
$^{+}CH_2C^{-}{=}CHCH_2OH$ (E)	G;D,K,R	164
$^{+}CH{=}CH(CH_2)_2OH$	G	165
$^{+}CH{=}CH(CH_2)_nOAc$ $n = 2$ or 3	G	166
$^{+}CH{=}CH(CH_2)_3OH$ (E)	G	167
$^{+}CH{=}CH(CH_2)_nOTHP$ $n > 5$ (E)	G	206
2		
$^{-}(CH_2)_2OH$	L	168
$^{-}CH_2{-}CHR{-}OH$	A,D,K,Si	169
	A	170
$^{-}CH_2{-}CRR'{-}OH$ R′ also H	A,D,K,R,S	171
$THPOCH_2CH^{-}CH{=}CH_2$ (→ *threo*)	A	172
$^{-}C({=}CH_2){-}CH_2OH$	A,K	173
	O	174
$R{-}CHOH{-}C^{-}{=}CH$	A,C,D,S,W	175
$CH_2{=}C{=}C^{-}CH_2OR$ R also H	A	176
$^{-}CH{=}CHOEt$ (Z)	A,K,L,R	177
$^{-}C{\equiv}C{-}OSiMe_2Bu{-}t$	K,Si	178
3		
$^{-}(CH_2)_3OH$	A	179
	K,L	180
	R	181
$^{-}CH_2CH_2{-}CH{-}OCH_2Ph$	C	182
$^{-}(CH_2)_nOPh$ $n = 3$–6	D	183
$^{-}CH_2CH_2CH_2OR$	C,D,Si,W,Z	184
	Review	65
$^{-}CHMe{-}CH_2CR_2{-}OH$	D,S,W	185
$^{-}CH_2C({=}CH_2)CH_2OH$	O	186
	A,K	187
$^{-}CH{=}CHCH_2OH$ (E)	L,R	188
	O	189
	A	179
$^{-}CH{=}CH(CH_2)_nOH$ (Z) $n = 1$–3	A,K,R	190
$^{-}CH{=}CHCH_2OTHP$ (E or Z)	A,K	191
$^{-}CHR'{=}CHCH_2OH$ (Z)	R	192
$^{-}CH{=}CHCR_2OH$ (Z)	D,K,R,Si	193
$^{-}C{=}C{-}CH_2OH$	R	194

Synthon	Method	Reference
$^{-}CH{=}CHCH_2OH$ (*Z*)	R	195
	Review	65
$^{-}C{\equiv}C{-}CH_2OH$	Review	65
≥4		
$^{-}(CH_2)_nOH$ $n = 4\text{–}9$	Z	196
$^{-}(CH_2)_3{-}CHMe{-}OSiR_3$	Z	197
$^{-}(CH_2)_3CR_2OH$ R = allyl	A,D,K,S,Si	198
$^{-}CH_2CH{=}CHCH_2OR$ (*Z*)	A	199
$^{-}CH_2{-}CMe{=}CHCH_2OH$ (*Z*)	R	200
$^{-}CH_2CH{=}CR{-}CH_2OH$	R	201
$^{-}CH{=}CMe{-}CH_2CH_2OH$ (*E*)	R	200
$^{-}C{\equiv}C({-}C{-})_nOH$ $n = 2\text{–}4$	R	202
$^{-}C{\equiv}C(CH_2)_3OH$	R	203,204
$^{-}(CH_2)_6OTHP$	T	205

ALDEHYDES

Synthon	Method	Reference
$^{+}CH_2CH(OCOPh){-}CHO$ (*R* or *S*)	G	207
$^{+}CH{=}CHCH{=}CHCHO$ (2*Z*,4*Z*)	G	208
$^{-}CH_2CH_2CH_2CHO$ (acetal)	A	301
	L	302
$^{-}CH_2CH{=}CH{-}CHO$ (*E*)	R	209
$^{-}CH_2CH{=}CMe{-}CHO$	A,K	210
$^{-}CH{=}CH(CH_2)_2CHO$ (*E*)	R	211
$^{-}(CH_2)_nCHO$	R	212

AMINES

Synthon	Method	Reference
$^{+}CH_2NH_2$	E	213
	G	214
	Review	215
$^{+}CH_2NHR$	G	216,217
$RCH_2CH^{+}NHR'$	G	218
$^{+}(CH_2)_nNR_2$ $n = 2$ or 3	E,N,P,Q	219
$^{+}CH_2CH_2NHCOPh$	E	220
$^{+}CHMe{-}CH_2NR_2$	E	221
$^{+}CH{=}CPh{-}CH_2NMe_2$ (*Z*)	G	222
$^{+}CH_2{-}NR{-}CH_2^{-}$	L(Y)	223
$^{-}CH_2{-}CH{-}NHPh$	D,Z	224
$^{-}CH_2{-}CHR{-}NHPh$	A,C,K,Si	169
	R,S,Z	225

$R_2NCH_2C^-{=}CHR'$	V	226
$PhNH{-}CH(CH_2^-)_2$	R,S,Si,W	227
$^-C{=}C{-}NR_2$	A,R	228
$^-(CH_2)_3NH_2$	C	229
$^-(CH_2)_3NMe_2$	K	230
$^-CH_2CH_2CH_2N{<}$	Review	65
$^-CH_2{-}C({=}CH_2)CH_2NMe_2$	A,K,R	231
$^-C{\equiv}CCH_2N(SiMe_3)_2$	A,C,K,R,Si	232
$^-C{\equiv}C{-}C{\equiv}C{-}NR_2$	C,D,K,R,Z	233

HALIDES

$^+CH{=}CHCl$ (E or Z)	G	234
$^+CH{=}CCl_2$	G	235
$^+C{\equiv}CCl$	E	236
$^+CH_2CH_2CH_2Cl$	Review	65
$^-CR(Cl){-}CH{=}CH_2$	R,Z	237
$^-CH_2CBr{=}CH_2$	A,K	54,238
$^-CH_2CH{=}CHBr$ (Z)	K	239
$^-CH_2{-}C{=}CHX$	A,C,T	240
$^-CH_2CH{=}CRCl$	R,Z	237
$^-CH{=}CHCH_2I$ (Z)	R	241,242
$^-(CH_2)_nCl$ $n > 3$	D,T	205

KETONES

$MeCO{-}C({=}CH_2){-}C^+$	E	243
$MeCOCH_2CH{=}CH{-}CH^+Me$	G	244
$MeCO{-}C{-}C{=}C{-}C^+$	E	243
$^-CH_2COCH_2CH_2CHO$	R	245
$^-CH_2CH{=}CHCOMe$ (E)	R	209
$^-C{-}C{=}C{-}COPh$	R	145
2-cyclohexenone 4-anion	R	246

SILANES

$Me_3SiCH^+CH{=}CH$ (E)	G	247
$Me_3SiCH_2CH_2^+$	G	248
$Me_3SiCH{=}CHCH_2^+$ ($E+Z$)	E,M	249
$Me_3SiCR{=}CHCH_2^+$ (E or Z)	E,G,Z	250
$Me_3SiCH_2CH{=}CHCH^+$ (E)	G	247

$Me_3SiCH_2^-$	A,K	251,252
$(Me_3Si)_2C^-$	A,K	253
$(Me_3Si)_3C^-$	R	254,255
$R_3SiC^-Cl_2$	R	256
$Me_3Si—CR^-$ (also cycloPr)	A,K	257
$Me_3SiC^-(CH_2)_2$	A,C,K,L	258
	A,K	259
$Me_3SiCH^-CH{=}CH_2$	A	260–264
$Me_3SiC^-{=}CH_2$	C	265
	A	266
	A,K	259
	L	267
$Me_3SiC^-{=}CHR$ (*E*)	C,R,Z	269
(*Z*)	R	270,271
	A,K	272
(*E* or *Z*)		273
$Me_3SiC^-{=}CR_2$	A	257
	A,D,R,V,W,X	274
$Me_3SiC^-{=}CHCH_2OH$ (*E*)	A,K	275
$Me_3SiC^-{=}CHCH_2OTHP$ (*E* or *Z*)	A,K	191
$Me_3Si—C^-{=}C{=}CH_2$	A	80
$Me_3Si—C^-{=}C{=}CH$	A,K	276
2		
$Me_3SiCH_2—C^-{=}CH_2$	L	54
	A	277
$Me_3SiCH_2CH_2^-$	A,K	278
	K	279
$Me_3SiCH{=}CH^-$ (*Z*)	A	280–282
	K	283
	V	268
	D,R,Si,Z	284
	A,K,O	285
	L	267
	K,R,Z	286
$Me_3SiCR{=}CH^-$ (*E*)	R	287
$Me_3SiCH{=}CR^-$ (*Z*)	L	267
$Me_2PhSiCH{=}CR^-$ (*E*)	C,D,L,O,R,X	288
$CH_2{=}C(SiMe_3)—CH_2^-$	A,C,K,Z	289
$^-CH—C(SiMe_2Ph){=}CH$	A,R,Si,W	290

3		
$Me_3SiCH_2—C=C^-$	R	291
$Me_3SiCH=CHCH_2^-$ (*E*)	A,K	264,292,293
	K	294
	R	295
	A,C,L,R	57
$iPr_3SiC≡CCH_2^-$	A,K,L,O,R	296
$CH_2=CHC(CH_2CH_2SiMe_3)=CHCH_2^-$	A,K	279
$Me_3SiCH^{\pm}$	A,K(Y)	297
$Me_3SiCMe^{\pm}$	A,K(Y)	298
$Me_3SiC^-=CH^+$ (*Z*)	R,W;G	299
	C,R,Z;G	269

2

FORMYL AND ACYL ANIONS: $^{-}CH{=}O$ and $^{-}CR{=}O$

David J. Ager

CONTENTS

2.1. INTRODUCTION

The electronegativity of oxygen naturally polarizes the carbonyl group so that the carbon atom is electrophilic ($^+$C—O$^-$). This property has made the carbonyl group one of the most important and versatile groups available for organic synthesis.[303–309] Many reactions utilize this phenomenon to introduce a moiety at the carbon atom—for example, addition of a Grignard reagent[310,311] or an alkyllithium[312,313] and reduction with an aluminium hydride[314–316] or hydroborane.[317,318] Other reactions take advantage of the polarity induced at

both the carbon and oxygen atoms; examples of these are the Wittig reaction,[319–326] hydroboration,[317,318] and hydrosilylation.[327–330]

The polarity has been enhanced by the addition of a Lewis acid, which has proven particularly useful in the context of olefin cyclizations.[331–334] Reactions of this type provide an alcohol as the product. The utility of the carbonyl group has been extended by the use of an acylium ion ($^{+}CR{=}O$) so that the functionality is introduced intact. The classic example is the aromatic Friedel–Crafts acylation reaction.[335–338] Although some purely aliphatic examples of this reaction are known,[339,340] the best results were obtained when the outcome of the intermediate carbocation was controlled, for example, by use of a silicon-substituted alkene.[331,341–345] To overcome some of the problems associated with this approach, various protected forms of the acylium ion have been used to prepare functionalized carbonyl compounds. These protected $^{+}C{=}O$ synthons invariably rely on groups that have been developed as acyl anion synthons.

The 1,3-dithiane group has been used as a masked $^{+}C{=}O$ synthon, and reaction occurred with nucleophilic carbon–carbon double bonds, such as enol ethers and related compounds,[346–350] allylsilanes,[351] and aromatic compounds[352,353] (Scheme 1). The removal of hydrogen from 1,3-dithiane **(1)** by

(1) $\xrightarrow[CH_2Cl_2]{Ph_3CBF_4}$ BF_4^- $\xrightarrow{R\text{–}C(OSiMe_3){=}CHR^1}$ (1)

the trityl cation is equivalent to removal of hydride from formaldehyde to give the $^{+}CH{=}O$ cation. Substituted derivatives ($^{+}CR{=}O$)[354] have been prepared by protonation of a ketene thioacetal[355] (see Sect. 2.2.1.1.).

Alternative reactions that employ a formyl cation equivalent are derived from 2-chloro-1,3-dithiane and its reactions with nucleophiles.[356–359] Other analogous systems that have also been used in this context are 1,3-dithiolanes,[360] benzo-1,3-dithiolanes,[361–363] and α-(trimethylsilyl)-phenylthioalkanes.[364–367]

Reversal of this natural polarity of the carbonyl group leads to formyl ($^{-}CH{=}O$) and acyl ($^{-}CR{=}O$) anions. These high-energy species are known (see Sect. 2.15.2.1.) but at present the high reactivity imposes many limitations on their direct use in synthetic methodology. However, many masked forms of these synthons have been introduced and used in synthesis.[11,368–374] This chapter discusses the comparative merits and limitations of these reagents. In addition to simple substitution patterns in these acyl anion equivalents, the presence of various functional groups within the synthon will be considered in the later sections.

Reactions that involve modification of the electrophilic moiety during the

condensation have not been included. A useful reaction of this type is the homologation of a carbonyl compound by a Wittig or related reaction (Scheme 2) where the original carbonyl group becomes a methylene group rather than an alcohol.[375–395]

$$\mathrm{RCHO} \xrightarrow[\text{2) Hydrolysis}]{\text{1) } Ar_3P{=}CR'{-}Y} \mathrm{RCH_2{-}CO{-}R'} \qquad (2)$$

$$Y = SR'', OR'', SeR'', \text{etc.}$$

Some common functional groups have been used as acyl anion equivalents, but only those reactions directly relevant to this usage have been included in the appropriate section. Examples are the oxidative cleavage of an olefin,[396–398] nitrile,[399–409] and carboxylic acid derivatives,[400,410–412] which are usually employed as $^{-}C{-}C^{+}{=}O$ synthons, and the hydration of acetylenes.[413–420]

2.2. 1,3-DITHIANES

2.2.1. Formyl Anion Equivalents

1,3-Dithianes are used widely as formyl and acyl anion equivalents.[368–370,373,421] The availability of the substituted derivatives **(2)** by the condensation of an aldehyde with 1,3-propanedithiol in the presence of a Lewis acid catalyst[421,422] has detracted from the use of 1,3-dithiane **(1)** as a formyl anion equivalent.

1,3-Dithiane **(1)** was easily deprotonated[423] by *n*-butyllithium in THF at −30° and alkylated smoothly with alkyl halides, in particular iodides[422,424] (Scheme 3), by an S_N2 reaction.[425]

(1) —[1) n-BuLi / THF/−23°C; 2) RX]→ (2) (3)

This alkylation has also been achieved with *p*-toluene- and benzenesulfonates.[426] 2-Lithio-1,3-dithiane is a relatively good nucleophile and gave the substitution product with the diethyl acetal of bromoacetaldehyde,[424,427,428] a system that often undergoes elimination. Allyl halides underwent substitution readily.[429,430] Some success has been achieved in the preparation of **2** by alkylation, when a silyl group was displaced to yield the carbon nucleophile.[431]

Elegant studies with 4,6-dimethyl-1,3-dithiane have shown that lithiation occurred equatorially, and reaction with iodomethane or a carbonyl compound resulted in equatorial substitution,[432–435] the thermodynamically controlled products. Kinetic control has been observed with 1,3-dithiane itself.[436]

The lithium anion of **1** opened epoxides to give the β-hydroxyaldehyde

derivative.[422,424] Attack occurred at the least hindered position,[437–441] and, in cyclic systems, in an axial manner.[442] This approach has been used for the preparation of intermediates in the synthesis of maytansine[443] and phyllanthocin.[444]

In a similar manner, carbonyl compounds have been condensed with 2-lithio-1,3-dithiane to provide masked α-hydroxycarbonyl compounds[422,424]; this approach has been used to synthesize the carbohydrate derivative, L-streptose.[445]

β-Dicarbonyl compounds have been synthesized by reaction of 2-lithio-1,3-dithiane with acylating reagents. Reaction with *N,N*-dimethylformamide gave the dimer **3**, which was cleaved thermally to provide the monomer **4** (Scheme 4).[446,447]

1) n-BuLi
2) DMF
Δ
rt (slow)
CHO
(1) (3) (4)

(4)

If **4** is required for further elaboration, it is conveniently prepared[448] *in situ*. By contrast, *N,N*-dimethylacetamide underwent deprotonation.[449] Addition occurred in the expected manner with ethyl chloroformate or carbon dioxide,[421,422] or with esters.[389,421,422,424,450] α-Dicarbonyl compounds have also been prepared by condensation of the anion derived from **1** and nitriles[421,422,451,452] with subsequent hydrolysis of the imine. Iminium salts provide a route to α-aminoaldehydes.[453,454] While reaction with *N*-acylaziridines gave protected α-ketoaldehydes,[455] the corresponding reaction with the aziridenecarbamate was used to illustrate the nucleophilicity of the dithiane system.[456] Reaction with vinyl azides gave the masked formamide derivative,[457] whereas aryl organochromium compounds provided a route to masked aromatic aldehydes.[458,459] Reaction with metal carbene complexes caused ring cleavage.[460]

Much attention has been given to the condensation reaction of 2-lithio-1,3-dithiane with α,β-unsaturated carbonyl compounds. The normal mode of addition with conjugated enones and enals was 1,2-.[370,422,461,462] In the presence of hexamethylphosphorictriamide (HMPA), α,β-unsaturated compounds unhindered at the β-position react by 1,4-addition[463–468] (Scheme 5). Care must be exercised, however, as chalcone[469] and enones with large groups adjacent to the carbonyl group[470] have shown a reversal of this latter trend. Conjugate addition has been put to further use by reaction of the enolate formed during the addition.[464,471] Nitroalkenes also underwent conjugate addition.[472–474]

2-Lithio-1,3-dithiane has been reacted with silyl,[475–477] germyl,[476] and

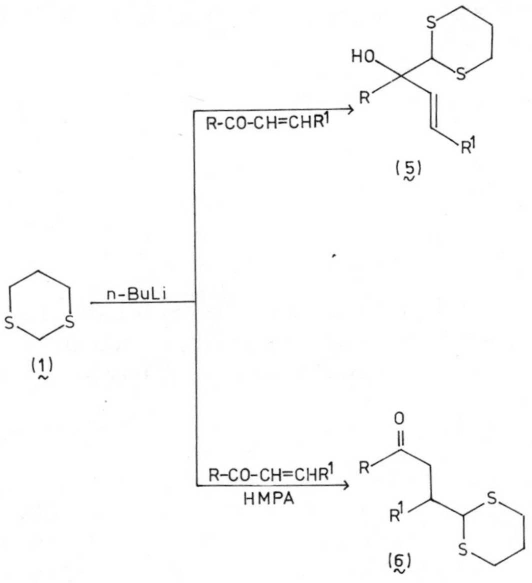

(5)

stannyl[476,478] chlorides. Other heteroatoms that have been used as electrophiles are phosphorus,[479] sulfur,[480–482] and boron.[483]

Other methods that have been used to substitute 1,3-dithiane **(1)** utilize formation of an ylide with an allylic compound followed by a [2,3] sigmatropic rearrangement under basic conditions.[484,485] The substituted 1,3-dithianes **(2)** have been converted to the corresponding carbonyl compound by hydrolysis, particularly in the presence of a heavy metal ion. The unmasking is discussed in Section 2.2.3.

A system closely related to **1** is 4,5-dihydro-5-methyl-1,3,5-dithiazine (**7,** Scheme 6).

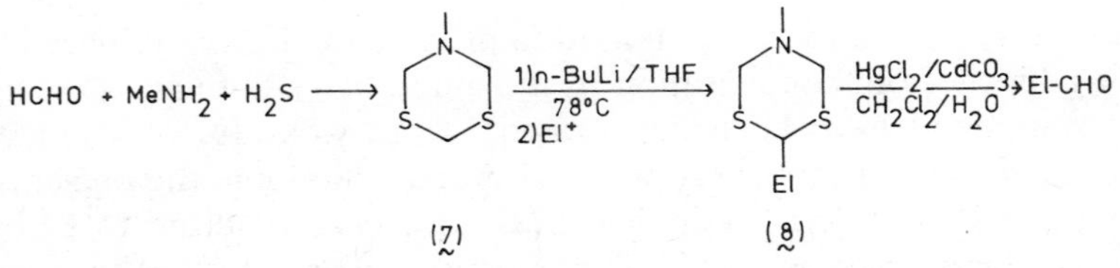

(6)

The anion derived from **7,** although stabilized by complex formation with the nitrogen at the 5-position, reacted with most electrophiles in a manner similar to 2-lithio-1,3-dithiane.[486,487] Mercury-assisted hydrolysis of **8** was more rapid than for **2.** The adduct **8** could not, however, be deprotonated to provide the acyl anion equivalent.

A system that also suffers from a similar drawback is that derived from 1,3,5-trithiane **(9)**,

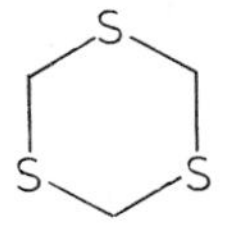

(9)

because preferential deprotonation occurs at an unsubstituted carbon atom. Reaction of the parent anion does, however, proceed smoothly, and aldehydes have been prepared by subsequent mercury(II)-assisted hydrolysis.[488–490]

2.2.1.1. Ketene Thioacetals

This aspect of 1,3-dithiane chemistry has been considered separately because of the wide utility of this class of compounds.[491] The alkylation of a ketene thioacetal provides an acyl anion equivalent (C=C—CO$^-$) (see Sect. 2.17.1.3.1.) or $^-$C—C—C=O equivalent (see Chapt. 6), depending upon the position at which reaction takes place. The addition of an organometallic reagent to the double bond provides an acyl (RCO$^-$) equivalent (see Sect. 2.2.2.). With regard to the synthesis of aldehydes, ketene thioacetals **(10)** have been reduced by triethylsilane in the presence of acid to afford the dithiane (**2,** Scheme 7).[492,493] The ketene derivatives **(10)** have been converted to α-haloesters by reaction with *N*-halosuccinimide.[494]

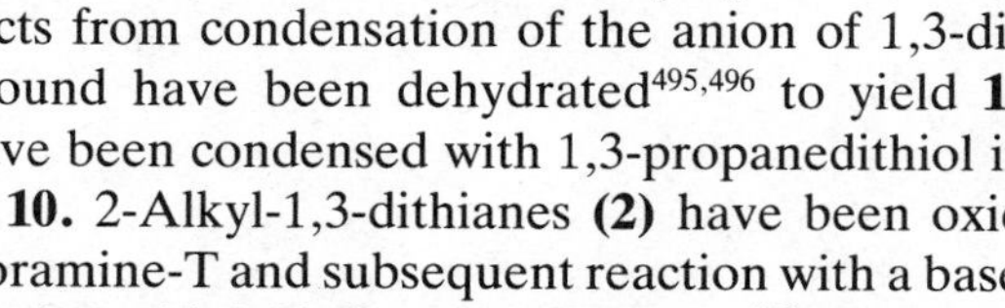

$$\textbf{(10)} \xrightarrow[\text{TFA}]{Et_3SiH} \textbf{(2)} \qquad (7)$$

A wide variety of routes have been used to prepare the ketene thioacetals **(10).**[491] The adducts from condensation of the anion of 1,3-dithiane **(1)** and a carbonyl compound have been dehydrated[495,496] to yield **10.** *N,N*-Dimethylthioamides have been condensed with 1,3-propanedithiol in the presence of a base to give **10.** 2-Alkyl-1,3-dithianes **(2)** have been oxidized to **10** by reaction with Chloramine-T and subsequent reaction with a base[497] or by anion formation and reaction with 2,2′-dipyridyl disulfide.[498] Derivatives of **2** bearing a γ-substituent have been isomerized or undergone elimination to provide the keteneacetal **10**[446,499] (see also Sect. 2.17.1.3.1.). Esters provide **10** when reacted with bis(dimethylaluminium)-1,2-ethanedithiolate[350,500] while 3-chloropropyldithioesters have been cyclized to **10** in the presence of lithium amide.[501] The most commonly used approaches rely on the Peterson olefination reaction,[355,461,489,494,502–505] its tin analog,[478] or the Wittig reaction.[446,504,506–512]

These ketene derivatives have been used to synthesize carbonyl compounds,[513] as illustrated in the synthesis of aromatin[514] and 17-oxoellipticine[515] (Scheme 8).

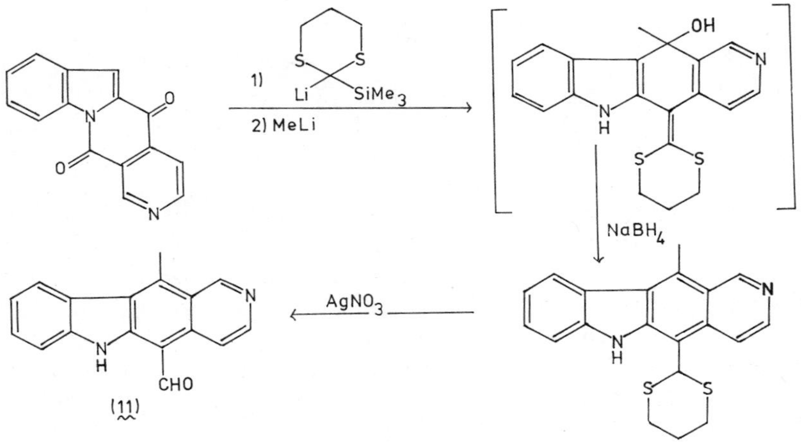

(8)

2.2.2. Acyl Anion Equivalents

2-Substituted-1,3-dithianes **(2)** were readily deprotonated[423] in a manner similar to the parent system **(1)** to give the equatorial 2-lithio-1,3-dithiane[433,435,516] existing as a tight ion pair in THF solution.[517] 2-Lithio-2-phenyl-1,3-dithiane (**12,** R = Ph) has been reacted with deuterium oxide[518] and tritium oxide[519] to provide the labeled benzaldehyde. Treatment of 2,2-diaryl-1,3-dithianes with *n*-butyllithium led to carbon-sulfur bond cleavage.[520]

Alkylation of the anion (**12,** Scheme 9) proceeded smoothly by an S_N2 reaction;[519] best results were obtained with primary iodides, bromides,[421,422,424,425,427,448,521–524] and tosylates,[525] as well as allyl halides.[422,430]

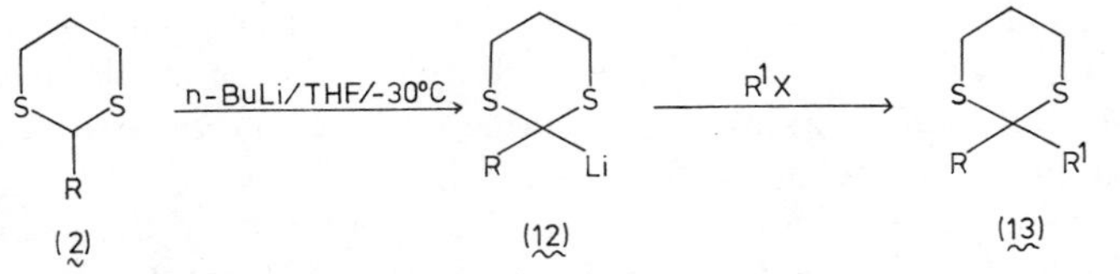

(9)

Selective reactions may, therefore, be carried out with ω-dihalocompounds.[427] Masked 1,*n*-dicarbonyl compounds have been prepared by sequential alkylation of 1,*n*-bis(1,3-dithianyl)alkanes[526,527] (see Sect. 2.17.2.2., 2.17.2.3.). The masked ketone derivatives **(13)** have been prepared by sequential alkylations in a "one-pot" reaction.[424] Reaction with dihaloalkanes has led to the synthesis

of cyclic ketones[346,427,477,528–530] (Scheme 10). Examples of the use of the 1,3-dithianyl group are illustrated by the syntheses of multistriatin,[531] prostaglandin analogs,[529,532] combretastatin,[533] and talaromycin B.[430]

(1) — 1) n-BuLi/THF/−30°C; 2) $Br(CH_2)_{n-1}Cl$; 3) n-BuLi → spiro dithiane, n = 4–8 (10)

An alternative approach to the masked ketones **(13)** has been to add an alkyllithium to a ketene thioacetal **(10)** and alkylation of the resultant anion[506,513,534,535]; cuprates did not add to **10.**[536] Ketene thioacetals **(10)** have provided ketones by ozonolysis.[514] Displacement of a silyl group[431] allowed reaction to occur in systems where a strong base would have reacted within the molecule (Scheme 11).[537] Indeed, care has to be exercised when other functional groups are present within the molecule because fragmentation can occur.[538–540]

$(CH_2)_{n-2}SiMe_3$ / CHO dithiane — N^nBu_4F, THF → spiro dithiane with OH, n = 4–7 (11)

2-Lithio-2-alkyl-1,3-dithianes **(12)** have been reacted with epoxides[422,427,437,541–546] and oxetane.[422] α-Hydroxycarbonyl compounds have been synthesized by condensation of the anion **(12)** with aldehydes and ketones.[422,424,443,444,447,546–549] The reverse of this condensation has been shown to occur with a strong base in a polar solvent,[550,551] and an α-group has been found to interact with the dithiane when sequential condensation of 1,3-dithiane with aldehydes was followed by reaction with methanesulfonyl chloride.[549] α-Diketones have been synthesized by use of aryl nitrile oxides as the electrophile.[552] Cyclic α-diketones, with one of the carbonyl groups protected as the dithiane, underwent an oxidative cleavage reaction leading to ω-ketoesters.[553,554] α-Dicarbonyl compounds have resulted from the use of carbon dioxide,[422,522] *N,N*-dimethylformamide,[422,555,556] ethyl formate and chloroformate,[421,422,522,555,557,558] lactones,[548] and acid chlorides, esters or nitriles[421,422] as electrophiles. A striking example is the use of dimethyl oxalate as the electrophile which gave a protected trione system in the synthesis of the antibiotic aplasmomycin.[559] Care has to be exercised with these condensations to minimize side reactions.[560]

The usual mode of addition to α,β-unsaturated carbonyl com-

pounds[421,422,439,561–563] is 1,2-, but this trend has been reversed by the addition of HMPA.[463,466,468,537] Examples of unsaturated systems that have undergone conjugate addition are enones,[463,465,557,564] butenolides,[565,566] enol ethers of β-diketones,[562,567,568] enals,[466–468] α,β-unsaturated esters[468] and amides,[569] and nitroalkenes.[472,474] In some systems, the 1,2-adduct has been isomerized to the 1,4-product by reaction with potassium hydride in HMPA.[550,564,570] The enolate initially formed has been used for further reaction,[565,566,569] the approach leading to an annulation procedure[571,572] (Scheme 12).

(1)
1)n-BuLi/THF/HMPA
2) Br
1)n-BuLi/THF
HMPA/-78°C
2)RCH=CHCOR^1
HCl/MeOH
H_2O

(12)

S_N2' displacement of an allyl alcohol ester has been achieved.[573] Addition to aziridines gave the amine derivative,[574] whereas aromatic nitro compounds gave coupling products.[575]

Heteroatoms that have been used as electrophiles are sulfur,[480,576,577] silicon,[475,476,521,578–580] germanium,[476] and tin.[476,478] While the sulfur adducts have been used to prepare esters,[576] the silyl derivatives have provided acylsilanes.[475,476]

2.2.3. Hydrolysis

The conversion of the 1,3-dithianes (**2** or **13**) to the carbonyl compound is closely related to that of other bisthioacetals. Many of the conditions drive the reaction to completion by complex formation with a metal ion or by making the sulfur less nucleophilic through oxidation or thioether formation.

The metal ion that has found particular favor for the conversion is mercury(II)[421,425,427,475,476,518,521,525,555,581–585]; this has been buffered by the addition of calcium or cadmium carbonate.[355,477,521,526,547,586–588] Boron trifluoride has been used to promote reaction in this system.[438,546,566,589] Reaction of the bis-

thioacetal with an alcohol in the presence of a mercury(II) salt has led to ketal formation.[490,590] A similar reaction has been observed in the presence of an acid catalyst.[591–594]

While the conversion has been achieved by the halogens,[584,595–598] *N*-bromosuccinimide has found wider use[421,462,504,546,574,599,600]; the addition of silver nitrate has also been found to be advantageous.[532,546,561,600,601] In a similar manner, the conversion has been carried out with Chloramine-T,[454,602,603] 1-chlorobenzotriazole,[604] and sulfuryl chloride.[605] The latter reagent has been used to accomplish a 1,2-carbonyl transposition.[606] Alkylation,[513,529,540,607–618] acylation,[619] or oxidation[563,620–627] have provided intermediates that were hydrolyzed easily. The transformation has also been carried out by thallium(III),[628] cerium(III),[629] or electrolysis.[630]

2.3. BISTHIOACETALS

2.3.1. Formyl Anion Equivalents

Many of the properties and reactions of this class of formyl anion equivalents are very similar to those of 1,3-dithiane **(1)** (see Sect. 2.2.1.). In a similar manner, attention has mostly been directed toward the synthesis of ketones because the bisthioacetals **(14)** are readily available by condensation of an aldehyde and a thiol in the presence of an acid catalyst.

2.3.1.1. Bis(alkylthio)acetals

Bis(ethylthio)methane **(15)** was deprotonated by lithium or sodium amide in liquid ammonia.[631] Alkylation of the resultant anion was accomplished with primary, and in some cases secondary, alkyl halides (Scheme 13).[631,632] Reaction of the anion with diethyl disulfide gave tetrakis(ethylthio)methane rather than the orthothioformate;[482,632] anionic reactions of orthothioformates have resulted in carbene formation.[480,633–636]

$$\underset{\mathbf{15}}{\text{RS—CH}_2\text{—SR}} \xrightarrow[\text{2) R'X}]{\text{1) NaNH}_2\text{/NH}_3} \underset{\mathbf{14}}{\text{RS—CHR'—SR}} \qquad (13)$$

The thioacetals **(14)** have also been prepared by substitution of an α-chlorosulfide by alkylthiolate[637,638] and addition of a Grignard reagent to the dithioester.[639,640] They have been formed by a cleavage reaction of an alkene activated by electron-withdrawing groups in the presence of a Lewis acid and ethanethiol.[641]

2.3.1.1.1. Ketene Thioacetals. Ketene thioacetals[491] **(16)**

$$\underset{\mathbf{16}}{R_2C{=}C(SR')_2}$$

have been prepared by the Peterson olefination[503] and Wittig[508] reactions, or by condensation of a Grignard reagent with carbon disulfide followed by treatment with LDA and an alkyl halide.[642,643] Variations on the last method have used a 1,1-dithiol[644] or dianion of a carboxylic acid[645] as starting material.

2.3.1.2. Bis(arylthio)acetals

Bis(phenylthio)methane (**15,** R = Ph) has been deprotonated by a metal amide[632] or *n*-butyllithium.[580] In addition to alkyl halides,[580,632,646,647] the anions have been reacted successfully with carbonyl compounds[482,580,646,648,649] and epoxides.[650] Conjugate addition to α,β-unsaturated ketones and esters has been achieved by the use of HMPA or the cuprate.[468,469,651] Boron electrophiles have led to the preparation of alcohols by rearrangement of the borate[483,652] and ketene thioacetals from the boronic esters.[653] Ketene bis(phenylthio)acetals (**16,** R = Ph) have been prepared by the Peterson[503] and Wittig[508] methods, condensation of an ester with aluminium thiophenoxide,[654] or elimination from a β-hydroxy-tris(phenylthio)alkane.[655] The masked aldehyde derivatives (**14,** R = Ph) have also been synthesized by a [2,3] sigmatropic rearrangement of a sulfur ylide[656] and by substitution of an α-chlorosulfide.[657]

The thioacetals (**14,** R = Ph) eliminate thiophenol in the presence of a copper(I) salt to give the vinylsulfide,[658] while the related reaction of **(17)** gave the masked aldehyde **(18)** (Scheme 14).[650]

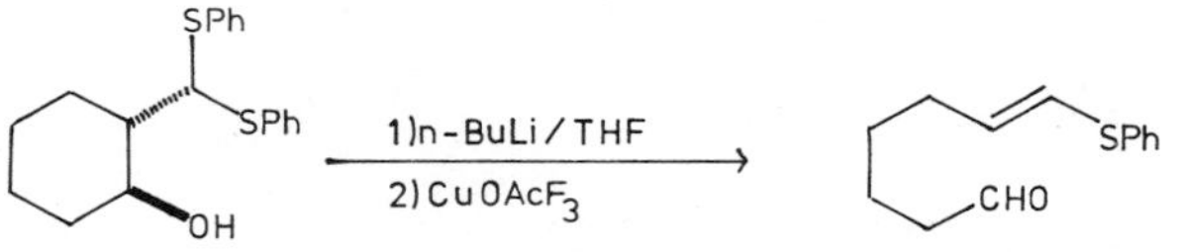

(17) —— 1) n-BuLi/THF; 2) CuOAcF$_3$ ——→ (18) (14)

2.3.1.3. Other Cyclic Thioacetals

1,3-Dithiolanes **(19)** have received little attention for their use as formyl and acyl anion equivalents. This is, no doubt, due to the observation that treatment of **(19)** with a base led to fragmentation of the dithiolane ring,[659] although this has been put to use for a 1,2-carbonyl transposition.[659] The system has found some use for the synthesis of ketene thioacetals by the Wittig reaction[660] and for investigation of new synthetic methods for the conversion of thioacetals to carbonyl compounds[592,614,661–666] (see Sect. 2.2.3.).

1,3-Benzodithiole **(20)** has been alkylated by reaction with *n*-butyllithium

followed by an alkyl halide;[667] this system has been hydrolyzed in the presence of a mercury(II) salt (see Sect. 2.2.2.). In a completely analogous manner 7,8-dimethyl-1,5-dihydro-2,4-benzodithiepin **(21)** has been alkylated[668] and reacted with epoxides.[669] The hydrolysis was accomplished by mercury(II)[669] or copper(II)[668] catalysis (Scheme 15).

$$\textbf{(21)} \xrightarrow[2)\,El^+]{1)\,n\text{-BuLi/THF/-60°C}} \text{2-El-benzodithiepin} \xrightarrow[Me_2CO/H_2O]{CuO/CuCl_2} \text{ElCHO} \quad (15)$$

2.3.1.4. Other Thioacetals

α-Thiodithiocarbamates have provided a method to prepare aldehydes (Scheme 16).[670] Methylenebis(*N,N*-dimethyldithiocarbamate) reacted in an analogous manner.[671]

$$MeSCH_2{-}SCS{-}NMe_2 \xrightarrow[2)\ RX]{1)\ BuLi/THF,\ -55°} MeS{-}CHR{-}SCSNMe_2 \xrightarrow{Hg^{2+}/MeOH} RCH(OMe)_2 \quad (16)$$

2.3.2. Acyl Anion Equivalents

2.3.2.1. Bis(alkylthio)acetals

The bis(ethylthio)acetals (**14,** R = Et) have been alkylated when an alkali metal amide was used as base in liquid ammonia, but the yields were not particularly good.[632] The alternative approach of addition of a Grignard reagent to a dithioester[672,673] followed by reaction with an electrophile is however a useful synthetic method (Scheme 17).[639,674] Reaction of the enolate of the dithiolate ester with a carbonyl compound provided a RO—C—CH—CO$^-$ synthon[639,641] (see Sect. 2.17.4.).

$$R{-}CS{-}SEt \xrightarrow[2)\ El^+]{1)\ EtMgI} El{-}CR(SEt)_2 \quad (17)$$

The thioketals have also been prepared from ketones by condensation with a thiol[675] or thiosilane[676,677] in the presence of a Lewis acid or with a disulfide in the presence of tri-*n*-butylphosphine.[678]

Cyclopropane dithioketals **(22)** have been synthesized (Scheme 18); they are useful precursors to ring-expanded α-hydroxyketones or α,β-unsaturated esters.[679,680]

$$MeSCH_2SMe \xrightarrow[\text{2) cyclohexanone}/-78°C;\ \text{3) TsCl/THF}/-78°C;\ \text{4) n-BuLi}]{\text{1) n-BuLi/hexane}/-50°C} \text{bicyclo[4.1.0]heptane-7,7-(SMe)}_2\ (\mathbf{22}) \quad (18)$$

In addition to the reaction of the acyl anion equivalents with alkyl halides, carbonyl compounds, and carboxylic acid derivatives,[632,639] enones usually gave the 1,2-adduct unless a third anion-stabilizing group was present[681,682] or HMPA was used as an additive.[468,469,683]

2.3.2.2. Bis(arylthio)acetals

The use of phenylthio rather than alkylthio acetals, with attendant increase in anion stabilization, improves the performance of this system's use as an RCO^- equivalent. Initial studies employed alkali metal amides in liquid ammonia.[632,647] An alternative has been the use of *n*-butyllithium–TMEDA complex in hexane with strict temperature control.[684] Care must be exercised with these alkylations because carbon–sulfur bond cleavage may occur.[520,685] The condensation with carbonyl compounds has proven less problematic and the adducts **23** have been used to synthesize ketones (Scheme 19),[646,686,687]

$$PhS\text{—}CHR\text{—}SPh \xrightarrow[\text{2) R'COR''}]{\text{1) BuLi/THF}} \underset{\mathbf{23}}{(PhS)_2CR\text{—}CR'R''\text{—}OH} \xrightarrow[(R''=H)]{\text{TFA}} R\text{—}COCH_2R' \quad (19)$$

α-phenylthioketones,[646] protected α-diketones,[646] 2-phenylthiobutadienes,[688] α-phenylthioenones,[689] butenolides,[690] γ-phenylthioacrylic esters,[690] and α,β-unsaturated ketones.[691] Other electrophiles have included ethyl chloroformate and acid chlorides,[646] boranes,[483] and α,β-unsaturated carbonyl compounds, the position of attack following other thioacetals.[465,468,469,651,683] This class of acyl anion equivalents has been used to prepare 2-phenylthiocyclobutanone (see Scheme 20)[692] and cyclopropanone dithioketals.[693–695]

$$\text{2,5-dimethoxytetrahydrofuran} \xrightarrow{\text{PhSH/HCl}} (PhS)_2CH\text{—}CH_2CH_2\text{—}CH(SPh)_2 \xrightarrow[-78°C]{\text{s-BuLi/TMEDA/THF}} \text{1,2-bis(phenylthio)cyclobutene} \xrightarrow[\text{AcOH}]{CuCl_2/TiCl_4/H_2O} \text{2-(phenylthio)cyclobutanone} \quad (20)$$

Alternative procedures[676–678] for the synthesis of the ketals are described in Section 2.3.2.1. The bisthioketals have been used to synthesize phenylvinylsulfides by base[696] and copper(I) catalyzed[648,658,697] elimination (see Sect. 2.14.2.).

2.3.2.3. Other Cyclic Thioacetals

7,8-Dimethyl-1,5-dihydro-2,4-benzodithiepin **(21)** has been used to prepare ketones by alkylation[668,698,699] or reaction with an epoxide.[669] In a similar manner, the lithium anion derived from 2-alkyl-1,3-benzodithioles **(24)** has been reacted with alkyl iodides, carbonyl compounds and epoxides.[667] Reaction with boranes leads to ketones (Scheme 21).[700] Addition of an alkyllithium to the trithiocarbonate derivative gave the thioorthoformate compound after alkylation.[701,702]

1) n-BuLi/THF; 2) $BR^2_2R^3$ → $\bar{B}R^2_2R^3$ adduct → (24) BR^2R^3 → R^1COR^2 (21)

2.4. OXIDIZED THIOACETALS

2.4.1. Formyl Anion Equivalents

2.4.1.1. Methyl Methylthiomethyl Sulfoxide (25)

Oxidation of a dithioacetal sulfur atom results in a system that can easily generate a carbanion. These give adducts **(26)** that are hydrolyzed under mild acidic conditions. Alkylation (Scheme 22) was facile[703,704]; omission of an electrophile resulted in formation of methyl bis(methylthio)methyl sulfoxide.[705]

$$\underset{\mathbf{25}}{MeSO{-}CH_2SMe} \xrightarrow[RX]{NaH/THF} \underset{\mathbf{26}}{MeSO{-}CHR{-}SMe} \xrightarrow{H_3O^+/THF} RCHO \quad (22)$$

25 ↓ 1) BuLi/THF 2) ArCHO 3) H_3O^+

$$\underset{\mathbf{27}}{HO{-}CHAr{-}CH(SMe)(SO{-}Me)} \xrightarrow{HCl/THF} \underset{\mathbf{28}}{Ar{-}COCH_2OH}$$

27 ↓ NaH/DMSO/MeI (23)

$$MeO{-}CHAr{-}CH(SMe)(SO{-}Me) \xrightarrow{CuCl_2/DME} MeO{-}CHAr{-}CHO$$

In addition to alkyl halides,[703,706] condensation has been achieved with aldehydes and ketones,[707] esters,[708] 2-bromopyridines,[709] and α,β-unsaturated ketones where 1,2-addition was the preferred mode of attack.[710] When Triton B was used as the base, aromatic aldehydes gave the ketene dithioacetal derivative.[711–715] Although the required adduct **27** was formed with *n*-butyllithium as base, acid-catalyzed hydrolysis gave the α-hydroxyketone **(28)** rather than the α-hydroxyaldehyde (Scheme 23).

This problem was circumvented by *O*-alkylation and reaction with copper(II) chloride or, alternatively, by treatment with triethyl orthoformate.[707]

The anion of **25** has been condensed with nitriles which led to substituted carboxylic acid derivatives,[711,716] and iminium salts which gave α-aminoaldehydes.[453] Substituted aldehydes have been prepared by addition of an anion to a ketene thioacetal monoxide.[717–720] Methyl methylthiomethyl sulfoxide **(25)** afforded aldehydes through the thioacetal when reacted with Grignard reagents.[721] Thermolysis of the adducts **(26)** gave the vinylsulfide.[722]

2.4.1.2. Ethyl Ethylthiomethyl Sulfoxide (29)

Use of this system rather than **25** has been advocated.[723] The anion derived from **29** has been reacted with alkyl halides,[723] and with aldehydes, ketones, esters, and acid chlorides.[724] Conjugate addition occurred with α,β-unsaturated esters, a contrast to enones that underwent 1,2-addition.[725] The aldehyde was unmasked by acid treatment in the presence of a mercury(II) salt (Scheme 24).[723,725]

$$\underset{\mathbf{29}}{\text{EtSO—CH}_2\text{SEt}} \xrightarrow[\text{2) RX}]{\text{1) BuLi or LDA}} \text{EtSO—CHR—SEt} \xrightarrow[\text{H}_3\text{O}^+]{\text{Hg}^{2+}} \text{RCHO} \qquad (24)$$

2.4.1.3. 1,3-Dithiane-1-oxide (30)

As with the analogous compounds discussed previously (Sect. 2.4.1.1., 2.4.1.2.), this formyl anion equivalent has been reacted[726] with alkyl halides, carbonyl compounds and esters; the yields, however, were lower than for **25** and **29.** Although deprotonation occurred in very high yield (96%), as detected by deuteration, with *n*-butyllithium as base, yields with other electrophiles were increased when LDA was used as base.[727] The sulfoxide oxygen induces chirality, and diastereomeric mixtures resulted when **30** was reacted with carbonyl

S S O

30

compounds.[727,728] The adducts have been prepared by oxidation of the appropriately substituted 1,3-dithiane.[729]

2.4.1.4. *p*-Tolyl *p*-Tolylthiomethyl Sulfoxide (31)

The stereochemical consequences of thioacetal monosulfoxides have been taken to their logical conclusion with **31** as a chiral formyl anion equivalent. The pathway illustrated in Scheme 25 resulted in the α-methoxyaldehyde **(32)** with a 70% ee.[730]

(31) 1) n-BuLi/THF/−78°C; 2) PhCHO/−78 → 0°C

1) $N^{n}Bu_4OH/Me_2SO_4$, H_2O/CH_2Cl_2; 2) $NaI/I_2/Ph_3P$

I_2/H_2O, $NaHCO_3$/dioxan → (32)

Ar = $pMeC_6H_4$ (25)

The anion of **(31)** in racemic form added in a 1,2-manner to enones, or 1,4- if HMPA was used as an additive.[731,732] Optically active α-hydroxyaldehyde derivatives were prepared by acylation followed by reduction[733,734] and unmasking of the aldehyde.[735]

2.4.1.5. Methylthiomethyl *p*-Tolyl Sulfone (31a)

The sulfone **(31a)** was alkylated under phase-transfer conditions to give **31b.** The latter was resistant to hydrolysis, and conversion to the aldehyde had to be performed by photolysis.[736]

$$MeS{-}CHR{-}SO_2{-}C_6H_4Me{-}p \qquad \textbf{31a}\ R = H \qquad \textbf{31b}\ R = \text{alkyl}$$

2.4.2. Acyl Anion Equivalents

2.4.2.1. Methyl Methylthiomethyl Sulfoxide (25)

In a manner similar to that already described for the synthesis of aldehydes (see Sect. 2.4.1.1, Scheme 22), **25** has been used for the preparation of ketones by reaction with excess base and alkyl halide.[706] In addition to symmetrical ketones, cyclic ketones have been prepared from the appropriate dihalocompound (Scheme 26).[737–739]

$$MeS\text{-}CH_2\text{-}SOMe \xrightarrow[X(CH_2)_{n-1}X]{\text{xs Base}} \text{(cyclic, ring size n) C(SMe)(S(O)Me)} \xrightarrow{H_3O^+} \text{(cyclic, ring size n) C=O} \quad (26)$$

(25)

The sulfone has also been used in this context.[740] The system has been employed to synthesize functionalized ketones (Scheme 27).[718,719]

$$CH_2{=}C(S(O)Me)(SMe) \xrightarrow[THF/-78°C]{Y\text{-}CH_2CH_2\text{-}Li} \left[Y\text{-}CH_2CH_2CH_2\text{-}\bar{C}(S(O)Me)(SMe) \right] \xrightarrow[HMPA]{RX}$$

$$Y\text{-}CH_2CH_2CH_2\text{-}C(S(O)Me)(SMe)\text{-}R \xrightarrow[MeCN]{HClO_4/H_2O} Y\text{-}CH_2CH_2CH_2\text{-}C(=O)\text{-}R \quad (27)$$

This procedure gave good results with ester enolates, but when more stable anions were used, equilibration occurred and alkylation took place[718] α to the anion-stabilizing group Y.

2.4.2.2. Ethyl Ethylthiomethyl Sulfoxide **(29)**

This reagent has been used as an acyl double anion equivalent $^-CO^-$ for sequential alkylations (Scheme 28; cf. Scheme 24).[723]

$$\underset{\mathbf{29a}}{EtSO{-}CHR{-}SEt} \xrightarrow[2)\ El^+]{1)\ LDA\ or\ BuLi} EtSO{-}CR(El){-}SEt \xrightarrow[THF/H_2O]{HgCl_2/HCl} R{-}CO{-}El \quad (28)$$

In addition to alkyl halides, successful electrophiles include aldehydes and acid chlorides.[729] With α,β-unsaturated ester and ketones, conjugate addition was the preferred mode of attack.[725]

2.4.2.3. 1,3-Dithiane-1-oxide **(30)**

2-Substituted 1,3-dithiane-1-oxides were readily deprotonated by LDA. Subsequent methylation with iodomethane gave the masked ketones in good yield.[727]

2.4.2.4. Methylthiomethyl *p*-Tolyl Sulfone **(31a)**

The alkylation of the masked aldehyde derivative **(31b)** was achieved after anion formation with sodium hydride in DMF. The ketone was obtained by photolytic or acid hydrolysis.[736]

2.5 HEMITHIOACETALS: α-ALKOXYSULFIDES

2.5.1. Formyl Anion Equivalents

As oxygen and sulfur are in the same periodic group, it has been possible to replace one sulfur atom in a bisthioacetal by oxygen. 1,3-Oxathianes **(33)**[741–743] have been alkylated by reaction with *s*-butyllithium followed by an alkyl halide (Scheme 29). The yields were high with primary alkyl iodides but surprisingly low with bromides.[744,745]

1) s-BuLi / THF / −78°C; 2) RX (33) (29)

Carbonyl compounds, nitriles, silicon, sulfur, and tin halides have also been successfully used as electrophiles. 1,2-Addition occurred with conjugated systems.[745]

The conformation of the ring and the presence of a chiral center have been

(+)-pulegone; $PhCH_2SH$, NaOH, 90%; Na/NH_3, MeOH, 80%; 89% $(CH_2O)_n/H^+$; (34); 1) n-BuLi / THF, 2) RCHO; DMSO, $(F_3Ac)_2O/NEt_3$; R^1MgX; NCS/$AgNO_3$/MeCN; OHC, OH (30)

used to prepare α-hydroxyaldehydes in a stereoselective manner. Although the initial studies were carried out with 2-acyl- and 2-aroyl-4,4,6-trimethyl-1,3-oxathianes and the 4,6,6-trimethyl analogs[746,747] the final asymmetric synthesis was based on **34** (Scheme 30).[748] This type of approach was used in the synthesis of the methyl ester of atrolactic acid where the oxathiane was derived from (+)-camphorsulfonic acid.[749]

The anion derived from methoxyphenylthiomethane[750,751] has been used as a formyl anion equivalent and condensed with carbonyl compounds,[699,752,753,1382] lactones,[754] and alkyl halides;[753,754] in some cases reaction with epoxides caused fragmentation of the reagent.[753] Good yields of adducts were obtained with a nitrile and *N,N*-dimethylamides.[753,755] The silicon adduct has been used in homologation reactions[756] and to prepare acylsilanes.[757] The sulfone derivative has also been employed and has the advantage of mild acid hydrolysis to unmask the aldehyde.[758]

1,3-Oxathianes have been converted to the corresponding carbonyl compounds by methods analogous to those used for bisthioacetals[600,754,759,760] (see Sect. 2.2.3.) and by electrochemical means.[761]

2.5.2. Acyl Anion Equivalents

The system has not been used extensively in this context. One interesting example introduced two substituents when only one equivalent of base was used (Scheme 31).[762] The 2-trimethylsilyl derivative of 1,3-oxathiane reacted with electrophiles in the expected manner when the anion was generated with *s*-butyllithium.[763]

SiMe3
1) s-BuLi / THF
2) PhCN
SiMe3
N
Ph
NSiMe3
Ph
3) MeI
Ph
45%

(31)

2.6. α-THIOSILANES

2.6.1. Formyl Anion Equivalents

To alleviate some of the problems associated with the hydrolysis of bis(thioacetals) (see Sect. 2.2.3.), alternative systems have been sought. One

of these is the use of phenylthiotrimethylsilylmethane **(35)** as a formyl anion equivalent. The silane **35** has been prepared by deprotonation of thioanisole[580,764] with *n*-butyllithium and reaction with chlorotrimethylsilane.[765,766] Subsequent deprotonation was achieved by *n*-butyllithium in THF[766,767] or TMEDA-hexane.[768] The anion (**36,** Scheme 32) reacted, in high yield, with primary alkyl halides to give the adducts **37.**[765,766,768]

$$PhSCH_2SiMe_3 \xrightarrow{BuLi} PhS{-}CHLi{-}SiMe_3 \xrightarrow{RX}$$

35 **36**

$$PhS{-}CHR{-}SiMe_3 \xrightarrow[3)\ H_3O^+]{1)\ MCPBA\quad 2)\ \Delta} RCHO \qquad (32)$$

37

The sulfoxide analog of **36** could not be alkylated.[769] In addition to alkyl halides, the anion **36** has been reacted with epoxides,[184,766,770] chlorotrimethylsilane, tri-*n*-butyltin chloride, and sulfur electrophiles.[770,771] The presence of the silyl group modifies some of the reactions with carbonyl compounds. The anion **36** reacted with aldehydes and ketones to give the vinylsulfides **38**[767,770,772] by the Peterson reaction.[773,774]

$$\mathbf{36} \xrightarrow{RCOR'} PhS{-}CH{=}CRR' \qquad (33)$$

38

Although vinylsulfides may be hydrolyzed to the carbonyl compound, **36** is not acting as a true formyl anion equivalent so that an α-hydroxyaldehyde would result. In a similar vein, reaction of **36** with carboxylic acid derivatives resulted in α-(phenylthio)ketones.[770] Reaction of phenylthiotrimethylsilylmethyllithium **(36)** with conjugated enones gave either the 1,4- or 1,2-adduct, depending on reaction conditions;[770] the 1,2-adducts were unstable and gave the 1-phenylthio-1,3-diene by the Peterson reaction.[773]

α-Silylsulfides have also been prepared from benzylthiol,[775] or by a base-induced ring opening of 2-(trimethylsilylmethyl)-1,3-dithianes,[776] a sulfur-ylide rearrangement,[777] substitution of a phenylthio group in bisthioacetals,[257,778] and addition of an alkyllithium to 1-phenylthio-1-trimethylsilylethene.[765,779]

The adducts **(37)** were converted to the aldehyde by oxidation to the sulfoxide, which then underwent the sila-Pummerer rearrangement[765,766,768,769,780] to give the *O*-trimethylsilylthioacetal; this was then hydrolysed by acid or base.

Reaction of the adducts **(37)** with Chloramine-T gave the vinylsilane.[465]

2.6.2. Acyl Anion Equivalents

1-Phenylthio-1-trimethylsilylalkanes **(37)** could not be used as acyl anion equivalents; deprotonation did not occur[771] unless an additional anion-stabi-

lizing group was present. Phenylketones were, however, prepared by the method shown[781] in Scheme 34.

$$\underset{(\mathbf{37},\ R = Ph)}{PhS{-}CHPh{-}SiMe_3} \xrightarrow[\text{2) RX}]{\text{1) BuLi/TMEDA}} PhS{-}CRPh{-}SiMe_3 \xrightarrow[\text{2) }\Delta\text{ 3) }H_3O^+]{\text{1) MCPBA}} Ph{-}COR \quad (34)$$

The required anion was prepared by indirect methods (Scheme 35),[771,782] of which the most useful was the sulfone route.[771,783]

$$CH_2{=}C(SPh)SiMe_3 \xrightarrow{R^2Li} RC(Li)(SPh)SiMe_3$$

$$RC(SPh)_2SiMe_3 \xrightarrow{LiC_{10}H_8} RC(Li)(SPh)SiMe_3 \xleftarrow{n\text{-}BuLi} RC(Sn^nBu_3)(SPh)SiMe_3$$

$$RC(Li)(SPh)SiMe_3 \xrightarrow{R^1X} RR^1C(SPh)SiMe_3 \xrightarrow[\text{2) }\Delta\text{ 3) }Hg^{2+}/H_2O]{\text{1) MCPBA}} RC(=O)R^1 \quad (35)$$

$$RCH(SO_2Ph)SiMe_3 \xrightarrow[\text{2) }R^1X]{\text{1) n-BuLi}} RR^1C(SO_2Ph)SiMe_3 \xrightarrow{\text{Redn}} RR^1C(SPh)SiMe_3$$

where $R^2CH_2 = R^1$

The presence of the silyl group gave rise to the Peterson reaction with carbonyl compounds.[774] A further complication with the system was the sila-Pummerer rearrangement giving vinylsulfides in addition to the *O*-silylthioacetal; the product ratio was dependent on substituents and mercury-catalyzed hydrolysis had to be employed to unmask the ketone.[771]

2.7. α-FUNCTIONAL SULFONES

2.7.1. Formyl Anion Equivalents

Several α-functionalized sulfones have been used as formyl anion equivalents. The cyclic sulfone **(39)**, prepared as shown in Scheme 36, was reacted with a wide variety of electrophiles but only the alkyl derivatives **(40)** were converted to the aldehydes by pyrolysis.[784]

$$^{i}PrCHO \xrightarrow[\substack{2)\,LiAlH_4\\3)\,CH_2O/H^+\\4)\,KMnO_4}]{1)\,S_2Cl_2} \mathbf{39} \xrightarrow[2)\,RX]{1)\,n\text{-}BuLi/THF/-80°C} \mathbf{40} \xrightarrow{\Delta} RCHO \qquad (36)$$

The acyclic analog **(41)** has been deprotonated with LDA as illustrated in Scheme 37, but KDA had to be used for condensation to occur with a carbonyl compound.[785]

$$\underset{\mathbf{41}}{PhSO_2{-}CH_2O{-}CHMe{-}OEt} \xrightarrow{1)\,LDA/THF/HMPA,\ -78°\ \ 2)\,RX} PhSO_2{-}CHR{-}OCHMe{-}OEt \xrightarrow{1)\,H^+\ \ 2)\,HO^-} R\dot{C}HO \qquad (37)$$

The methyl ether **(42)** analog of **41** has also found use as a formyl anion equivalent.[758]

Although phenylsulfonylnitromethane **(43)** was easily deprotonated and alkylated, the adduct could not be converted to an aldehyde.[786] Phenylsulfonylmethyl chloride **(44)** has been used as a nucleophilic agent to prepare aldehydes. The method, however, was limited to condensation with carbonyl compounds which gave the α,β-epoxysulfone. Reaction with a nucleophile gave the α-substituted aldehyde.[787,788]

$$PhSO_2{-}CH_2{-}Y$$

42 Y = OMe
43 Y = NO_2
44 Y = Cl

2.7.2. Acyl Anion Equivalents

The α-alkoxysulfone (**45,** Scheme 38) has been used as an acyl anion equivalent. Symmetrical ketones, including a cyclic example, have been prepared[785] from the parent compound **41** by treatment with an alkyl halide and two equivalents of LDA.

$$\underset{\mathbf{45}}{PhSO_2\text{—}CHR\text{—}OCHMe\text{—}OEt} \xrightarrow[\text{2) R'X}]{\text{1) LDA/THF/HMPA, }-78^\circ} PhSO_2\text{—}CRR'\text{—}OCHMe\text{—}OEt \xrightarrow[\text{2) }HO^-]{\text{1) }H_3O^+} RCOR' \quad (38)$$

2.8. NITROGEN-CONTAINING SULFIDES

2.8.1. Benzothiazoles

Although benzothiazoles **(46)** have been used in many synthetic methods as a masked aldehyde group,[789–792] the derived vinyl anion has not been exploited as a formyl anion equivalent to any great extent.[457,793] A silyl group has been displaced by fluoride ion to provide an alternative method to the adducts.[794]

(46)

2.8.2. α-Thionitriles

This class of compounds[795] has not been used as a direct formyl anion equivalent, although alkylations[796–798] and condensations with carbonyl compounds were facile.[799,800] They have been prepared from nitriles[801] while desulfurization[799] gave the nitriles (see Sect. 2.12.2.4.).

By contrast, the dithiocarbamate **(47)** has been used to prepare ketones (Scheme 39). The stability of the acyl anion equivalent was demonstrated by the use of phase-transfer conditions.[802] 1,*n*-Dihalocompounds led to cyclic ketones.[802]

$$\underset{\mathbf{47}}{NC\text{—}CH_2S\text{—}CSNMe_2} \xrightarrow[\text{PTC}]{\text{RX, }HO^-} NC\text{—}CHR\text{—}SCSNMe_2 \xrightarrow[\text{PTC}]{\text{R'X, }HO^-} NC\text{—}CRR'\text{—}SCS\text{—}NMe_2 \xrightarrow[\text{or 1) NBS or NCS; 2) NaOH/}H_2O\text{/MeCN}]{\text{NaOH/}H_2O\text{/EtOH/}\Delta} R\text{—}CO\text{—}R' \quad (39)$$

2.8.3. α-Thioisonitriles

TosMIC (**48**) has been alkylated under phase-transfer conditions.[803–806] A second alkylation, however, required the use of more vigorous conditions with sodium or potassium hydride as the base.[804,806–808] In addition to alkyl halides, condensation occurred with carbonyl compounds, in the presence of thallium ethoxide, to give an oxazoline; aqueous acid treatment led to the α-hydroxyaldehyde[809–811] (Scheme 40). Other electrophiles have included acid chlorides[812] and 1-phenylsulfonyl-1,3-butadiene.[813]

$$\text{Ts}\diagup\!\!\diagdown\text{NC}\ (\mathbf{48}) \xrightarrow[\text{RR}^1\text{CO}]{\text{TlOEt/EtOH/DME}} \text{oxazoline (R, R}^1\text{, EtO)} \xrightarrow{H_3O^+} \text{RR}^1\text{C(OH)CHO} \quad (40)$$

2.9. α-THIOCARBOXYLIC ACID DERIVATIVES

This class of umpolung reagents relies upon the natural polarity of the carboxylic acid moiety ($^{-}$C—COOR) and provides a method for this group to be used as an acyl anion equivalent[814] (see Sect. 2.13.2.1). The dianion of the α-thiocarboxylic acid (**49,** Scheme 41) was generated with two equivalents of LDA. Condensation occurred readily with alkyl halides[815,816] or carbonyl compounds[817] and a second alkylation could also be accomplished.[816] 1,2-Addition occurred with α,β-unsaturated carbonyl compounds.[817] Reaction with esters or lactones led to α-thioketones.[817] The carbonyl compound was unmasked by treatment with NCS.[816,818,819]

$$\underset{\mathbf{49}}{\text{RS—CH}_2\text{COOH}} \xrightarrow[\text{2) El}^+\ \text{3) LDA}\ \text{4) El}'^+]{\text{1) 2LDA/THF/HMPA, }-78^\circ} \text{RS—C(El)(El}'\text{)—COOH} \quad (41)$$

$$\xrightarrow[\text{2) H}_3\text{O}^+]{\text{1) NCS/NaHCO}_3\text{, MeOH}} \text{El—CO—El}'$$

Alternative procedures were electrolysis[820] or reduction of the acid followed by elimination and hydrolysis of the vinylsulfide (see Sect. 2.14.2).[815]

2.10. OTHER SULFUR COMPOUNDS

2.10.1. Sulfoxides

As a sulfoxide stabilizes an α-carbanion, this class of compound has found limited use as a formyl anion equivalent.[368,369,373,421] The anion is usually derived

from a methyl sulfoxide. The carbonyl group has been unmasked by a Pummerer rearrangement[821–824] or iodine in methanol.[825,826]

2.10.2. Sulfones

Sulfones[369,821,827,828] may be regarded as acyl anion equivalents. An α-anion may be reacted with a variety of electrophiles. Conversion to the ketone has been realized by anion formation followed by reaction with molybdenum[829] or bis(trimethylsilyl)[830] peroxides (Scheme 42). A related method has been to react the anion with a carbon tetrahalide followed by hydrolysis of the halide with TFA in the presence of silver perchlorate.[831] The transformation has also been accomplished through a boronic ester.[832]

$$RCH_2\text{—}SO_2Ar \xrightarrow[\text{2) R'X}]{\text{1) BuLi}} R\text{—}CHR'\text{—}SO_2Ar \xrightarrow[\text{2) MoO}_5\text{/py/HMPA or (Me}_3\text{Si)}_2\text{O}_2]{\text{1) LDA/THF, }-78^\circ} RCOR' \quad (42)$$

2.10.3. α-Chlorosulfur Compounds

In addition to the α-chlorosulfones (see Sect. 2.10.2), α-chlorosulfides[833,834] have been converted to aldehydes. α-Chlorosulfoxides have been converted to vinylsulfides, which in turn have been hydrolyzed to the carbonyl compounds[835] (see Sect. 2.14.2). These transformations open up the possibility of sulfides and sulfoxides being used as formyl anion equivalents.

N,N-Dimethylbenzylamine reacted with phenylthiomethyl chloride in the presence of base to give the rearranged product, dimethylamino-phenylthio-*o*-tolylmethane, which was hydrolyzed to *o*-tolualdehyde; benzyl phenyl sulfide underwent a similar reaction.[836]

2.10.4. Miscellaneous

Thermolysis of the sodium salt of *S*-allyl-*S*′-methyldithiocarbonate tosylhydrazone followed by methylation gave the ketene thioacetal (Scheme 43).[837]

$$MeSCS\text{—}NHNHTs \xrightarrow[\text{CH}_2\text{=CHCH}_2\text{X}]{\text{NaH/MeCN}} (MeS)(CH_2\text{=}CHCH_2)C\text{=}N\text{—}NHTs \xrightarrow[\text{2) MeI}]{\text{1) NaH/THF/}\Delta} CH_2\text{=}CHCH\text{=}C(SMe)_2 \quad (43)$$

1-Phenylthiocyclopropyltriphenylphosphonium tetrafluoroborate has been used as a $^{+}C—C—CO^{-}$ synthon, as illustrated in Scheme 44.[838]

(44)

2.11. SELENIDES

2.11.1. Formyl Anion Equivalents

2.11.1.1. α-SILYLSELENIDES

In a manner similar to α-silylsulfides (see Sect. 2.6.1.), phenylselenotrimethylsilylmethane **(50)** has been used as a formyl anion equivalent (Scheme 45).[839,840]

$$\mathrm{PhSeSePh} \xrightarrow[\text{2) } Me_3SiCH_2Cl]{\text{1) } NaBH_4/MeOH} \underset{\mathbf{50}}{\mathrm{PhSeCH_2SiMe_3}} \xrightarrow[\text{2) } RCH_2X]{\text{1) LDA/THF, } -78°} \underset{\mathbf{51}}{\mathrm{RCH_2CH(SePh)(SiMe_3)}} \xrightarrow{30\%\ H_2O_2,\ 0°\text{ to rt}} \mathrm{RCH_2CHO} \quad (45)$$

The advantages of this system are: LDA rather than *n*-butyllithium was used as the base and the Pummerer-type rearrangement occurred below 25°. The presence of the silicon group led to vinylselenides when carbonyl compounds were employed as the electrophile.[772] The vinylsilane was obtained by treatment of the adduct **51** with Chloramine-T.[780]

2.11.1.2. SELENOACETALS

This class of compounds has found very little use as a formyl anion equivalent.[841]

2.11.2. Acyl Anion Equivalents

2.11.2.1. α-SILYLSELENIDES

The acyl anion equivalents **(52)** have been prepared by deprotonation of the appropriate α-silylselenide **(51).** An alternative method was to displace a selenium group from an acetal by treatment with *n*-butyllithium (Scheme 46).[842,843] Condensation with carbonyl compounds gave the β-hydroxysilane in addition to the vinylselenide.[374,842] The Pummerer reaction, however, was not clean and resulted in elimination of selenium as well as the required rearrangement.[843]

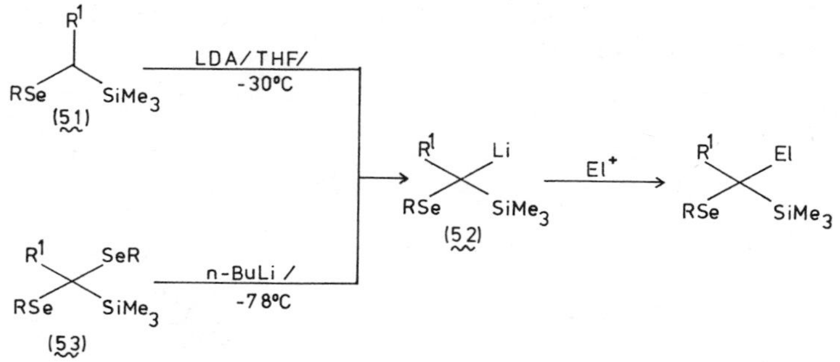

(46)

2.11.2.2. SELENOKETALS

Bisselenoketals **(54)** have been prepared by reaction of the anion **55** with an electrophile, which has included alkyl halides, epoxides, and carbonyl compounds.[844] 1,2- versus 1,4-addition could be controlled by the use of HMPA, as for the sulfur compounds.[465,468,469,683] Again, the anion **55** was prepared by direct deprotonation or displacement of a selenium group.[844,845] The ketone was unmasked by hydrolysis in the presence of a mercury(II) or copper(II) salt, or by oxidation (Scheme 47).[846]

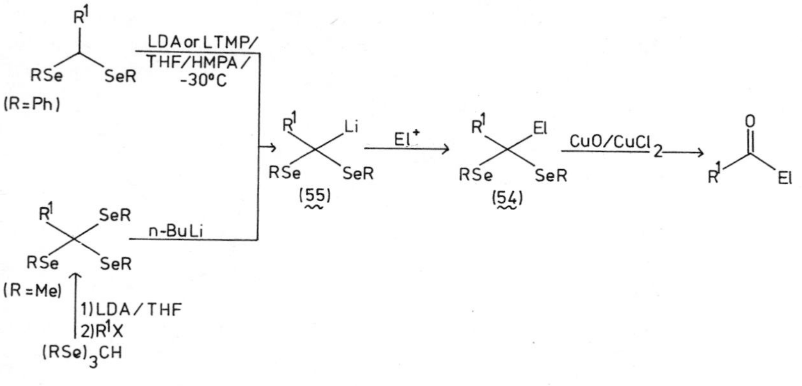

(47)

2.12. NITROGEN COMPOUNDS

2.12.1. Formyl Anion Equivalents

Although *N*-alkyl imines have been deprotonated and reacted with a wide variety of electrophiles, the allyl anion was formed, which resulted in the system acting as a $^{-}C{-}CHO$ synthon.[847–852] The substituted imine **56** was, however, successfully employed as a HCO^{-} synthon (Scheme 48).[853] The anion derived from the nitro compound **(57)** underwent conjugate addition with α,β-unsaturated ketones to give, after unmasking with hydroxylamine, 1,4-dicarbonyl compounds.[854] In a similar manner, pyridinium *p*-toluenesulfonylmethylide **(58)** underwent conjugate addition with *N*-substituted maleimides in the presence of an alcohol to give the vinyl ether, which was hydrolyzed to the aldehyde.[855]

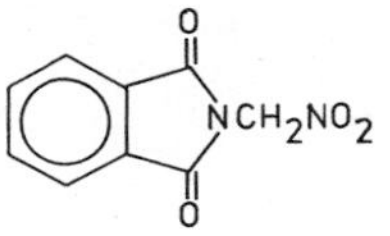

NCH_2NO_2

(57)

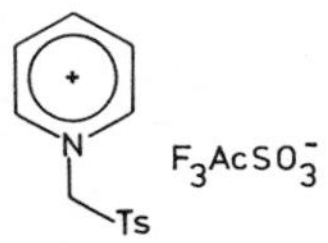

$F_3AcSO_3^-$ Ts

(58)

$$PhCH{=}NCH_2COOEt \xrightarrow[\text{2) RX}]{\text{1) LDA/THF/HMPA, }-78^\circ} PhCH{=}N{-}CHRCOOEt \xrightarrow[\text{2) NaIO}_4]{\text{1) LAH}} RCHO \quad (48)$$

56

3-Methylthio-1,4-diphenyl-*s*-triazolium iodide **(59)** was deprotonated by sodium hydride in DMF. Subsequent reaction with an alkyl halide and reduction gave the aldehyde (Scheme 49).[856] Although reaction occurred with benzaldehyde, hydride loss resulted in a change of oxidation levels and cleavage. The system is, therefore, limited to a formyl anion equivalent with alkyl halides only. The related carbene **(60)** obtained by thermolysis of the trichloromethyl compound did not provide a HCO^{-} equivalent.[857–859]

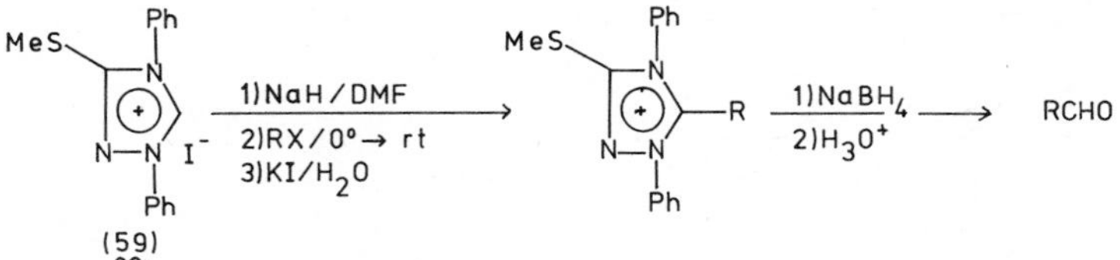

$$\mathbf{59}\ (\text{MeS, Ph, N, N–N, Ph, }I^-) \xrightarrow[\text{2) RX/0}^\circ\rightarrow\text{rt; 3) KI/H}_2\text{O}]{\text{1) NaH/DMF}} (\text{MeS, Ph, R}) \xrightarrow[\text{2) H}_3\text{O}^+]{\text{1) NaBH}_4} RCHO \quad (49)$$

(59)

Ph
N
N
Ph
(60)

2.12.2. Acyl Anion Equivalents

2.12.2.1. α-AMINONITRILES

The first use of α-aminonitriles[409,860–862] as an acyl anion equivalent was for the synthesis of aryl ketones.[863] Anions of α-dialkylaminoarylacetonitriles (**61,** Scheme 50)[864,865] have now been reacted with a wide range of electrophiles such as deuterium oxide,[866] alkyl halides,[863,864,867–872] also under phase-transfer conditions,[873] benzyl halides,[868,873–876] allyl halides,[873,877,878] and with aryl halides, epichlorohydrin, acid chlorides, and ethyl bromoacetate.[877] Conjugate addition occurred with α,β-unsaturated nitriles,[877,879–882] esters,[877,879] and ketones.[883–885] Typical bases are sodium hydride, sodium or potassium amide, and LDA.

$$\mathrm{ArCHO} \xrightarrow[\mathrm{R_2NH}]{\mathrm{KCN}} \underset{\mathbf{61}}{\mathrm{R_2N{-}CHAr{-}CN}} \xrightarrow[\text{2) El}^+]{\text{1) base}} \mathrm{R_2N{-}CAr(El){-}CN} \xrightarrow{\mathrm{H_3O^+}} \mathrm{ArCO{-}El} \qquad (50)$$

$$\mathrm{R''_2NCH_2CN} \xrightarrow[\text{2) RX}]{\text{1) LDA/THF/HMPA, }-78^\circ} \underset{\mathbf{62}}{\mathrm{R''_2N{-}CHR{-}CN}} \xrightarrow[\text{2) R'X}]{\text{1) LDA/THF/HMPA, }-78^\circ} \mathrm{R''_2N{-}CRR'{-}CN} \xrightarrow[\text{or HCl/H}_2\text{O}]{\text{Cu}^{2+}\text{/H}_2\text{O/EtOH}} \mathrm{R{-}CO{-}R'} \qquad (51)$$

Aliphatic ketones have also been synthesized by this strategy (Scheme 51). The required anion was generated either by deprotonation of **62** with LDA[867,886,887] or KDA[887] or by addition of an alkyllithium to an α-amino-α,β-unsaturated nitrile;[888] not all alkyllithiums underwent conjugate addition.[888] Reactions with alkyl halides[886] and aldehydes[887,889] proceeded smoothly, and 1,4-addition occurred with α,β-unsaturated esters[890,891] and ketones.[891] The parent compound (**62;** R = H) gave α-amino-α,β-unsaturated nitriles with carbonyl compounds, but the yields were higher when a Peterson olefination reaction was used.[892]

The above reactions have been carried out with a wide variety of amino groups. The most common, however, has been the morpholine derivative, although diethylamino has been advocated for aliphatic derivatives.[886] Some success has been achieved in the enantioselective synthesis of α-hydroxyketones by use of chiral amines.[893]

The system has been used to prepare aldehydes and ketones by a [2,3] sigmatropic rearrangement (Scheme 52).[894,895]

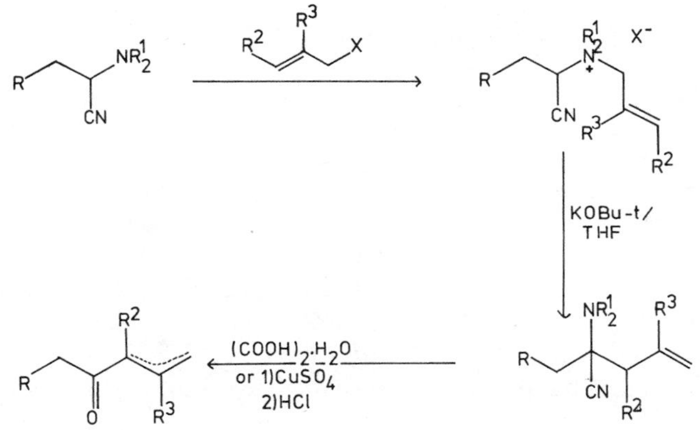

(52)

This approach has been used in the synthesis of α-sinensal,[896] artemisia ketone,[897] γ-cyclocitral,[898] and helminthosporic acid derivatives.[899,900]

$$\underset{\mathbf{63}}{RCH_2NO_2} \xrightarrow[2)\ El^+]{1)\ 2BuLi/THF/HMPA,\ -65^\circ} \underset{\mathbf{64}}{El{-}CHR{-}NO_2} \longrightarrow R{-}CO{-}El \quad (53)$$

2.12.2.2. Nitroalkanes

The ability of a nitro group to stabilize an α-carbanion has been exploited in the synthesis of ketones.[901,902] For a nitroalkane **(63)**[403,903] to be a latent acyl anion equivalent, it has to be primary. The dilithium derivatives of **63** reacted with a wide variety of electrophiles, which included carbonates, acid anhydrides, esters, aldehydes, and ketones.[904] The dianion was also prepared by addition of an alkyllithium to nitroethylene. Conjugate addition occurred between α,β-unsaturated ketones[905] and esters and nitroalkanes[906,907] in the presence of an amine.[908] The adducts **(64)** have been converted to ketones by methods such as electrolysis,[907,909] treatment with base followed by acid hydrolysis (the Nef reaction),[905,910–913] silica,[914,915] singlet oxygen,[916] persulfate oxidation,[917] nitrite oxidation,[918] potassium permanganate,[919,1632] chromium oxidation,[920,921] ozonolysis of the anion,[922] and titanium trichloride.[906,923–925]

2.12.2.3. Imines

The substituted imine **56** has been used to prepare ketones in a manner analogous to that described for aldehydes (Sect. 2.12.1; Scheme 48).[853] A related system consists of *N*-benzylidenebenzylamines $ArCH{=}NCH_2Ar'$ **(65);** these have been alkylated, but the position of reaction depended upon the substituents.[926] The *t*-butylhydrazones **(66)** were deprotonated by *n*-butyllithium and reacted cleanly with carbonyl compounds (Scheme 54). Alkylation also occurred at carbon, with the exception of iodomethane which gave predominately *N*-alkylation;[927] conjugate addition took place with α,β-unsaturated esters and nitriles.[928]

$$t\text{BuNHNH}_2 + \text{R'CHO} \xrightarrow{\text{AcOH}} \underset{\textbf{66}}{t\text{BuNHN{=}CHR'}} \xrightarrow[\text{3) BuLi 4) H}_2\text{O}]{\text{1) BuLi/THF, 0° 2) R}_2\text{CO}} \tag{54}$$

$$t\text{BuNH—N{=}CR'—CR}_2\text{—OH} \xrightarrow{\text{H}_3\text{O}^+} \text{R'CO—CR}_2\text{—OH}$$

The usual mode of deprotonation of an imine provides the allyl anion. The vinyl anion has been prepared by the addition of alkyllithiums to isocyanides (Scheme 55).[929] This anion has been reacted with a variety of electrophiles: deuterium oxide,[930] carbon dioxide,[930,931] boron compounds,[932–934] and alkyl halides, epoxides, ethyl chloroformate, and aldehydes.[931] Most studies have employed *t*-butyl isocyanide.

$$\text{R—Met} + \text{R'NC} \longrightarrow \text{R'N{=}CR—Met} \xrightarrow{\text{El}^+} \tag{55}$$

$$\text{R'N{=}CR—El} \xrightarrow{\text{H}_3\text{O}^+} \text{RCO—El}$$

2.12.2.4. Cyanohydrins

Cyanohydrins[409,795,861,935,936] have found widespread use as acyl anion equivalents. The presence of an aryl group increased the acidity of the "aldehyde" proton and allowed weaker bases to be used than for the alkyl analogs. The masked carbonyl compound **67** has usually been prepared from the aldehyde,[937] and various groups have been used to protect the hydroxyl group. Examples of protecting groups include ethoxyethyl,[938–939] tetrahydropyranyl,[940] ethyl,[458,941–943] and esters.[944–947] The groups that have found most favor are ethoxyethyl and trimethylsilyl. These latter compounds (**67;** R′ = Me_3Si) have been prepared from the cyanohydrin, but the most common method was treatment of the aldehyde with trimethylsilylcyanide.[948–955]

$$\underset{\textbf{67}}{\text{R'O—CHR—CN}} \xrightarrow[\text{2) El}^+]{\text{1) base}} \text{R'O—}\overset{\text{El}}{\underset{\text{CN}}{\text{C}}}\text{—R} \xrightarrow{\text{H}_3\text{O}^+} \text{R—CO—El} \tag{56}$$

Also, hydrogen cyanide may be added to a silyl enol ether.[956,957] *O*-Silylcyanohydrins have only been used to prepare aryl ketones, and electrophiles that have been successful are alkyl halides,[953–955] tosylates,[954,955] allyl halides,[958] an aminating reagent,[959] and carbonyl compounds.[960,961] Conjugate addition occurred with enones,[962–964] although the solvent[962] and electronic properties of the aryl group[963] had a significant influence on the reaction outcome. Similar results were obtained with enals and α,β-unsaturated esters.[965] A variation has been to use the triscyanohydrin from tris(cyano)methylsilane.[966]

Various derivatives have been used to prepare cyclic ketones;[939,945] the majority of examples involve the ethoxyethyl compound (**67**; R′ = —CHMe—OEt). The ring size can range from three[939] to macrocyclic[967–974] (Scheme 57). The ethoxyethyl group allowed the synthesis of alkyl ketones, and electrophiles have included carbonyl compounds,[975] alkyl halides,[872,976–978] and conjugate additions to enones[979–981] and unsaturated sulfoxides.[982]

EtO O CN X n —Base→ EtO O CN n —H_3O^+→ O n (57)

Ketones have also been prepared from cyanohydrins by *O*-alkylation followed by a [2,3] sigmatropic rearrangement.[983–985] In addition, cyanohydrins have provided a means for using the natural polarity of the nitrile group as an acyl anion equivalent.[400–402,404] The transformation nitrile → cyanohydrin was achieved by reaction of the anion with an oxygen[406–408] or halogen[405] electrophile.

2.12.2.4.1. Cyanide- and Thiazolium-Catalyzed Condensations. The classic benzoin reaction may be considered as an *in situ* formation of a cyanohydrin and subsequent acyl anion formation.[986–989] The acyloin reaction can be regarded in a similar manner.[990]

Aromatic aldehydes added to a variety of α,β-unsaturated compounds, including nitriles, esters, and ketones, in the presence of cyanide ion in a polar solvent.[991–1000] The same reaction has been accomplished with a thiazolium salt, such as 3-benzyl-5-(2-hydroxyethyl)-4-methyl-1,3-thiazolium chloride, as catalyst.[1000–1002] The advantages of this catalyst are: the method is not limited to aryl derivatives,[1003–1006] it can be polymer supported,[1007] and saturated aldehydes have been used as the electrophile.[1008–1019] Intramolecular cyclizations have been achieved with this catalyst.[1020–1024] α-Ketoacids have been employed in place of aldehydes and provided good yields of the same adducts as decarboxylation occurred.[1025]

2.12.2.5. Other Methods

2-Oxazolin-5-ones (**68**) have been used as acyl anion equivalents as shown in Scheme (58).[1026–1027]

$$\mathrm{R{-}CH(COOH)(NHCOR^1)} \xrightarrow{Ac_2O} \mathbf{(68)} \xrightarrow[\text{i-PrNEt}]{R^2X/DMF} \xrightarrow[\text{2) Pb(OAc)}_4\text{, THF/HMPA}]{\text{1) } H_2O} \mathrm{RCOR^2} \quad (58)$$

Symmetrical ketones have been prepared from the hippuric acid analog (**68; R = H**).[1028] In addition to reactive halides, the system has been condensed with α,β-unsaturated compounds; although reaction occurred at a different position, an acyl anion equivalent was still produced (Scheme 59).[1029–1032]

$$\xrightarrow[NEt_3]{CH_2{=}CHY} \xrightarrow[H_3O^+]{\text{NaOH or}} \mathrm{RCOCH_2CH_2Y} \quad (59)$$

Reaction of a 1,1-dichloroalkane with two equivalents of pyridine, followed by hydrolysis with deuterium oxide, gave the deuterioaldehyde.[1033] The same product resulted from treatment of an α-ketoacid with an anhydride in pyridine followed by deuterium oxide.[1034,1035]

2.13. OXYGEN COMPOUNDS

2.13.1. Formyl Anion Equivalents

1,3-Dioxolane **(69)** underwent addition to activated alkenes by a radical reaction in the presence of oxygen.

(69)

The outcome, however, was the same as if **69** had acted as a formyl anion equivalent.[1036] An acetal anion has been prepared by transmetallation of a tin compound **(70)** (Scheme 60).[1037,1038] Although alkylation gave the aldehyde, condensation with benzaldehyde led to methyl phenylacetate upon hydrolysis.

$$\mathrm{Bu_3SnH} \xrightarrow[(EtO)_2CHOPh]{iPrMgCl} \underset{\mathbf{70}}{\mathrm{Bu_3SnCH(OEt)_2}} \xrightarrow[\text{2) El}^+ \text{ 3) } H_3O^+]{\text{1) BuLi/THF, }-78^\circ} \mathrm{El{-}CHO} \quad (60)$$

2.13.2. Acyl Anion Equivalents

Although not generally applicable as acyl anion equivalents, the 1,3-dioxolane derived from *p*-oxazolinylbenzaldehyde was deprotonated by *n*-butyllithium and condensed with alkyl halides or carbonyl compounds.[1039]

2.13.2.1. Carboxylic Acid Derivatives

The natural polarity of a carboxylic acid, $^{-}$C—COOH, has been put to use as an acyl anion equivalent by first reacting the dianion from RCH_2COOH with R′X, followed by oxidation of the α-carbon atom and decarboxylation. This has been achieved by reaction of the acid dianion with oxygen (Scheme 61).[410,412,1040–1042] a sulfur electrophile,[816,819,1043] or formylation.[411,1044] β-Diesters may be converted to the ketone *via* the diacid by electrolysis.[1045]

$$RR^1CH\text{—}COOH \xrightarrow[THF/0°C]{2LDA} RR^1C^{-}\text{—}COO^{-} \xrightarrow[2) Et_2O/O_2]{1) \rightarrow -78°C} RR^1C(O_2^{-})COO^{-} \xrightarrow{Me_2NCH(OMe)_2/CH_2Cl_2} [\text{cyclic peroxide, NMe}_2] \longrightarrow RR^1C{=}O \qquad (61)$$

An alternative method was to convert the acid derivative to an alkene and to unmask the ketone by ozonolysis.[396,397]

The lithium acyloin enediolate **(71)** has been used to prepare ketones (and aldehydes) by the route shown in Scheme 62.[1046,1047] As the authors state,[1046] "this method constitutes a detour but facile way to the synthesis of ketones."

$$RCOOR' \xrightarrow[Me_3SiCl]{Na} Me_3SiO\text{—}CR{=}CR\text{—}OSiMe_3 \xrightarrow[DME,\ rt]{MeLi} {}^{-}O\text{—}CR{=}CR\text{—}O^{-}\ (\mathbf{71})$$

$$\mathbf{71} \xrightarrow{R''X} RCO\text{—}CRR''\text{—}OH \xrightarrow[MeOH]{NaBH_4} HO\text{—}CHR\text{—}CRR''\text{—}OH \xrightarrow[Et_2O]{PbTA} R\text{—}CO\text{—}R''\ (+\ RCHO) \qquad (62)$$

2.13.2.2. Acylsilanes

Acylsilanes,[771] sometimes referred to as α-silylketones **(72)**, have enjoyed some success as acyl anion equivalents. They have been prepared by a variety

of methods, including silylation of 1,3-dithiane[475,476] (see Sect. 2.2.1), but were susceptible to hydrolysis to give the aldehyde.[1048] Arylacylsilanes have been condensed with alkyl halides and carbonyl compounds in the presence of fluoride ion (Scheme 63).[1049–1051]

$$\underset{\mathbf{72}}{ArCOSiMe_3} \xrightarrow[\text{18-crown-6 or DMSO or HMPA}]{\text{RX, KF}} ArCOR \qquad (63)$$

2.14. VINYL COMPOUNDS

2.14.1. Vinyl Ethers

Methyl vinyl ether (**73,** R = Me; Scheme 64) was deprotonated with *t*-butyllithium at $-65°$ to give the acyl anion equivalent (**74,** R = Me).

$$\underset{\mathbf{73}}{CH_2{=}CH{-}OR} \xrightarrow{t\text{BuLi}} \underset{\mathbf{74}}{CH_2{=}C(OR)(Li)} \xrightarrow{El^+}$$

$$CH_2{=}C(OR)(El) \xrightarrow{H_3O^+} CH_3CO{-}El \qquad (64)$$

The vinyllithium **74** has been reacted with a wide variety of electrophiles; a clean reaction was observed with alkyl halides, carbonyl compounds, and an allyl bromide.[1052] Benzonitrile and benzoic acid gave 1-phenyl-1,2-propanedione, after acidic hydrolysis, but the major product obtained from methyl benzoate was derived from the addition of two equivalents of **74.** Reaction of **74** with boranes led to ketones after oxidative work-up.[1053] Although 1,2-addition was observed with α,β-unsaturated ketones,[1052] 1,4-addition occurred with the cuprate.[1054] The vinyl ether was hydrolyzed to the carbonyl compound by protic acid; reaction with a peracid gave the α-hydroxyketones.[1055]

Ethyl vinyl ether (**73;** R = Et) has been employed in a manner analogous to that described for methyl vinyl ether.[1056,1057] The addition of the cuprate to enones was, however, more susceptible to steric effects in the α,β-unsaturated ketone.[267,1058] The reaction with an epoxide was promoted by boron trifluoride etherate.[1059]

An attempt to use phenyl vinyl ether (**73;** R = Ph) resulted in metallation of the aromatic *ortho*-position, although the vinyl anion was formed by a second deprotonation.[1060]

In addition to vinyllithiums, other coupling reagents have been employed. The zinc anion of **73** (R = Et) coupled with aryl and vinyl halides[1061,1062] in

the presence of a palladium catalyst. This approach has been extended to propargylic ethers which are $C{=}C{-}CO^-$ equivalents (see Sect. 2.17.1.2.1.). Allyl ethers gave the vinyl anion when treated with butyl potassium.[1063] Reaction of the vinylstannane **(75)** with an acid chloride in the presence of a palladium(II) catalyst gave, after acid hydrolysis, the dione (Scheme 65).[265]

$$CH_2{=}C(OMe)(SnMe_3) \xrightarrow[2)\ H_3O^+]{1)\ RCOCl,\ Pd^{2+}} R{-}CO{-}CO{-}Me \qquad (65)$$

75

Cyclic vinyl ethers have also been used as acyl anion equivalents.[1064–1066] Hydrolysis in these cases resulted in the formation of ω-hydroxycarbonyl compounds, the position of the hydroxyl group being determined by the ring size of the cyclic vinyl ether[1067,1068] (see Sect. 2.17.4.). The method was used as a basis for the synthesis of *endo*-brevicomin[1069] and spiroketals.[1070]

2.14.2. Vinylsulfides

The property of sulfur to stabilize an α-anion led to initial problems with the addition of the alkyllithium across the double bond of the vinylsulfide.[1071,1072] Careful choice of base and solvent systems allowed formation of the required vinylanion **(76)**; systems that have been successful are *s*-[1073] and *n*-butyllithium,[1074,1075] or LDA[1076,1077] in THF-HMPA and LDA in hexane.[1078] The vinyl anion **76** has also been prepared by tin–lithium exchange,[1079] addition of an alkyllithium to a thioketene,[1080] displacement of a sulfur group from a ketene thioacetal,[695,1081] and addition to an acetylene.[1082–1085]

$$CH_2{=}CHSPh \xrightarrow{Base} CH_2{=}C(SPh)(Li) \xrightarrow{El^+} CH_2{=}C(SPh)(El) \xrightarrow[H_2O]{Hg^{2+}} CH_3CO{-}El \qquad (66)$$

76

Vinylsulfides have been prepared from α-silylsulfides by the Peterson reaction[253,767,771,773,774] and by elimination of thiophenol from bisthioacetals.[658,697,722,1086–1090] Other methods have been based on the Wittig reaction (see Sect. 2.1.),[383,387,688,1091–1097] *N*-tosyl-hydrazones,[1098–1100] and α-thioesters.[1098,1101]

In a manner similar to vinyl ethers, the sulfur analogs have been coupled with vinyl and aryl halides.[1062] A silyl group has been displaced to give the carbanion adduct when treated with fluoride in the presence of an aldehyde

(Scheme 67).[1102] A 1,2-dithioalkene has been used as a type of acyl anion equivalent, but hydrolysis was not carried out.[1103]

$$CH_2{=}C(SPh)(SiMe_3) \xrightarrow[PhCHO]{Bu_4NF} CH_2{=}C(SPh)(CHOH{-}Ph) \quad (67)$$

Vinylsulfides have been hydrolyzed to the carbonyl compounds by methods that have been used for thioacetals (see Sect. 2.2.3.),[495,1104–1109] such as mercuric chloride in aqueous acetonitrile.[1073]

A chiral vinylsulfoxide has been employed in a synthesis of α-tocopherol.[1110] Although 1-bromo-1-phenylthio-2-ethoxyethene underwent halogen–metal exchange and reacted as a vinylsulfide, the oxygen group converts this to a $^{-}$C—CHO synthon.[1111]

2.14.3. Vinylsilanes

Unlike vinyl ethers and sulfides, vinylsilanes are not readily hydrolyzed. In addition to preparing carbonyl compounds, vinylsilanes have found widespread use for the regio- and stereochemical control of many electrophilic reactions.[327,345,1112] As acyl anion equivalents, the required anion has been prepared by metal–halogen exchange (Scheme 68).[259,1113–1114]

$$R_2C{=}CBr{-}SiR'_3 \xrightarrow[2)\ El^+]{1)\ Mg\ or\ n\text{-}\ or\ t\text{BuLi}} \underset{\mathbf{77}}{R_2C{=}C(El){-}SiR'_3} \xrightarrow[2)\ H_3O^+]{1)\ MCPBA} R_2CH{-}CO{-}El \quad (68)$$

The vinyllithium or Grignard reagents have been reacted with alkyl halides,[255,284,1079,1115–1118] often with copper(I) catalysis, carbonyl compounds,[1115,1119–1125] α,β-unsaturated carbonyl compounds,[267,1118,1126,1127] acid anhydrides,[1128] and epoxides.[1113,1129] The vinylanions have also been prepared by the addition of organometallic reagents to acetylenes.[271,274,287,1062,1130–1133] The use of boron and aluminium reagents to achieve this allows coupling reactions not readily accessible to the transmetallation procedure. Other anionic methods for the preparation of vinylsilanes involved the elimination of selenium from β-hydroxyselenides,[1134] the Peterson elimination[773,774] from bissilylanions,[253,534,772,1135] and the use of allylsilanes.[57,249,293,294,1136,1137]

The conversion of the vinylsilane **(77)** to the carbonyl compound was achieved by oxidation with a peracid and acid hydrolysis of the intermediate α,β-epoxy silane.[267,327,771,774,1112,1120,1122,1138–1140] The stereochemical constraints of the elimination limit this approach to acyclic cases.[1141–1144] Oxidative procedures have been used with alkoxysubstituted silyl groups.[1145]

2.14.4. Vinylselenides

The vinylanion **(78)** has been prepared by reaction of the vinylselenide **(79)** with LDA[1146,1147] or KDA.[1148] An alternative procedure is the transmetallion of a selenide with an alkyllithium[1079,1149] (Scheme 69).[1150] Reaction occurred in high yield with alkyl halides, epoxides, carbonyl compounds, and DMF.[1148,1149] The conversion to the carbonyl compound was accomplished by mercury(II) catalyzed hydrolysis.[374,394,1150]

$$\underset{\mathbf{79}}{R_2C{=}CH{-}SeR'} \xrightarrow[\text{THF, }-78^\circ]{\text{LDA or KDA}} \underset{\mathbf{78}}{R_2C{=}C(SeR')Li(K)} \xrightarrow{El^+} R_2C{=}C(SeR')El \quad (69)$$

2.14.5. Nitrogen Compounds

One aspect of these compounds has already been discussed in Section 2.12.2.3. (imines). The position of deprotonation in an enamine was influenced by other substituents,[1151–1154] but formation of the vinylanion was not the usual course of reaction. This problem has been overcome by use of a hydrazone (Scheme 70).[1155] Yields of carbonyl compounds were, however, often no more than moderate.

$$RCH_2CO{-}NHNHSO_2Ar \xrightarrow[\text{2) }HNR'_2]{\text{1) }PCl_5} RCH_2C(NR'_2){=}NNHSO_2Ar \xrightarrow[\text{2) }10^\circ\text{ 3) }El^-\text{ 4) }H^+]{\text{1) }t\text{BuLi/THF, }-78^\circ} RCH_2CO{-}El \quad (70)$$

2.15. ORGANOMETALLICS

2.15.1. Formyl Anion Equivalents

The addition of the formyl anion equivalent HCO^- from carbon monoxide has been accomplished with various transition metals, in particular iron (Scheme 71),[1156–1162] palladium,[1163] nickel,[1164] and manganese.[1383]

$$Fe(CO)_5 \xrightarrow[\text{or 1) Na—Hg/THF 2) RBr, }Ph_3P\text{ 3) AcOH}]{\text{1) KOH 2) Amberlyst A-26 3) RX/THF}} RCHO \quad (71)$$

2.15.2. Acyl Anion Equivalents

2.15.2.1. Acyl Anions

These important species[1165] have been prepared by the addition of an alkyllithium to carbon monoxide, but their high reactivity has made their use

fraught with difficulties, as many side reactions can result.[1166,1167] The reaction depended upon the alkyllithium employed; *t*-butyllithium has often given the best results. These acyl anions have been reacted with chlorotrimethylsilane,[1168–1170] carbonyl compounds,[1171–1173] carbon disulfide,[1174] and isocyanates.[1175] Grignard reagents have also led to carbonyl compounds by reaction with carbon monoxide.[1176]

2.15.2.2. Other Systems

Many of these methods have been based on the addition of two alkyl halides to carbon monoxide by use of a transition metal catalyst[1177] as shown for iron in Scheme 72.[1178–1185] In addition to alkyl halides, alkenes have been used[1186] in the condensation and have given a route to cyclic ketones.[1187] Palladium provided a route to α-ketoamides.[1381]

$$Na_2Fe(CO)_4 \xrightarrow[2)\ R'X]{1)\ RX} R\text{—}CO\text{—}R' \qquad (72)$$

Nickel carbonyl has been reacted with an alkyllithium to give, after conjugate addition to an α,β-unsaturated ketone,[1188,1189] or an acetylene,[1190] a 1,4-dicarbonyl compound (Scheme 73).

$$RLi \xrightarrow{Ni(CO)_4} [RCONi(CO)_3]^- \xrightarrow{R'CH{=}CHCOR''} R''COCH_2\text{—}CHR'\text{—}COR \qquad (73)$$

Metallic nickel has been used to couple benzyl and acid halides[1191]; benzyl halides led to the 1,3-diarylpropan-2-one with nickel and oxalyl chloride monoester.[1192] In a similar manner, aryl and alkyl halides coupled under palladium catalysis in the presence of a zinc–copper couple,[1193] organotin,[1194,1195] or organozinc compounds.[1196–1198]

α-Mercural acetates reacted with enones under reducing conditions to yield the γ-acetylketones. As the mercural was derived from a ketone, umpolung had occurred but an alcohol derivative resulted.[1199]

Reaction of beryllium with an acid halide leads to an intermediate that reacted with carbonyl compounds and acid halides as an acyl anion.[1200] Boron has been used in analogous reactions.[1201,1202]

2.16. MISCELLANEOUS METHODS

2.16.1. Formyl Anion Equivalents

As with the sulfur and selenium compounds, tellurium acetals have been used to prepare aldehydes; unmasking was achieved with iodine.[1203] Even the boron analog of this type of compounds has been used as a formyl anion equivalent.[1204] Phosphorus ylides have been reacted with 1,*n*-dihaloalkanes and then

oxygen to produce the cyclic ketone in a moderate yield.[1205] Tetraethyl methanediphosphonate has been used as a formyl anion equivalent with carbonyl compounds relying on oxidation of the intermediate vinylphosphonate.[1206]

Dichloromethyllithium has been reacted with carbonyl compounds to give, after reaction with base or lithium chloride, the α-hydroxy- or α-chloroaldehydes,[1207–1212] respectively. Electrolysis of chloroform and carbon tetrachloride in the presence of an aldehyde gave the same adducts.[1213]

As mentioned previously (see Sect. 2.1.), the Wittig reaction and ozonolysis[398] may be used to prepare aldehydes by formal umpolung reagents. The Peterson reaction has been used to prepare vinylboranes, which were oxidized to give the aldehyde.[1214,1215]

Two systems that do not act as true formyl anion equivalents, but have proven useful, are α-halosilanes[255,1216–1219] and methoxymethyltrimethylsilane.[1216,1220]

One reaction that has not received much attention for umpolung is the addition of an organometallic reagent to a thioketone (C=S), as the addition occurs in the opposite manner to a ketone[1221] (see Sect. 2.3.2.1.).

2.16.2. Acyl Anion Equivalents

The *gem*-diboron system has been extended from a formyl (Sect. 2.16.1.) to an acyl anion equivalent.[1204] Diethyl α-trimethylsilyloxy phosphonates **(80)** have been deprotonated with LDA. Alkylation proceeded smoothly with primary alkyl halides;[1222–1224] the ketone was unmasked by treatment with base (Scheme 74). Carbonyl compounds also reacted with the acyl anion equivalent, and the α-trimethylsilyloxyketone could be obtained directly by silicon migration.[1225,1226]

$$\underset{\mathbf{80}}{Me_3SiO—CHR—PO(OEt)_2} \xrightarrow[2)\ El^+]{1)\ LDA/THF,\ -78°}$$

$$Me_3SiO—CR(El)—PO(OEt)_2 \xrightarrow[EtOH]{NaOH/H_2O} R—CO—El \qquad (74)$$

Sequential reaction of electrophiles with the anion derived from diethyl (trimethylsilylethoxymethyl)phosphonate, followed by desilylation and dephosphorylation, led to carbonyl compounds.[1227]

Geminal dihalocompounds have found some use as acyl anion equivalents[1228] (see Sect. 2.16.1.).

Acyloins have been deprotonated by reaction with sodium hydroxide in DMSO. Condensation occurred with various halides[1229–1231] and with Michael acceptors.[1231] The carbonyl group was unmasked by oxidative cleavage.

Although they do not involve anions, two interesting reactions are the photochemical condensation of aldehydes with electron-deficient alkenes to

give the ketone.[1232] The transformation has been achieved[1233] from the acid anhydride rather than the aldehyde in the presence of Vitamin B_{12}.

2.17. FUNCTIONALIZED ACYL ANION EQUIVALENTS

This section discusses acyl anion equivalents that contain another functional group. Many variations are, of course, possible and only acyl anion equivalents where the two functional groups under consideration interact in this moiety have been included.

2.17.1. α,β-Unsaturated Acyl Anions (C=C—CO⁻ and C≡C—CO⁻)

2.17.1.1. By Protection

The acetylenic 1,3-dioxolane **(81)** was deprotonated by *n*-butyllithium in THF at the acetal carbon. Reaction with an electrophile followed by acid hydrolysis gave the substituted ynone (Scheme 75).[1234,1235]

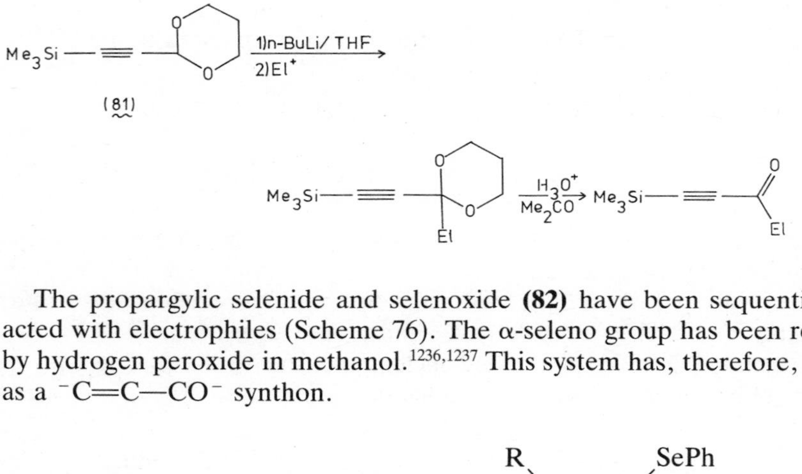

$Me_3Si—≡—$(dioxane) $\xrightarrow[2) El^+]{1) n\text{-BuLi/THF}}$ $Me_3Si—≡—C(El)$(dioxane) $\xrightarrow[Me_2CO]{H_3O^+}$ $Me_3Si—≡—C(=O)El$ (75)

(81)

The propargylic selenide and selenoxide **(82)** have been sequentially reacted with electrophiles (Scheme 76). The α-seleno group has been removed by hydrogen peroxide in methanol.[1236,1237] This system has, therefore, reacted as a ⁻C=C—CO⁻ synthon.

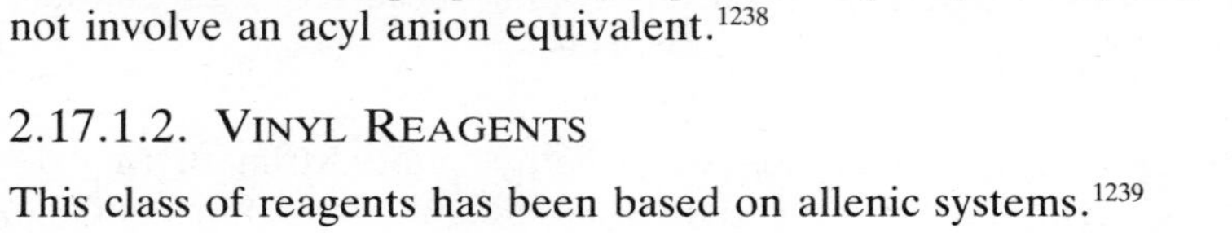

$$Ph—Se(O)_n—CH_2C{\equiv}CH \xrightarrow[2)\ RX\ 3)\ El^+]{1)\ 2LDA/THF,\ -78^\circ} RCH{=}C(SePh)(CO—El) \quad (76)$$

82

n = 0 or 1.

Enals have been prepared by a protected phosphorus ylide, but this does not involve an acyl anion equivalent.[1238]

2.17.1.2. Vinyl Reagents

This class of reagents has been based on allenic systems.[1239]

2.17.1.2.1. Allenic Ethers. Protected allenic alcohols **(83)** were deprotonated α to the oxygen by *n*-butyllithium; alkylation and deprotection with acid gave the enone (Scheme 77).[1240,1241]

$$HC{\equiv}CCH_2{-}OR \xrightarrow{KOBu\text{-}t} \underset{\mathbf{83}}{CH_2{=}C{=}CH{-}OR} \xrightarrow[2)\ El^+]{1)\ BuLi/Et_2O,\ -30^\circ} CH_2{=}C{=}C(OR)(El) \xrightarrow{H_3O^+} CH_2{=}CH{-}CO{-}El \qquad (77)$$

The alcohols were available from the propargylic ether by reaction with potassium *t*-butoxide.[1240] In addition to alkyl halides,[1242] carbonyl compounds[1241,1243] and α,β-unsaturated ketones[1243] (1,2-addition) have been used as electrophiles. Coupling with aryl halides was achieved in the presence of a palladium catalyst.[79,1061]

The presence of an alkyl[1244,1245] or silyl group[1246] on the acetylene led to alkylation α to this group. A second alkylation did, however, occur α to the oxygen atom (Scheme 78), the system having reacted as a $^{-}C{=}C{-}CO^{-}$ synthon.[1247,1248] The system has been extended to the cumulated triene.[1249,1250]

$$R{-}C{\equiv}CCH_2{-}OR' \xrightarrow[2)\ R''X]{1)\ BuLi/TMEDA} RR''C{=}C{=}CH{-}OR' \xrightarrow[2)\ El^+]{1)\ BuLi} RR''C{=}C{=}C(El){-}OR' \qquad (78)$$

2.17.1.2.2. Allenic Sulfides. This class of compounds reacted in a similar manner to the oxygen ethers. The notable difference was that the presence of a group on the γ-carbon did not affect the position of reaction; this was α to the sulfur.[1251–1254] The sulfoxide gave similar but sluggish reactions.[1255] Allenic thioacetal monoxides **(84)** added nucleophiles to give the enal derivative (Scheme 79) in a manner analogous to ketene thioacetals.[1256,1257]

$$\underset{\mathbf{84}}{R_2C{=}C{=}C(SO{-}Ph)(SPh)} \xrightarrow[2)\ H_2O]{1)\ Nu} R_2C{=}C(Nu){-}CH(SO{-}Ph)(SPh) \qquad (79)$$

Allenic sulfides carrying an additional alkoxy group may lead to some interesting variations; reaction with an electrophile occurred α to the sulfur,[1258–1260] producing $CO{-}C{-}CO^{-}$ equivalents.

2.17.1.2.3. Other Systems. Reaction of 1-diethylamino-3-trimethylsilyl-1-propyne with *n*-butyllithium and a carbonyl compound gave, after aqueous

hydrolysis, a mixture of products, mainly from addition α to the silicon.[1261] The unsubstituted allene **(85)** did, however, react in the required manner to give C=C—CO⁻ and ⁻C=C—CO⁻ equivalents (Scheme 80).[1262]

$$(EtO)_2PO\text{—}NHMe \xrightarrow[\text{2) } BrCH_2C\equiv CH \text{; 3) } NaH/\Delta]{\text{1) NaH}} \underset{\mathbf{85}}{(EtO)_2PO\text{—}NMe\text{—}CH=C=CH_2} \xrightarrow[\text{2) RX}]{\text{1) BuLi}}$$

$$(EtO)_2PO\text{—}NMe\text{—}CR=C=CH_2 \xrightarrow{H_3O^+} R\text{—}CO\text{—}CH=CH_2 \qquad (80)$$

$$\downarrow \text{1) BuLi; 2) R'X}$$

$$(EtO)_2PO\text{—}NMe\text{—}CR=C=CHR' \xrightarrow{H_3O^+} R\text{—}CO\text{—}CH=CHR'(E)$$

Enals have been synthesized by condensation of a vinyl Grignard reagent and carbonyl compound followed by thermal rearrangement of the derived thiocarbamate.[1263]

2.17.1.3. Allyl Reagents

The majority of systems that have been used in this context rely on the position of alkylation of an allyl anion. As this can vary between electrophiles with any one system, there is a large degree of overlap between the C=C—CO⁻ and ⁻C—C—CO synthons (see Chapt. 6).

2.17.1.3.1. 1,3-Dithianylketene Thioacetals. These acetals **(10)** (see Sect. 2.2.1.1.) were readily deprotonated to yield the allyl anion. Reaction with alkyl halides (Scheme 81) at the α-position gave, after hydrolysis, the enone.[489,499,500,1264–1266]

(10) —[1) n-BuLi/THF; 2) El⁺]→ 2-El-2-(R¹-vinyl)-1,3-dithiane —[Hg^II, H₂O]→ El—CO—CH=CH—R¹ (81)

By contrast, carbonyl compounds reacted at the γ-carbon.[502] The regioselectivity can be influenced greatly by the presence of a second anion-stabilizing group within the system.[446,1267–1269] The ratio has been correlated to the hardness of both the leaving and alkyl groups of the alkylating agent.[1270] Addition to enones resulted in 1,4-addition through the γ-position,[1271,1272] but this trend was reversed, giving the 1,4-α-adduct when the cuprate or HMPA was used.[1273] Reaction with a vinylsulfone gave the α-adduct.[1274] By contrast, 2-(2-trimethylsilylethynyl)-1,3-dithiane reacted with electrophiles α to the dithiane providing the acetylene product.[1275]

An alkyllithium was added to 1,1-dithio-1,3-diene, followed by reaction with an electrophile to provide the masked enone.[1276] In this case the diene acted as a $^+C{-}C{=}C{-}CO^-$ synthon.

2.17.1.3.2. Other Sulfur Reagents. A Wittig variation utilizing ketenethioacetals (Sect. 2.3.1.1.1.) has been used to synthesize enals.[1277] Phenyl allyl sulfide was deprotonated with *n*-butyllithium and alkylation occurred at the α-position.[1278,1279] The Pummerer reaction could then be used to afford the carbonyl compound (see Sect. 2.10.1.). The allyl *N,N*-dimethyldithiocarbamate was also alkylated at the α-position, but its removal involved a rearrangement that led to a $^-C{-}C{-}CHO$ synthon.[1280] Reaction of the allyl anion derived from 1-phenylthio-1-trimethylsilyl-1-propene occurred at the γ-position with a carbonyl compound but subsequent deprotonation and alkylation occurred at the α-position.[1281] Alkylation of 1-methoxy-1-phenylthio-2-propene **(86)** occurred at the α-position (Scheme 82); a subsequent rearrangement reaction allowed this system to act as a $^-C{=}C{-}CO^-$ or $^{2-}C{=}C{-}CO$ synthon.[1282–1284]

$$\underset{\mathbf{86}}{CH_2{=}CH{-}CH(OMe){-}SPh} \xrightarrow[\text{2) RX}]{\text{1) LDA/THF, }-78°} CH_2{=}CH{-}CR(OMe){-}SPh \qquad (82)$$

The anion derived from 1,3-bis(methylthio)propene acted as a $^-C{=}C{-}CHO$ synthon[1279,1285,1286] as did a selenium analog.[1287]

2.17.1.3.3. Cyanohydrins. The *O*-trimethylsilylcyanohydrins **(87),** prepared from the enal and trimethylsilylcyanide,[948,953,1288] were readily deprotonated by LDA. Reaction with alkyl halides[953,1288] or carbonyl compounds[965,1289–1291] occurred at the α-position (Scheme 83). Reaction with a silicon electrophile gave α- and/or γ-adducts.[1292–1294] The α-adduct from an allyl halide has been converted to the γ-adduct by thermal rearrangement.[1295]

$$RCH{=}CH{-}CHO \xrightarrow[\text{ZnI}_2]{\text{Me}_3\text{SiCN}} \underset{\mathbf{87}}{RCH{=}CHCH(CN)(OSiMe_3)} \xrightarrow[\text{2) El}^+]{\text{1) LDA/THF, }-78°} RCH{=}CH{-}C(CN)(El)(OSiMe_3) \qquad (83)$$

Alkyl- and ethoxyethyl-protected cyanohydrins also showed the tendency to undergo α-reactions with alkyl halides,[938,1296] epoxides,[1297] and enones.[1298] The approach has been used in the synthesis of macrolides.[970,973]

2.17.1.3.4. α-Aminonitriles. Allyl anions derived from unsaturated α-aminonitriles have been alkylated. The position of alkylation depends upon the size of the alkylating agent and groups on nitrogen.[1299] Although steric requirements normally favored γ-alkylation, α-alkylation was achieved in a few isolated cases.[1300,1301] Reaction at the α-site was promoted for carbonyl compounds by the presence of zinc chloride.[1302] The problem has been circumvented by the use of 4-phenylthio-2-*N,N*-dimethylaminobutyronitrile.[1303]

2.17.1.3.5. Other Systems. The 1,1-dichloroallyl anion was prepared by displacement of a lead group; reaction with unhindered carbonyl compounds occurred at the α-position.[1304]

2.17.2. Dicarbonyl Compounds

2.17.2.1. 1,2-Dicarbonyls

In these systems the two functional groups work together to stabilize the α-anion. Indeed, many syntheses of these systems rely on reaction of an enolate,[1305,1306] with subsequent predictable reactions.[389,1307] The required CO—CO⁻ equivalents were generated under mild conditions and were alkylated readily[448,474,1308] when the masking group was a bisthioacetal. 1,3-Oxathianes have already been discussed in this context[762] (see Sect. 2.5.2).

The acetal **(88)** underwent conjugate additions to enones and enoic esters as described in Section 2.12.2.4.1. (see Scheme 84).[1309] Enolates of α,α-dichloroketones have been condensed with aldehydes, although conversion to an α-dicarbonyl compound was not carried out.[1310]

$$\underset{\mathbf{88}}{(EtO)_2CH{-}CHO} \xrightarrow[CH_2{=}CHCOR,\ dioxan]{thiazolium\ salt/Et_3N} (EtO)_2CH{-}CO{-}CH_2CH_2{-}COR \quad (84)$$

2.17.2.2. 1,3-Dicarbonyls

Protection[1311] has been used to synthesize 1,3-dicarbonyl compounds by sequential alkylation of the bis-1,3-dithiane **(89)** (Scheme 85).[427]

1-Methylthio-3-methoxy-1-propene (see Sect. 2.17.1.2.2.) was used as a CO—C—CO⁻ equivalent *via* formation of the vinyl anion and alkylation α to the sulfur; hydrolysis with mercury(II) catalysis gave the dione.[1258]

2-Isopropylthio-1-decalones **(90)**[1312] were deprotonated by LTMP to provide the vinyl anion. Conjugate addition of this anion to methyl acrylate gave a new pentannulation procedure (Scheme 86).[1313] The α-oxoketene dithiacetal system[1314–1317] has provided monoacetal protected 1,3-diketones by reduction[1318] and a ⁺C—C=C—CO⁻ synthon by Wittig reaction followed by addition of an alkyllithium.[1319]

1)LiTMP/THF/-78°C; 2)H_2C=CHCOOMe; SPr-i; MeOOC; (90) (86)

Condensation of a β-ketoaldehyde with a secondary amine gave the β-aminoenone, further converted to the vinyl anion with *t*-butyllithium (Scheme 87).[1320]

$$RCO—CH_2CHO \xrightarrow{R'_2NH} RCOCH{=}CHNR'_2 \xrightarrow[El^+]{tBuLi/THF,\ -115°} RCOCH{=}C(El)—NR'_2 \qquad (87)$$

2.17.2.3. 1,4-Dicarbonyls

Again the appropriate bis(1,3-dithiane) has been sequentially alkylated to provide 1,4-dicarbonyl compounds.[526,527] Furan has provided an extremely versatile method to these diones because it can be sequentially reacted at the 2- and 5-positions, providing a ⁻CO—C—C—CO⁻ synthon. In addition to alkyl halides,[1321,1322] electrophiles have included carbon dioxide,[1321,1323] carbonyl compounds,[1321] epoxides,[1324–1327] esters,[1321] acid chlorides,[1321] acids,[1328] and nitriles.[1321] Reaction with oxygen caused dimerization.[1329] The 1,4-dicarbonyl moiety has been unmasked by acid[1322] or cerium ammonium nitrate[1330] (Scheme 88).

1)n-BuLi/THF; 2)RBr; AcOH; H_2SO_4/H_2O; R (88)

The presence of a group that can complex lithium has been shown to modify the position of reaction in the furan ring under certain conditions.[1331,1332] 1,2,5-

Trione systems were also available from protected furfuraldehyde derivatives.[557,1333,1334] Oxidative cleavage of 2,5-disubstituted furans provided a route to conjugated ene-1,4-diones.[1335–1338]

The acyloin from a protected γ-ketoester has been employed as a RCO—C—CO⁻ synthon.[1047] ⁺CO—C—C—CO⁻ chemistry is available via the cyclobutene acyloin as shown in Scheme 89.[1339]

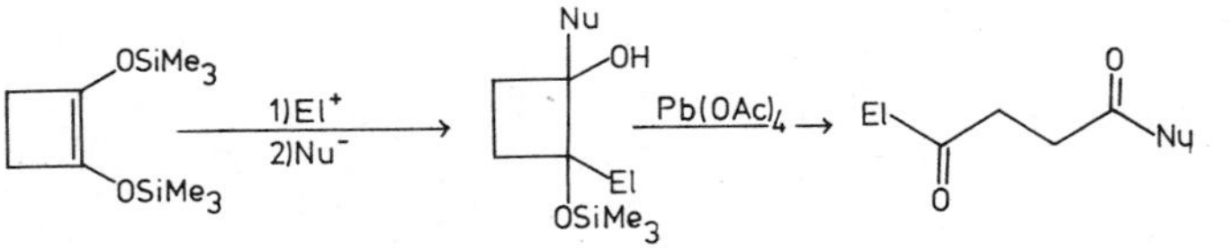

(89)

2.17.2.4. Other Dicarbonyls

These are available from the appropriate 1,*n*-bis-(1′,3′-dithian-2′-yl)alkane.[577]

2.17.3. Ketocarboxylic Acid Derivatives

2.17.3.1. α-Ketocarboxylic Acid Derivatives

Ethyl 1,3-dithianyl-2-carboxylate **(91)** was deprotonated by sodium hydride in DMF; alkylation followed by reaction with NBS gave the α-ketoester (Scheme 90).[1340,1341]

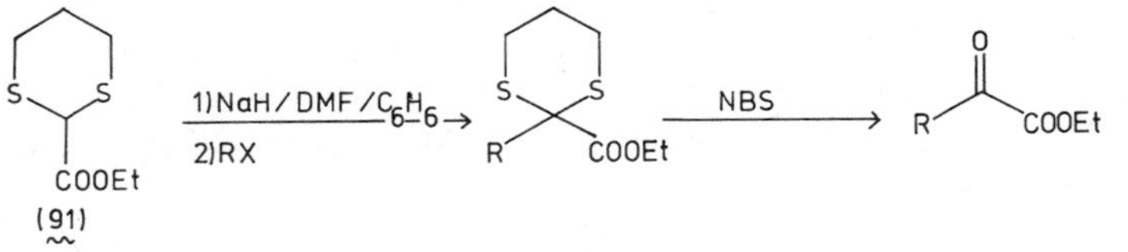

(90)

Alkylation was also achieved under phase-transfer conditions.[1342,1343] Other electrophiles have included acetates,[1344] nitroalkenes,[474] aldehydes,[1345,1346] and α,β-unsaturated aldehydes,[1345] the position of attack being controlled by the counterion. The methyl ester gave 1,4-addition with Δ^2-butenolides.[1347] The approach has been used in a synthesis of brefeldin A.[1268]

The ketene derived from an acid chloride underwent cycloaddition reactions.[1348] The anion of 2-cyano-1,3-dithiane was reacted with alkyl halides[1349] and epoxides.[1350]

The dianion of 2,2-bis(ethylthio)acetic acid, formed with potassium hexamethyldisilazide, was alkylated by alkyl iodides, bromides, and tosylates in high yield.[1351,1352] Reaction also occurred with epoxides and aziridines. The anion of the methyl ester underwent conjugate addition with a wide variety of α,β-unsaturated substrates.[1353–1356] Methyl bis(phenylthio)acetate also gave these Michael adducts.[1357] Compounds of this type were also obtained by

addition of an enolate to methyl 2-methylthioacrylate, followed by reaction with LDA and dimethyl disulfide,[1358] while ethyl 2-lithio-2-carboxy-1,3-dithiolane underwent 1,4-addition with enones.[1359]

The presence of the α-ester group allowed the use of acetals for this class of umpolung reagents. The anion derived from methyl dimethoxyacetate **(92)** was reacted with alkyl halides[1360] or carbonyl compounds[1361] (Scheme 91). Conjugate addition was observed with butenolides.[1347]

$$\underset{\mathbf{92}}{(MeO)_2CH{-}COOMe} \xrightarrow[2)\ El^+]{1)\ LDA/THF,\ -78^\circ} (MeO)_2C(El){-}COOMe \qquad (91)$$

An EtOCO—CO$^-$ synthon was obtained by reaction of 1,2-diethoxy-1,2-ditrimethylsiloxyethene (**93,** Scheme 92) with carbonyl compounds in the presence of zinc chloride. 1,4-Addition occurred with enones.[1362]

$$EtOCO{-}COOEt \xrightarrow[Me_3SiCl]{Na/K} \underset{\mathbf{93}}{(EtO)(Me_3SiO)C{=}C(OSiMe_3)(OEt)} \xrightarrow[2)\ H_3O^+]{1)\ R_2CO/ZnCl_2} HO{-}CR_2{-}COCOOEt \qquad (92)$$

The enolate of *t*-butyl bis(trimethylsilyl)acetate was condensed with carbonyl compounds to provide the α-trimethylsilyl-α,β-unsaturated ester (see Sect. 2.14.3.).[1363] The anion derived from 2,2,2-trifluoroethyl tosylate was reacted with carbonyl compounds to provide, after hydrolysis, the α-keto-acid.[1364]

Esters of dichloroacetic acid were alkylated *via* the enolate,[1365] which was also prepared by reaction of the trichloroacetate with *i*-propylmagnesium bromide.[1366]

2.17.3.2. β-Ketocarboxylic Acid Derivatives

The anion derived from the diethylacetal of 1-methylthio-1-propyn-3-one was alkylated α to the thio group to provide, after mercury(II)-catalyzed hydrolysis, the β-ketoester.[1259]

The dianion of propiolic acid gave a similar synthon, which was reacted with epoxides.[1367,1368] Deprotonation of methyl 3-phenylthio-2-methyl-2-propenoate gave the anion α to sulfur; reaction occurred with aldehydes and acid chlorides and α,β-unsaturated esters underwent conjugate addition to give the substituted cyclopentenones.[1369] This approach was used in a synthesis[1370] of methylenomycin A.

2.17.3.3. Other Ketocarboxylic Acid Derivatives

δ-Ketoacids have been prepared from 1,2-cyclohexanedione (Scheme 93).[1371]

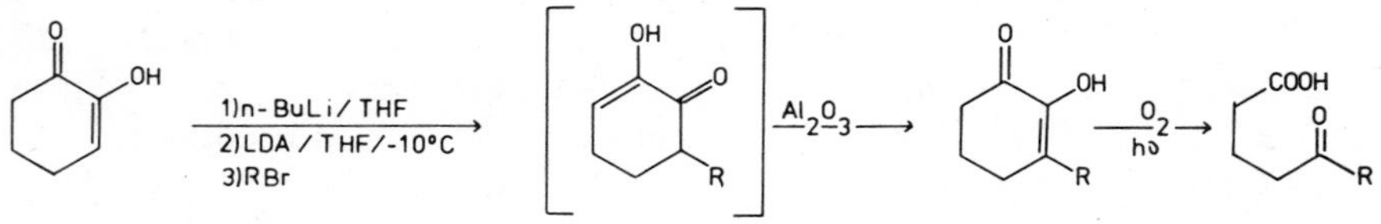

1) n-BuLi/THF; 2) LDA/THF/−10°C; 3) RBr → [ ] → Al_2O_3 → → O_2, hν →

(93)

2.17.4. Oxygen-Substituted Acyl Anions

The dialkoxy ether **(94)**, which contains both an allyl and vinyl ether moiety, gave the vinyl adduct when alkylated (Scheme 94).[1372] An alternative procedure to prepare the anion was displacement of a stannyl group.[1373]

$$\underset{\mathbf{94}}{BuOCH{=}CHCH_2OBu} \xrightarrow[2)\ MeI]{1)\ tBuLi/THF/HMPA} BuO{-}CMe{=}CH{-}CH_2OBu \quad (94)$$

Vinyl ethers derived from cyclic ethers (Sect. 2.14.1.) are sources for the HO—(C)$_n$—C=O$^-$ synthons, n depending upon the ring size.[1066–1070] The dihydropyran route has been used to prepare spiroketals.[1374]

Many members of this class of reagents depend on modified acyl anion equivalents such as α-alkoxycyanohydrins,[1375] conjugate additions of α-alkoxyaldehydes catalyzed by thiazolium salts,[1376] and 1-lithio-1-trimethylsilyl-1-propen-3-ol tetrahydropyranyl ether,[191] and 1-lithio-1-*t*-butylthio-3-methoxy-1-alkenes.[1377] In a similar manner, addition of a Grignard reagent to β-hydroxydithioesters provided the HO—C—C—CO$^-$ equivalent.[639,640]

Condensation of the vinyl sulfide **(95)** with benzaldehyde allowed stereoselective acylation (Scheme 95).[1378]

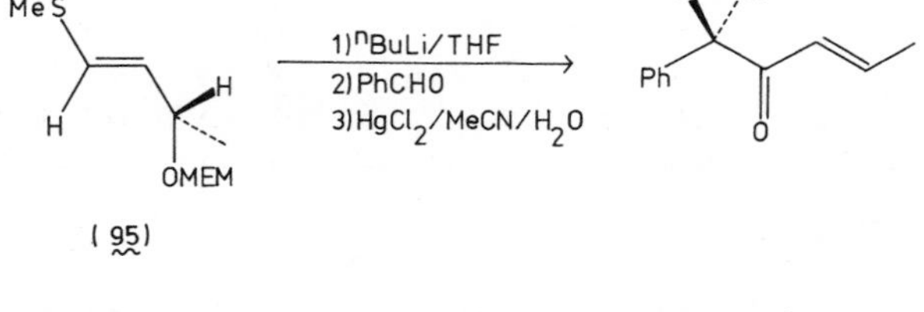

MeS, H, H, OMEM (95) — 1) ⁿBuLi/THF; 2) PhCHO; 3) $HgCl_2$/MeCN/H_2O → H, OH, Ph, O

(95)

2.17.5. Acyl Anion Equivalents With ω-Cation Centers

2.17.5.1. $^+$C—CO$^-$

The most common members of this class of acyl anion equivalents are ketene thioacetals (see Sect. 2.2.2.). Alkyllithiums add across the double bond to

provide the more stable anion, α to the sulfur atoms; the anion can then be reacted in the usual manner as an acyl anion equivalent. The addition of alkyllithiums[534,535] to 2-alkylidene-1,3-dithianes is susceptible to steric effects.[536] The reactivity of the ketene derivative toward the addition of the anion has been increased by oxidation at the sulfur, as with the ketene dithioacetal monoxide system,[717–719] or by formation of a sulfonium salt[513] (see Sect. 2.2.2.). 2-(*N*-Methylanilino)acrylonitrile has also been used to provide a $^+C{-}CO^-$ equivalent[888,1630] (see Sect. 2.12.2.1.), as has 1-phenylthio-1-trimethylsilylethene[782] (see Sect. 2.6.2.). Aryl vinyl sulfoxides[1631] are potential but unproven sources for the $^+CH_2CO^-$ synthon.

2.17.5.2. $^+CO{-}C{-}CO^-$

Methyl 3-lithio-2-methyl-3-phenylthio-2-propenoate has been reacted with aldehydes, acid chlorides, and α,β-unsaturated esters to provide routes to butenolides and cyclopentenones *via* the $^+CO{-}C{-}CO^-$ synthon (see Sect. 2.17.3.2.).

The cyclopentanone annulation synthon, 1-phenylthiocyclopropyltriphenylphosphonium tetrafluoborate, may also be considered as a member of this general class[838] as can β-thioenones.[1313]

2.17.5.3. Others

Conjugated ketene thioacetals have been used as $^+C{-}C{=}C{-}CO^-$ equivalents, whereas 1,2-bis(trimethylsilyloxy)cyclobutene provides a $^+CO{-}C{-}C{-}CO^-$ synthon[1339] (see Sect. 2.17.2.3.).

2.17.6. Acyl Anion Equivalents with ω-Anionic Centers

2.17.6.1. $^-C{-}CO^-$

Dithioesters have been used as a source for this synthon, the acyl anion equivalent being derived[639] from Grignard addition to the C=S double bond (Scheme 96).

$$\text{Me—CS—SEt} \xrightarrow[\text{3) } H_2O \text{ 4) protection}]{\text{1) LDA 2) RCHO}}$$

$$\text{EtO—CHMe—O—CHR—CH}_2\text{CSSEt} \xrightarrow[\text{2) El}^+]{\text{1) EtMgI}} \qquad (96)$$

$$\text{EtO—CHMe—O—CHR—CH}_2\text{—C(SEt)}_2\text{—El}$$

2.17.6.2. $^{-}C{=}C{-}CO^{-}$ AND $^{-}C{-}C{-}CO^{-}$

Many of these synthons are based on allenes and acetylenes; the acyl anion part of a number of these has already been considered (Sect. 2.17.1.2. and 2.17.1.3.).

The anion derived from methoxyallene was silylated α to the alkoxy group. Subsequent anion formation and alkylation led, after acid treatment, to the α,β-unsaturated acylsilane or, after reaction with fluoride and acid, to the enal.[1248] In a similar manner, 3-trimethylsilyl-2-propyn-1-ol *t*-butyl ether provided β-silylenones.[1246] This method has also been used to prepare β-alkyl- and β-phenylenones,[1244,1245] and has been employed in a synthesis of pyrenophorin.[1379] A nitrogen allene has been used successfully to provide enones by means of this synthon.[1262]

3-Methylthio-2-propyn-1-ol methyl ether was alkylated α to the thio group and, hence, led to β-thioenones.[1258] An example that also employs sulfur is outlined in Scheme 97; the second alkylation occurred α to sulfur with alkyl halides but α to oxygen when the electrophile was an aldehyde.[1282] Similar methodology led to β-silylenones.[1283]

$$CH_2{=}CR{-}CH(OMe){-}SPh \xrightarrow[R'X,\ -78^\circ]{LDA/THF} CH_2{=}CR{-}CR'(OMe){-}SPh \xrightarrow[Hexane,\ \Delta]{SiO_2}$$

$$PhSCH_2{-}CR{=}CR'{-}OMe \xrightarrow[2)\ R''X,\ -78^\circ]{1)\ BuLi/TMEDA/THF,\ 0^\circ} \quad (97)$$

$$PhS{-}CHR''{-}CR{=}CR'{-}OMe \xrightarrow[Dioxan/H_2O]{NaIO_4} R''{-}CH{=}CR{-}CO{-}R'\ (E)$$

The allyl anion derived from an enamine has been used to provide a $^{-}C{-}C{-}CO^{-}$ synthon, but the presence of another functional group modifies the umpoled nature of these reagents.[1153,1154]

Phenyl propargyl selenide has been used to prepare α-phenylseleno-enones (Scheme 98).[1236,1237]

$$PhSeCH_2C{\equiv}CH \xrightarrow[2)\ RX\ 3)\ El^+]{1)\ 2LDA/THF,\ -78^\circ} PhSe{-}CHR{-}C{\equiv}C{-}El$$

$$\xrightarrow{MCPBA/CH_2Cl_2,\ -78^\circ\ to\ -30^\circ} RCH{=}C(SePh)(CO{-}El) \quad (98)$$

Although the acetylenic 1,3-dioxolane **(81)** provides an excellent $C{\equiv}C{-}CO^{-}$ synthon, reaction with silicon, germanium, or tin halides in ether gave reaction α to the silyl group, which led to the α,β-unsaturated acids.[1235]

2.17.6.3. Others

γ,δ-Unsaturated dithioesters have been used to provide a $^{-}C{=}C{-}C{=}C{-}CO^{-}$ synthon.[1380]

3

HYDROXYCARBONYL ANIONS AND RELATED SYNTHONS

Michael Kolb

CONTENTS

3.1. INTRODUCTION

As the premier functionality in organic synthesis, the carbonyl group is intimately involved in many carbon–carbon bond-forming reactions. It merits this position mainly due to the versatile *electrophilic* nature of the carbonyl carbon atom. Actually, numerous hydroxycarbonyl synthons of the $^{+}COOH$ variety are known and serve to prepare carboxylic acids and related compounds,[1384,1385] the most prominent being CO_2, XCOOR, $(RCO)_2O$, etc.

This chapter will describe reaction sequences that rely on an alternative reactivity pattern of the carbonyl group. Formally at least, the carbonyl carbon atom in these reactions has abandoned its traditional electrophilic reactivity

and functions as an apparent *nucleophilic* center. The utility of this umpoled carbonyl reactivity in the synthesis of carboxylic acids and related functionalities will be discussed.

Conceptually, several possibilities exist to achieve an umpolung of the carbonyl reactivity (Chart 1). Generally the putative carbonyl carbon atom will be modified by substituents X, Y, and Z in a manner that permits not only the stabilization of a negative charge, but also incorporates enough nucleophilic character at the carbon atom to allow for the desired transformation. An additional requirement for the choice of X, Y, and Z obviously is that they can be manipulated easily to regenerate the carbonyl function at the desired point in the synthetic scheme. Within these constraints, a large number of substitution patterns for X, Y, and Z can be proposed, and actually many have been investigated. They will be discussed in detail below.

In general, two complementary approaches are apparent. The first, described by entries A, B, and C (Chart 1), considers the reaction of a carbon electrophile R^+ with an appropriately substituted methine, methylene, or methyl carbanionic species. The other approach, D and E, has the R group already incorporated in the carbanionic structure, and addition of an electrophilic substituent leads to the products. The carbanionic species in most cases is generated formally by proton abstraction from the corresponding precursor.

It should be mentioned at this point that the expressions "carbon electrophile," "carbanions," etc., are used in the above context in a formal sense only. Classification of reactions into entries A–E does not express any judgment of the reaction mechanisms involved in the sequences described.

Since much thought in synthetic chemistry is based on synthons that are obtained by a heterolytic disconnection of bonds, it is both convenient and illustrative to use charged synthons when writing a synthetic operation. It is for that reason only that formally charged species are used throughout in the description of reactions.

The possible alternative to entries A–E, which would introduce consecutively two Y^+ units onto a $RC{=}X^{2-}$ species, has been excluded from the scope of this chapter. Also, examples in which a carbonyl function is prepared

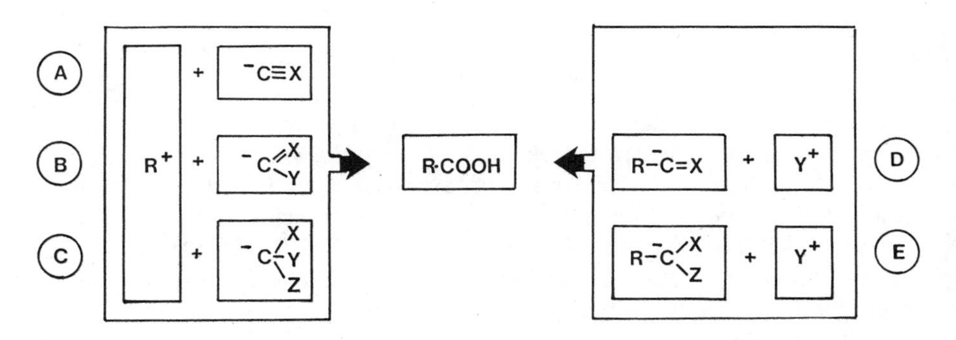

first, and only afterwards, in a separate reaction, transformed into a carboxyl group, for example, by oxidation, have in general been omitted. Only selected, useful synthetic reactions that result in a nucleophilic carboxylation of a carbonyl group concurrent with further elaboration of the final carboxyl carbon atom will be discussed.

Under these restrictions, the present chapter attempts to cover the literature on $^{-}$COOH chemistry that has appeared since 1950. The first review in this area was published in 1969 by Seebach,[11] followed then by several papers describing subsequent developments.[370,371,373,1386–1388] (I would like to acknowledge gratefully the help of Professor Seebach in compiling part of the literature presented here, and the Merrell Dow Strasbourg Research Center library staff who were very helpful in the search for umpolung references through their literature retrieval system.)

We now will consider briefly the classical solutions that exist for the title subject. Then we will discuss in detail the more recent developments on umpoled hydroxycarbonyl synthons. For this, reaction sequences involving chemistry with umpoled hydroxycarbonyl and related synthons, and which generate the same carboxyl derivative, will be discussed together in categories. Ten categories, as specified in the chapter contents list, will be distinguished.

3.2. CLASSICAL HYDROXYCARBONYL ANIONS AND RELATED SYNTHONS

The broadness of the problem of umpolung of reactivity was recognized some time ago,[1389–1395] and an excellent review[1388] discusses the matter in detail. In this context, four classical solutions to this problem must be mentioned.

The cyanide ion has been employed for decades in the conversion of halides to nitriles and thence to carboxylic acid derivatives upon hydrolysis[1396,1397] (Scheme 1).

$$\mathrm{RX} + {}^{-}\mathrm{CN} \longrightarrow \mathrm{RCN} \longrightarrow \mathrm{RCOOH} \qquad (1)$$

The reaction of carboxylic acids with isocyanides to give the corresponding formamides (Scheme 2) is an example of a classical umpoled aminocarbonyl synthon.[1398–1405] With acid chlorides, α-ketoamides are formed (Scheme 3) after hydrolysis.[1398–1400,1406]

$$\mathrm{RNC} + \mathrm{R'COOH} \longrightarrow \mathrm{RNH{-}CHO} \qquad (2)^{1401}$$

$$\mathrm{PhNC} + \mathrm{MeCOCl} \xrightarrow{\text{Hydrolysis}} \mathrm{PhNH{-}COCOMe} \qquad (3)^{1406}$$

Depending on the reaction conditions, the use of hydroxylamine as reaction partner gives access to formamidoximines, formamidines or ureas[1398–1400,1407,1408] (Scheme 4).

$$PhNC + H_2NOH \longrightarrow \begin{cases} PhNH—CH=NOH \\ \xrightarrow{ZnCl_2} PhNH—CH=NPh \\ PhNH—CONH_2 \end{cases} \qquad (4)^{1407,1408}$$

In this context also belongs the Passerini reaction.[1398–1400,1409–1411] Scheme (5) gives the outcome of this unique three-component condensation of isocyanides with aldehydes or ketones and carboxylic acids.

$$R'NC + R''COOH + R_2CO \longrightarrow R'NH—CO—CR_2—OCOR'' \quad (5)$$

Another example of classical $^{-}$COOH chemistry is the use of trihalomethanes.[1412] To synthesize α-substituted carboxylic acids, nucleophiles were reacted with trichloromethylcarbinols, prepared in turn from aldehydes or ketones and chloroform. An early example is the synthesis of α-phenyloxyisobutyric acid from acetone, chloroform, phenol, and KOH, described by Link[1413] (Scheme 6), and more recently by others.[1414]

$$CHCl_3 + Me_2CO + PhOH \xrightarrow{KOH} PhOCMe_2COOH \qquad (6)$$

Conversion of trichloromethylcarbinols, obtained from chloroform and benzaldehyde into α-chlorophenylacetic acid by aqueous KOH (Scheme 7),[1415] into α-methoxyphenylacetic acid by KOH/methanol,[1416–1418] or α-ethoxyphenylacetic acid by NaOEt in ethanol (Scheme 8)[1416–1418] has been described.

$$PhCHO + CHCl_3 \xrightarrow{KOH} Ph—CHCl—COOH \qquad (7)$$

$$PhCHO + CHCl_3 \xrightarrow{NaOEt} Ph—CHOEt—COOH \qquad (8)$$

Anions of acetylenes show umpoled carbonyl reactivity and can be used for a variety of carbon–carbon bond-forming reactions.[370,420,1419] However, despite the fact that oxidative cleavage of acetylene derivatives to carboxylic acids, for example, by ozone[1420–1424] or RuO_4,[1425] has been known for some time, this has only recently been exploited for the purpose of $^{-}$COOH equivalency (Scheme 9).[1426]

$$\begin{array}{ccccc} RCHO + HC\equiv CMgX & \longrightarrow & RCHOHC\equiv CH & \longrightarrow & RCHNPhth—C\equiv CH \\ & & \downarrow & & \downarrow \\ & & RCHOH—COOH & & R—CHNH_2—COOH \end{array} \qquad (9)$$

3.3. RECENT HYDROXYCARBONYL ANIONS AND RELATED SYNTHONS

3.3.1. Cyanide Anions $^{-}$CN

Since the early uses of the cyanide ion as an umpoled carboxyl function, described in the preceding section, a variety of new, versatile, mild, refined, and "sophisticated" procedures for the same purpose have appeared in the literature.

In a series of papers,[1427] Nagata and co-workers have described the addition of cyanide to α,β-unsaturated carbonyl systems in a conjugate fashion. Their methods largely avoid the side reactions encountered in traditional procedures. The reagents employed are either a combination of HCN and alkyl aluminium compounds or alkyl aluminium cyanides. Not only 1,4-additions to enones[937,1428–1436] (Scheme 10), but also epoxide opening (Scheme 11)[1437] and 1,2-addition reactions to carbonyl functions (Scheme 12)[937] were investigated (for base-catalyzed cleavages of epoxides with HCN, see Ref. 1438). The yields are generally good to excellent, exceeding 80% in most cases.

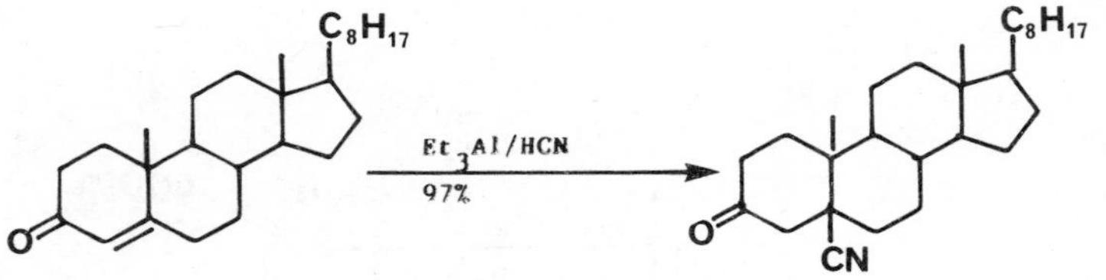

(10)

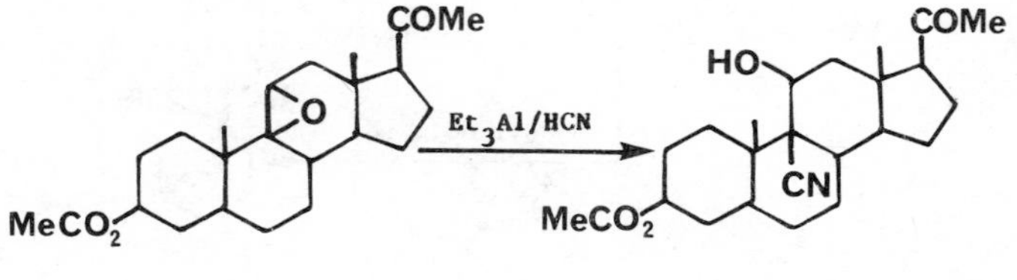

(11)

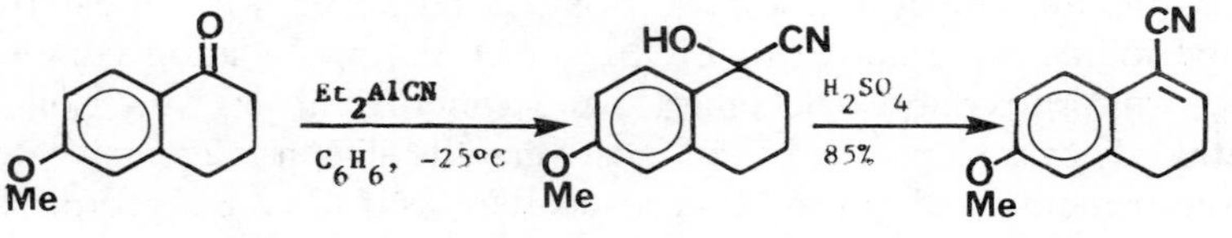

(12)

Aryl and vinyl halides can be converted in excellent yield to α,β-unsaturated nitriles by $NaCu(CN)_2$, a reagent described by House (Scheme 13)[1439] for use at elevated temperatures in DMF or HMPA, or by $K_4Ni_2(CN)_4$ at

ambient temperature following a procedure described by Corey (Scheme 14).[1440]

1-iodonaphthalene $\xrightarrow[\text{150–155°C, 4h, 98\%}]{\text{NaCu(CN)}_2\text{, DME}}$ 1-cyanonaphthalene (13)

$$\text{PhCH=CHBr } (E) \xrightarrow[\text{KCN, rt, 2 h, 78\%}]{K_4Ni_2(CN)_4\text{, MeOH}} \text{PhCH=CH—CN } (E) \quad (14)$$

Nitriles may also be prepared conveniently from ketones by means of hydrazone derivatives,[1441–1443] for example, methoxycarbonyl hydrazones,[1443] in excellent yields (Scheme 15).[1444]

(15)

Alternatively, when the diazene was added to LiOMe in DME containing dimethyl carbonate, the cyanoester was isolated. However, when the diazene and methyl iodide were added to LiOMe in DME, methylation competed favorably with carboxylation (Scheme 15) and α-methyl nitriles were isolated.

Trimethylsilylcyanide is one of the most versatile silicon reagents.[1445] Specifically the trimethylsilyl group often reacts here as if it were a proton, and therefore one might expect trimethylsilylcyanide to approximate the reactions (and toxicity) of hydrogen cyanide. For example, addition of trimethylsilylcyanide to a carbonyl group with[952] or without[1446] a catalyst gives protected α-hydroxy nitriles (Scheme 16). α,β-Unsaturated carbonyl functions react under these conditions cleanly in the 1,2-manner.[952,1445]

(fluorenone) —[$AlCl_3$, TMSCN; 3h, 73%]→ (9-cyano-9-(trimethylsilyloxy)fluorene: O–$SiMe_3$, CN) (16)

Dithioacetals[1447] and orthoesters or -amides,[1448] upon treatment with trimethylsilylcyanide and $SnCl_4$ catalysis, afford nitriles with quite an unusual spectrum of substitution patterns (Schemes 17, 18). Acid chlorides react with Me_3SiCN to afford 2-(trimethylsilyloxy)-2-propenonitriles, precursors of α-keto nitriles.[1293]

$$Et_2C(SPh)_2 + Me_3SiCN \xrightarrow[\text{rt, 6 h, 78\%}]{SnCl_3,\ CH_2Cl_2} Et_2C(CN)SPh$$

1) MCPBA, 2) Δ, 73% → (E)—MeCH=CEt—CN

Raney Ni, 81% → $Et_2CH—CN$ (17)[1447]

$$\left.\begin{matrix}(RO)_4C & \text{or}\\ (MeO)_2CHNR_2 & \text{or}\\ (EtO)_3CNMe_2 & \text{or}\\ (EtO)_2C(NMe_2)_2 & \end{matrix}\right\} \xrightarrow[63\text{–}86\%]{Me_3SiCN,\ SnCl_4} \left\{\begin{matrix}(RO)_3CCN\\ \text{or } MeOCH(NR_2)CN\\ \text{or } (EtO)_2C(NMe_2)CN\\ \text{or } EtOC(NMe_2)_2CN\end{matrix}\right. \quad (18)^{1448}$$

Another group of new and useful reagents that introduce a cyanide function into derivatized carbonyl groups consists of the cyanophosphates.[1449] Upon reaction with enamines, they offer access to protected α-amino nitriles under nonaqueous conditions[1450–1452] (Scheme 19), and so complement the procedures involved in the usual Strecker synthesis.

(steroidal pyrrolidine enamine, C_8H_{17}) —[3 $(EtO)_2P(O)CN$; 3 K_2CO_3, THF, reflux; 7h, 92%]→ (NC, N-pyrrolidinyl steroid, C_8H_{17}) (19)

With ketones as reaction partner in the presence of catalytic amounts of LDA, the sequence gives *O*-phosphorylated cyanohydrins.[1453]

Conjugate addition reactions of the cyanide ion[1427] have been applied to α,β-unsaturated aldehydes, ketones and esters by Saegusa.[1451] Catalysis with $TiCl_4$ promotes this transformation of enones with *t*-BuNC (Scheme 20).[1451] Acetals react under analogous conditions. This procedure has the merit of

combining mild reaction conditions with avoiding the use of toxic cyanides and with excellent yields of γ-keto nitriles.

$$\text{(cyclohexylidene ketone)} \xrightarrow[CH_2Cl_2,\ 0°C,\ 14h,\ 85\%]{TiCl_4,\ t\text{-}BuNC} \text{(1-cyano-cyclohexyl ketone)} \quad (20)$$

3.3.2. Chlorocarbonyl Anions ^-COCl

The addition of CO to olefins, catalyzed by $PdCl_2$, as described by Tsuji,[1452] gives acid chlorides in moderate yield (Scheme 21) in the absence of other nucleophiles. Asymmetric olefins, such as propene, react at the terminal carbon atom; those higher than pentene give other, lower boiling products.

$$CH_2{=}CH_2 + PdCl_2 + CO \xrightarrow[40\%]{rt,\ 50\ kg/cm^2} ClCH_2CH_2COCl \quad (21)$$

A curiosity is the reaction of CCl_4 with hydrocarbons in the presence of *t*-butyl peroxide, which also affords acid chlorides (Scheme 22).[2311]

$$\text{cyclohexane} + CCl_4 + CO + (t\text{-BuO})_2 \xrightarrow[16h,\ 46\%]{6000\ psi,\ 130°C} \text{cyclohexyl-}COCl \quad (22)$$

3.3.3. Aminocarbonyl Anions $^-CONH_2$

Routes to primary amides through a ^-COOH synthon different from ^-CN are fairly rare. Principally, all cases reported that do not rely on hydrolysis[1435,1436,1454] of nitriles are based on the treatment of an activated carboxyl derivative intermediate with ammonia. The carboxyl derivative, in turn, is obtained by a reaction sequence that then has involved an umpoled hydroxycarbonyl synthon.

A procedure reported by Tsuchihashi[823] describes the addition of α-sulfinyl carbanions to aryl aldehydes. Oxidation of the reaction product gives α-keto sulfoxides, which, after Pummerer rearrangement and quenching with ammonia, afford protected α-hydroxy amides (Scheme 23).[823]

$$PhCHO + TolSOCH_2Li \longrightarrow PhCHOH{-}CH_2SO{-}Tol \xrightarrow[84\%]{MnO_2,\ CH_2Cl_2}$$

$$PhCO{-}CH_2SO{-}Tol \xrightarrow[115°,\ 14\ h]{AcOH,\ AcONa} Ph{-}CHOAc{-}COSAr \quad (23)$$

$$\downarrow \text{conc. } NH_4OH,\ MeOH,\ 73\%$$

$$Ph{-}CHOAc{-}CONH_2$$

The corresponding α-keto derivatives can be obtained in moderate overall yield by treating nitriles with the sodium salt of methyl methylthiomethyl sulfoxide[711,1455] (Scheme 24).

$$iPrCN \xrightarrow[\text{NaH, 72\%}]{\text{MeSCH}_2\text{—SO—Me}} (H_2N)(iPr)C{=}C(SOMe)(SMe) \xrightarrow[87\%]{Ac_2O} iPr\text{—}C(NHAc)(SMe)\text{—}COSMe$$

$$\xrightarrow[84\%]{NH_3,\ iPrOH} iPr\text{—}C(NHAc)(SMe)\text{—}CONH_2 \xrightarrow[73\%]{\text{MCPBA, }H_2O,\ CH_2Cl_2} iPrCO\text{—}CONH_2 \qquad (24)^{1455}$$

For an example of the synthesis of primary amides by means of carbamoyllithium intermediates, see the procedure described in Scheme (40).[1456]

3.3.4. (Alkylamino)carbonyl Anions $^{-}$CONHR

Two reaction sequences dominate the use of umpoled hydroxycarbonyl synthons in the preparation of secondary amides. These are either from the realm of isonitrile chemistry[1402–1405] or CO addition reactions.[1457–1463]

Ugi,[1464–1468] and later McFarland[1469] and Hagedorn,[1470] describe in a series of papers the reaction of isonitriles with aldehydes[1464,1467–1470] (Scheme 25) and ketones[1464,1467,1470] (Scheme 26), acyl halides[1465,1471] (Scheme 27) (but see Ref. 1472 for a different outcome of this reaction), and imines or immonium salts of aldehydes or ketones[1464,1467,1473,1474] (Scheme 28) under hydrolytic conditions, to obtain α-substituted secondary amides. Interestingly, the procedure described in sequence (28) gives also access to β-lactams. The product yields are good—in fact, excellent in most cases.

$$NH_3 + iPrCHO + HCOOH + CX\text{—}NC \xrightarrow[54\%]{} CX\text{—}NHCO\text{—}CH(Pr\text{-}i)\text{—}NHCHO \qquad (25)^{1464}$$

where CX = cyclohexyl.

$$CX\text{—}NC + Me_2CO \xrightarrow[0°,\ 71\%]{H_2SO_4,\ H_2O} CX\text{—}NHCO\text{—}CMe_2\text{—}OH \qquad (26)^{1470}$$

$$AcCl + BuNC \xrightarrow[\text{2) 50° 10 min, 81\%}]{\text{1) 20° 45 min}} Ac\text{—}CCl{=}NBu \xrightarrow[84\%]{H_2O} Ac\text{—}CONHBu \qquad (27)^{1465}$$

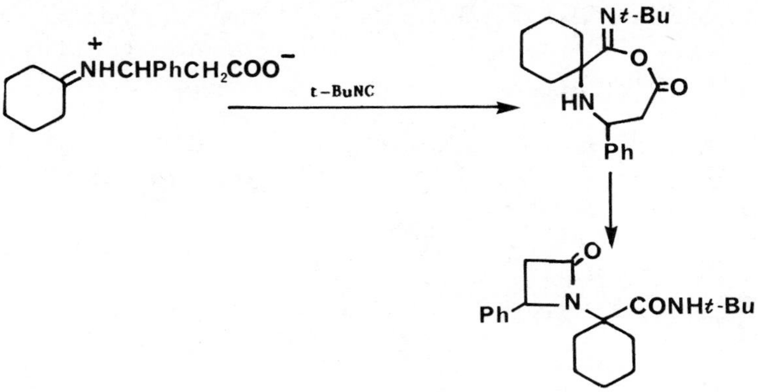

(28)

The conditions of the addition reaction have recently been refined by the use of $TiCl_4$ as a catalyst (Scheme 29)[1475] (for a review of Ti reagents in organic synthesis, see Ref. 1476).

$$\text{RCHO} + \text{MeNC} \longrightarrow \text{HO—CHR—CONH—Me} \qquad (29)$$

where R = C_5H_{11}, 96%; R = $C_6H_5CH{=}CH—$, 36%.

This also has allowed the extension of the reaction scheme to dimethyl acetals and ketals (Scheme 30)[1477] to yield α-alkoxy amides. α,β-Unsaturated aldehydes were found to react under these conditions in the 1,2-manner,[1475] which contrasts with their reactivity pattern in similar reactions described by Saegusa[1451] for the synthesis of nitriles (see Scheme 20).

$$\text{MeCH(OEt)}_2 + \text{CX—NC} \xrightarrow[\text{2) H}_2\text{O 90\%}]{\text{1) TiCl}_4\text{, CH}_2\text{Cl}_2} \text{EtO—CHMe—CONH—CX} \qquad (30)$$

Various mechanisms were proposed to explain the outcome of these reactions.[1475]

The alternative important route to secondary amides utilizing $^-$CONHR synthons relies on addition reactions of carbon monoxide.[1457–1463] Catalyzed by various Pd(II) complexes in the presence of amines, CO reacts with aryl and vinyl halides (Scheme 31)[1478] to give amides.

$$\underset{(Z)}{\text{PhCH}{=}\text{CHBr}} + \text{PhNH}_2 + \text{Br}_2\text{Pd(PPh}_3)_2 \longrightarrow \underset{(Z)}{\text{PhCH}{=}\text{CH—CONHPh}} \qquad (31)$$

Side products from *p*-bromonitrobenzene in this reaction are ureas, formed by attack of a second amine on the nitro group. This has led to a modification of the reaction conditions, which then allows the use of nitroarenes in place of aryl halides for the preparation of *N*-substituted ureas in moderate to good yields.[1478,1479]

Japanese authors describe a related sequence that either involves the reaction of $[RCOFe(CO)_4^-]$[1480] or of $RNHMgBr/Fe(CO)_5$ (suggesting an $RNHCOFe(CO)_4^- BrMg^+$ intermediate) with nitro compounds (Scheme 32).[1481] The same authors also report the use of $Na_2Fe(CO)_4$ in this context.[1482]

$$PhNHMgBr + p\text{-}MeOC_6H_4NO_2 + Fe(CO)_5 \xrightarrow[92\%]{H^+} p\text{-}MeOC_6H_4NHCONHPh \qquad (32)$$

The addition of RNH^- metal$^+$ to CO and subsequent quenching with electrophiles has been reported on two occasions[1483,1484] to yield secondary amides, however in fair yields only. The choice of electrophiles is limited, only protons[1483] (using *t*-BuNHLi) or allylic halides[1484] $[LiCu(NHR)_2]$ having been reported.

Recently, Rautenstrauch has described an improved version of this sequence, which affords secondary amides through the use of tripropylcarbazoyllithium (Scheme 33).[1485] The reaction of this carbamoyllithium derivative with methyl iodide (slow), aldehydes, ketones, and esters (to obtain α-keto amides) was investigated.

$$Pr_2N\text{-}NHPr \longrightarrow Pr_2N\text{-}NLiPr \xrightarrow{CO,\ -75^\circ C} Pr_2N\text{-}NCOLi \xrightarrow[\text{cyclohexanone}]{-75^\circ C,\ 80\%} \text{1-(PrN–NPr}_2\text{–CO)-cyclohexan-1-ol} \xrightarrow[85\%]{Li/NH_3,\ THF} \text{1-(NHPr–CO)-cyclohexan-1-ol} \qquad (33)$$

The reaction sequences described in the foregoing chapter for the preparation of primary amides from aldehydes involving a Pummerer rearrangement (see Scheme 23),[823] can, by a different choice in the amine component, also be employed for secondary amides.

Conversion of carbonyl derivatives into secondary amides was described[756] recently using methoxyphenylthiotrimethylsilylmethyllithium, prepared from phenylthiomethyl methyl ether.

For an entry to secondary amides by means of carbamoyllithium intermediates, see below (Scheme 40).[1456]

3.3.5. (Dialkylamino)carbonyl Anions $^{-}CONR_2$

Addition reactions of carbon monoxide[1457–1463] were shown in the preceding section on umpoled monoalkylaminocarbonyl synthons to be useful for the preparation of secondary amides. This approach can be adapted to the synthesis of tertiary amides, then involving umpoled dialkylaminocarbonyl synthons: $^{-}CONR_2$.

An unusual example is the addition reaction of diazenes to CO in the presence of catalytic amounts of Pd(II) salts giving access to a variety of 2-aryl indazolinones (Scheme 34)[1486]; for the carboxylation of azobenzene with $Co_2(CO)_8$, see Refs. 1487 and 1488.

Me N=N Me —— 1) $PdCl_2$ 99% ; 2) CO, 100°C, H_2O, 100 atm, 3 h, 97% ——→ Me, O, HN-N, Me (34)

Aminometalloorganic reagents, such as $LiNR_2$[1489–1492] (Scheme 35) or $LiCu(NR_2)_2$ (Scheme 36),[1484] react with CO to afford carbamoyllithiums.

NH —— n-BuLi, DME/THF, −75°C ——→ NLi —— CO, −75°C ——→ NCOLi —— O= (cyclohexanone), −75°C, 85% ——→ NHCO–C(OH) + NHCOCO–C(OH) (35)

4 : 1

$$CuCl + LiNEt_2 \xrightarrow[-20^\circ]{HMPA/THF} CuNEt_2 \xrightarrow{LiNEt_2} (Et_2N)_2CuLi \xrightarrow[rt]{CO}$$

$$(Et_2NCO)_2CuLi \xrightarrow[76\%]{CH_2{=}CHCH_2Br} Et_2NCO{-}CH_2CH{=}CH_2 \qquad (36)^{1484}$$

These intermediates can be trapped with electrophiles, such as primary alkyl iodides[1484,1489,1490] or ketones,[1489] or with allylic halides, aryl halides, acyl halides, or α,β-unsaturated ketones[1484] (1,4-addition), giving variously functionalized tertiary amides. In some cases α-keto amides may appear as side products; exceptionally, they are isolated as the main product.[1489]

The synthetic utility of the carbamoyllithiums is primarily limited by (1) the reaction of these organometallics with CO, (2) the incomplete conversion to amides observed in certain cases, and (3) secondary addition reactions of unhindered carbamoyllithium species with the primary adducts.

Nickel tetracarbonyl reacts with secondary amines and vinyl halides to yield α,β-unsaturated tertiary amides (Scheme 37).[1440]

$$\underset{(E)}{PhCH{=}CHBr} + Ni(CO)_4 + HN(CH_2)_4 \xrightarrow[5\ h,\ 82\%]{MeOH,\ 60^\circ}$$

$$\underset{(E)}{PhCH{=}CHCO}{-}N\begin{matrix} CH_2CH_2 \\ | \\ CH_2CH_2 \end{matrix} \qquad (37)$$

Upon treatment with $LiNR_2$, nickel tetracarbonyl can be reacted with vinylic, allylic, benzylic, and phenylic halides or acid chlorides to afford, in good-to-excellent yield, tertiary amides (Scheme 38).[1493,1494]

$$LiNMe_2 + Ni(CO)_4 \xrightarrow{Et_2O} Li(Me_2NCONi(CO)_3)$$

Then (Et₂O, 10°C): with BuCOCl → $Bu{-}COCO{-}NMe_2$ (97%); with 3-bromocyclohexene → 3-cyclohexenyl-$CONMe_2$ (99%) (38)

On reaction with benzaldehyde, or aceto- or benzophenone, the corresponding α-hydroxy amides were obtained. The possible use of $Na_2Fe(CO)_4$, in place of the nickel complex, for these purposes should be mentioned.[1158]

Methoxyphenylthiotrimethylsilylmethyllithium, previously described as an umpoled aminocarbonyl synthon in the synthesis of secondary amides, has also been used for tertiary amides.[756]

The reaction of the potassium salt of $NCCH_2NMePh$ with aryl esters gives α-keto amides in moderate yields (Scheme 39),[1495] and thus complements other methods reported earlier for the synthesis of α-keto amides.[1496–1503]

$$\text{PhMeN}\frown\text{CN} \xrightarrow[\text{25°C, 3h, 81\%}]{\text{KH, THF; p-ClPhCOOEt}} \text{4-ClC}_6\text{H}_4\text{C(OH)=C(CN)NPhMe} \xrightarrow[\text{50°C, 5h, 80\%}]{\text{Cu(OAc)}_2\text{, 50\% EtOH}} \text{4-ClC}_6\text{H}_4\text{COCONPhMe} \quad (39)$$

The number of possible routes for the synthesis of tertiary amides by means of umpoled dialkylaminocarbonyl or related synthons is, compared with those reported for secondary amides, increased by the possibility of metallating dialkyl formamides directly. The resulting carbamoyllithium species can then be reacted with electrophiles to obtain the desired carbon skeleton. Pioneering work of Schöllkopf[1456,1504,1505] and others[1496,1506–1508] has demonstrated the scope of this approach. Dimethyl[1505] and diisopropyl[1496,1506–1508] formamides serve as starting materials, LDA[1456,1505,1506] or *t*-BuLi[1496,1507,1508] as bases. Good-to-excellent yields are achieved in reactions of the intermediate carbamoyllithium species with aldehydes and ketones in general[1456,1496,1507] (Scheme 40) and with esters[1496,1507] and benzylic halides[1496] in isolated examples (Scheme 41).

$$\text{OHC—NR—CH}_2\text{OMe} \xrightarrow[-75°,\ 88\%]{\text{LDA, THF}} \text{LiCO—NR—CH}_2\text{OMe} \xrightarrow[-75°,\ 85\%]{\text{Ph}_2\text{CO}} \text{HO—CPh}_2\text{—CO—NR—CH}_2\text{OMe} \quad (40)^{1456}$$

where (R = Me)

$$\text{LiCON(Pr-}i)_2 + \text{PhCOOEt} \xrightarrow[-95°,\ 70\%]{\text{THF/Et}_2\text{O/pentane}} \text{PhCOCON(Pr-}i)_2 \quad (41)^{1496}$$

Less reactive alkylating reagents give complex mixtures of products, suggestive of decomposition of the carbamoyllithium species. In the case of α,β-unsaturated aldehydes or ketones, the reaction leads to 1,2-addition products. By a judicious choice of formamide substituents, a direct entry to both primary and secondary amides is available (Scheme 40)[1456] (R = CH_2OMe, removed by acid hydrolysis).

A very recent publication describes the reaction of diaryl chloroformamides with aldehydes and ketones, giving tertiary amides. SmI_2 was used to promote the transformation (Scheme 42).[1509]

$$\text{Ph}_2\text{N—COCl} + \text{C}_7\text{H}_{15}\text{CHO} \xrightarrow[\text{2) H}^+,\ 67\%]{\text{1) SmI}_2\text{, THF, rt}} \text{Ph}_2\text{NCO—CHOH—C}_7\text{H}_{15} \quad (42)$$

$R_2NCOSmI_2$ is suggested as the intermediate reactive species in these reactions. This approach is different from the foregoing, which rely either on a deprotonation of formamides by strong bases, or organolithium carbamoylation. This "acyl anion" is the result of a reductive process, a direct carbonyl umpolung by way of electron transfer.

3.3.6. Hydroxycarbonyl Anions $^{-}COOH$

This category constitutes, together with the one on $^{-}COOR$ presented in the following section, by far the largest part of the present chapter on hydroxycarbonyl anion and related synthons. About one-half of all reaction sequences featuring umpoled hydroxycarbonyl or related synthons actually involve the preparation of carboxylic acids, substituted with a wide spectrum of functionalities. The strategies used in these approaches toward $^{-}COOH$ vary considerably and give an interesting picture of the ingenuity of chemists.

Nevertheless, an addition reaction[1457–1463] on CO, achieved by a variety of methods (and used since 1927),[1510,1513–1516] has been the most widely used approach until now.[1518]

Catalyzed by $Ni(CO)_4$,[1510–1512,1516,1517] aryl halides[1510,1512] (Scheme 43), olefins,[1510] vinyl halides,[1511] and allyl alcohols[1511] were reacted with CO to afford carboxylic acids, in many cases in excellent yields.

1-Chloronaphthalene $\xrightarrow[Ni(CO)_4,\ 110°C,\ 6\ h;\ 95\%]{DMF,\ CO,\ Ca(OH)_2}$ 1-Naphthoic acid (COOH) (43)

In chronological order, catalysis by Pd(II) salts[1457,1519,1520] and the use of $Na_2Fe(CO)_4$ (Scheme 44)[1158] or of catalytic amounts of Co(II) salts[1521–1526] (Scheme 45) or of $Fe(CO)_5$[1518,1527] (Scheme 46) under phase-transfer conditions are also on record.

$$Na_2Fe(CO)_4 + C_5H_{11}Br \xrightarrow[87\%]{O_2,\ THF} C_5H_{11}COOH \qquad (44)^{1158}$$

$$p\text{-}Br—C_6H_4—COMe \xrightarrow[Bu_4NBr,\ h\nu,\ 65°;\ 2)\ H^+,\ 90\%]{1)\ Co_2(CO)_8,\ PhH/NaOH,\ CO} p\text{-}HOOC—C_6H_4—COMe \qquad (45)^{1526}$$

$$PhCH_2Br \xrightarrow[NaOH,\ (Bu_4N)_2SO_4,\ 75\%]{Fe(CO)_5,\ CO,\ PhH/H_2O} PhCH_2COOH \qquad (46)^{1527}$$

Related to these addition reactions, which fall into the general category of entry D (Chart 1), is the reaction of organolithium compounds with methyl isocyanide.[1528] The concept behind entries B and C (Chart 1) has also led to a series of $^{-}$COOH synthons that have been investigated intensively for the synthesis of carboxylic acids. One of the earlier approaches is the use of a carbanionic species, which can formally be described as RS—C(metal)=NR, derived either from the corresponding halo derivative[1529,1530] or by deprotonation[1531–1533]; a recent publication takes advantage of a desilylation procedure (Scheme 49, below).[794] Typical electrophiles are aldehydes and ketones[1529,1533] (Scheme 47), carbonates,[1531] or carbon dioxide[1530,1532] (Scheme 48).

1) EtBr, Mg 2) PhCHO 85%; 1) n-BuLi 2) PhCHO 97%; CHOHPh; R: H, Me (47)

PhLi, −60°C; Li; CO_2 40%; COOH (48)

Epoxides can also act as electrophiles,[1533] as can esters and α,β-unsaturated ketones (1,4-addition) (Scheme 49).[794]

$SiMe_3$; CsF, Bu_4NF/SiO_2; 50% (49)

The resulting thiazole derivatives are then ready for further hydrolytic manipulations toward carboxylic acids.[1534]

By analogy, imidazole derivatives have been metallated at C-2 and then reacted with electrophiles to obtain 2-substituted analogs.[1535] However no efficient hydrolysis procedure for the final imidazole products, in order to obtain carboxylic acids, was reported.

Tris-heterosubstituted carbanions are other prominent intermediates used in $^{-}$COOH chemistry. Ortho thioformyl carbanions,[632,635,636,701,845,1536,1537] generated from various sources by a variety of bases, have been treated with assorted electrophiles, such as primary and secondary alkyl iodides[635,636,845] (Scheme 50), bromides,[632,636] chlorides,[636] haloformates,[636,845] epoxides,[636,845]

(Scheme 51), aldehydes,[636,845] ketones,[845] disulfides,[636,845] allyl halides,[635] trialkylsilylhalides,[636] and α,β-unsaturated esters (1,4-addition) (Scheme 52).[1536] The thiocarboxylic acid orthoesters are hydrolyzed by treatment with mercury(II) oxide and boron trifluoride etherate.[538]

$$(PhS)_3CH \xrightarrow{BuLi} (PhS)_3CLi \xleftarrow{BuLi} (PhS)_3C{-}SPh$$

$$(PhS)_3CLi \xrightarrow{MeI,\ 88\%} (PhS)_3CMe \qquad (50)^{845}$$

$$(PhS)_3CLi + MeCH{-}CH_2\ (\text{epoxide, }O) \xrightarrow{30\%} (PhS)_3C{-}CH_2{-}CHMe{-}OH \qquad (51)^{636}$$

$(MeS)_3CLi$ + $C_{13}H_{27}$-butenolide $\xrightarrow{1)\ THF,\ -78°C;\ 2)\ H_2CO,\ -40°C}$ lactone bearing $(MeS)_3C$ and CH_2OH, $C_{13}H_{27}$ $\xrightarrow{HgO,\ BF_3\ Et_2O,\ 80\%}$ lactone bearing HOOC and CH_2OH, $C_{13}H_{27}$ (52)

1,3-Dithiane $\xrightarrow{SO_2Cl_2,\ -40°C \rightarrow 25°C}$ 2-chloro-1,3-dithiane

2-chloro-1,3-dithiane $\xrightarrow{P(OEt)_3,\ 85\%}$ 2-PO(OEt)$_2$-1,3-dithiane $\xrightarrow{n\text{-BuLi, cyclohexanone},\ 96\%}$ 2-cyclohexylidene-1,3-dithiane

2-chloro-1,3-dithiane $\xrightarrow{PPh_3,\ 78\%}$ 2-$\overset{+}{P}Ph_3\ Cl^-$-1,3-dithiane $\xrightarrow{KOH/EtOH,\ Ph{-}CH{=}CH{-}CHO,\ 96\%}$ 2-(3-phenylallylidene)-1,3-dithiane (53)

In related approaches, one of the thio substituents in the carbanionic species may be replaced by a phosphorus function, either as a phosphonate or a phosphonium group[357,1538] (Scheme 53). An alternative is to use a dithioalkyl-

substituted acetonitrile[1349,1539–1541] as illustrated (Scheme 54); for the bismethylthio and dithiane analogs, see Refs. 1349 and 1540–1541, respectively.

$$(EtS)_2CHCHO \xrightarrow[\text{2) MsCl 75–80\%}]{\text{1) NH}_2\text{OH, Et}_3\text{N}} (EtS)_2CH—CN \xrightarrow[\text{2) C}_8\text{H}_{17}\text{Br}]{\text{1) KH, THF}}$$

$$(EtS)_2C(CN)—C_8H_{17} \xrightarrow[\text{75\%}]{\text{NCS, MeCN/H}_2\text{O}} HOOC—C_8H_{17} \qquad (54)^{1539}$$

The cyano compound is accessible from the corresponding aldehyde.[1542] The potassium derivative generated in this sequence undergoes a clean 1,4-addition to α,β-unsaturated ketones at low temperatures; it reacts neither with esters nor aldehydes, probably due to a reversal of the initial addition step.

Trialkylsilyl substituted bisalkylthiomethanes upon metallation and reaction with aldehydes or ketones give ketene thioacetals[461,494] that can be hydrolyzed[495,538,1315,1543–1549] to carboxylic acids[503,506,515,1550,1551] (Scheme 55).

2-SiMe$_3$-1,3-dithiane $\xrightarrow{\text{n-BuLi}}$ 2-Li-2-SiMe$_3$-1,3-dithiane; with OHC-N(morpholino), 59% → 2-(morpholinomethylene)-1,3-dithiane; with Ph$_2$CO, 87% → 2-(=CPh$_2$)-1,3-dithiane (55)

As a modern extension of the classical haloform procedure,[1412] recent papers describe the use of trihalosubstituted methanes as a $^-$COOH equivalent,[1552–1555] mainly with the objective of preparing phenylacetic acid derivatives. A direct synthesis of amino acids has been reported (Scheme 56),[1555] the yields exceeding those obtained for these examples in the Strecker procedure. The main improvement in the efficiency of this reaction is due to new solvent/base conditions.

$$m\text{—Cl—}C_6H_4\text{—CHO} \xrightarrow[\text{2) HCl 3) NH}_4\text{OH 83\%}]{\text{1) CHBr}_3\text{, MeO(CH}_2)_2\text{OH, KOH/LiNH}_2\text{ cat.}} m\text{—Cl—}C_6H_4CH(NH_2)COOH \qquad (56)$$

Further examples of tris-heterosubstituted starting materials, useful for this aspect of $^{-}$COOH chemistry, include tris-selenosubstituted methanes (Scheme 57),[842,1556] giving excellent yields in alkylation reactions, and tetraethyl dimethylaminomethylene diphosphonate, obtained from dimethylformamide dimethyl acetal and diethyl phosphite, in reaction with aryl aldehydes (Scheme 58).[1557]

$$(MeSe)_3CH \xrightarrow[\text{MeI } 80\%]{\text{LDA, THF, } -78°} (MeSe)_3C\text{—}Me \quad (57)$$

The hydrolysis of the phosphonate intermediate (Scheme 58) occurs smoothly in strongly acidic media.[1558] Ketones do not react under the conditions described.

$$(MeO)_2CH\text{—}NMe_2 \xrightarrow{(EtO)_2POH} [(EtO)_2PO]_2CH\text{—}NMe_2 \xrightarrow[\text{2) PhCHO } 66\%]{\text{1) NaH}}$$
$$PhCH\text{=}C(NMe_2)\text{—}PO(OEt)_2 \xrightarrow{H^+,\ 75\%} PhCH_2COOH \quad (58)$$

The reaction of aryl aldehydes with thiocarbonates, mediated by trimethyl phosphite,[512] provides another interesting entry into ketene thioacetals, equivalents of carboxylic acids.[495,538,1315,1543–1549]

An α-sulfinyl carbanion, with further elaboration of the putative carbonyl carbon atom through a Pummerer rearrangement, has also been exploited as a $^{-}$COOH synthon[823,1559] (Scheme 59) (for a corresponding sequence for preparing amides, see Scheme 23).

$$PhCOCH_2\text{—}SO\text{—}Tol \xrightarrow{NaOAc,\ Ac_2O}$$
$$AcO\text{—}CHPh\text{—}CO\text{—}STol \xrightarrow[65\%]{NaOH} HO\text{—}CHPh\text{—}COOH \quad (59)^{823}$$

In the example of Scheme (60),[758] the lithium salt of phenylthiomethyl methyl ether is employed. A Jones oxidation of the intermediate to obtain the carboxylic acid is required. Alkyl and allyl halides, the latter without isomerization, afford by this procedure carboxylic acids in good yields; carbonyl electrophiles give synthetically useful α-acetoxy carboxylic acids.

$$MeOCH_2SPh \xrightarrow[\text{2) PhCHO } 72\%]{\text{1) BuLi, THF, } -78°} HO\text{—}CHPh\text{—}CH(OMe)\text{—}SPh \longrightarrow$$
$$AcO\text{—}CHPh\text{—}CH(OMe)\text{—}SPh \xrightarrow[60\%]{\text{Jones}} AcO\text{—}CHPh\text{—}COOH \quad (60)$$

Another useful preparation of α-hydroxycarboxylic acids by means of umpoled hydroxycarbonyl synthons consists of the reaction of aldehydes with cyanophosphates.[1449] Protected cyanohydrins are obtained that on hydrolysis give the desired hydroxycarboxylic acids.[1453]

3.3.7. Alkoxycarbonyl Anions $^{-}$COOR

As abundant as the references on $^{-}$COOH, covered in the foregoing section, are those in the literature that give access to carboxylic acid esters through umpoled hydroxycarbonyl or related synthons. In this section, however, addition reactions to carbon monoxide are by far the most frequently used, and cover about one-half of all reported sequences. An extensive series of papers by Tsuji describes the synthesis of esters from CO by Pd(II)-catalyzed addition of olefins,[1560,1564–1569] allyl halides, alcohols, ethers and esters,[1561,1563] acetylenes,[1562,1570,1571] allenes,[1572,1573] and propargyl halides[1574] in the presence of an alcohol. The yields obtained in these reactions vary considerably, depending on the reaction conditions used (Scheme 61).[1563,1565] Multifunctional systems were also investigated (Scheme 62),[1564,1571] however with limited success only.

$$CH_2{=}CHCH_2X + PdCl_2 \xrightarrow[5\text{–}80\%]{CO,\ EtOH} CH_2{=}CHCH_2{-}COOEt$$

$$X = Cl, OH, OTs, OAc, OEt, OPh \tag{61}$$

$$RCH{=}CH_2 + PdCl_2 \xrightarrow[5\text{–}41\%]{CO,\ R'OH} RCH_2CH_2COOR'$$

$$CH_2{=}CHCH_2COOEt \xrightarrow{PdCl_2,\ CO,\ EtOH} EtOCO{-}CH_2CH{=}CHCOOEt\ (E)$$

$$HC{\equiv}C{-}COOEt \xrightarrow[EtOH,\ 10\%\ HCl]{PdCl_2,\ CO} \tag{62}$$

$$\begin{array}{lr} EtOCO{-}CH{=}CH{-}COOEt\ (E) & 20\% \\ +\ EtOCO{-}CH{=}C(COOEt)_2 & 8\% \\ +\ (EtOCO)_2C{=}CHCH{=}C(COOEt)_2 & 10\% \\ +\ EtOCO{-}CH_2CH(COOEt)_2 & 32\% \end{array}$$

Analogous reaction sequences have been developed by others,[1575–1577] and more recently a paper by Heck describes experiments investigating a variety of reaction conditions employing several different Pd(II) complexes, such as

$Pd(OAc)_2$, $PhPdX(PPh_3)_2$ (with X = Br or Cl), $Pd_2Br_4(PPh_3)_2$ or $PdX_2(PPh_3)_2$ (with X = I, Br or Cl) (Scheme 63).[1578]

$$p\text{-BrC}_6\text{H}_4\text{CN} \xrightarrow[\text{CO 1 atm, BuOH, Bu}_3\text{N, 89\%}]{\text{PdBr}_2(\text{PPh}_3)_2\text{, 14 h, 100°}} p\text{-BuOCO—C}_6\text{H}_4\text{CN} \qquad (63)$$

Aryl halides react smoothly to give the corresponding esters. Carbobutoxylation of vinylic iodides and bromides is also feasible with these triphenylphosphine Pd complexes, although the stereochemistry of the starting material is not preserved under these conditions. However, with $PdI_2(PPh_3)_2$ as catalyst, the reaction occurs with retention of configuration.[1578] The yields in these reactions vary from 10% to 83%. In this context it is interesting to note that carbalkoxylation of vinylic halides to give β-substituted acrylates or crotonates is now reaching industrial maturity.[1579]

Described recently was an interesting application of Pd(II)-catalyzed addition reactions of CO in the presence of an alcohol. Silyloxycyclopropanes were quantitatively converted to β-acetoxy mercury ketones and then treated with $PdCl_2$/CO to afford γ-keto esters in reasonably good yields (Scheme 64).[1580]

$$\text{(cyclopropane: H}_2\text{C, H}_2\text{C, C(OSiMe}_3\text{)(Bu—}t\text{))} \xrightarrow[\text{rt, 30 min}]{\text{Hg(OAc)}_2\text{, ROH}} [t\text{-BuCOCH}_2\text{CH}_2\text{—HgOAc}] \xrightarrow[-20° \text{ to rt, 20 h, 52\%}]{\text{PdCl}_2\text{, EtOH, CO 1 atm}}$$

$$t\text{-BuCOCH}_2\text{CH}_2\text{—COOEt} \qquad (64)$$

Other metal derivatives used in this context are Ni-allyl complexes[1581] or $Ni(CO)_4$ in reactions with aryl,[1440,1511,1582] vinyl,[1440,1583] (Scheme 65) or allyl[1440,1511,1584–1593] (Scheme 66) halides. Secondary halides having the possibility of rearranging to the primary isomer will react in the latter form. The same is true for tertiary halides, which do not react as such, but rearrange first to secondary allyl halides.[1511]

$$\text{4-}t\text{-Bu-1-bromocyclohexene} \xrightarrow[\text{tert.-BuOH, 60°C, 16 h, 76\%}]{\text{Ni(CO)}_4\text{, tert.-BuOK}} \text{4-}t\text{-Bu-1-(COOEt)cyclohexene} \qquad (65)$$

$$\text{CH}_2\text{=CMe—CH}_2\text{Cl + CO + MeOH} \qquad (66)^{1511}$$

$$\xrightarrow{\text{Ni(CO)}_4\text{, 25°, 2–3 atm, 50\%}} \text{CH}_2\text{=CMe—CH}_2\text{COOMe}$$

$$\xrightarrow{\text{Ni(CO)}_4\text{, HC≡CH, 25°, 1 atm, 80\%}} \text{CH}_2\text{=CMe—CH}_2\text{CH=CHCOOMe } (Z)$$

Iron-carbonyl complexes such as $KHFe(CO)_4$,[1160,1594] $Na_2Fe(CO)_4$[1158,1482] (Scheme 67), or $Fe(CO)_5$[1518,1595] (Scheme 68) are another group of metal derivatives that have been used successfully in the synthesis of esters from a variety of electrophiles. The procedure described in Scheme (67) gives good yields with primary halides and tosylates; for secondary halides, the transformation is less satisfactory.

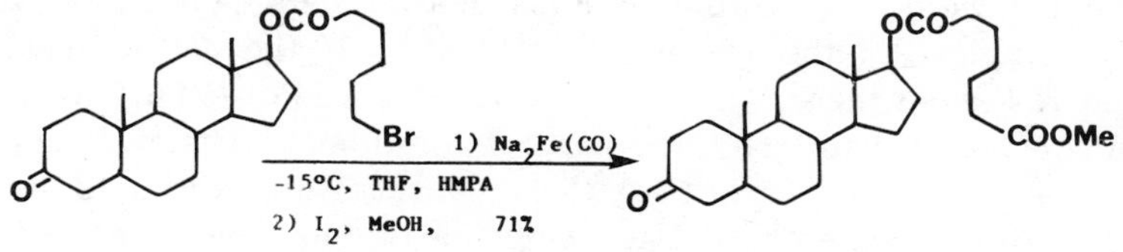

OCO, Br — 1) $Na_2Fe(CO)_4$, −15°C, THF, HMPA; 2) I_2, MeOH, 71% → OCO, COOMe (67)

Reaction conditions utilizing phase transfer catalysis are reported for the sequence with $Fe(CO)_5$ (Scheme 68).[1518]

$$PhCH_2Br \xrightarrow[K_2CO_3,\ rt,\ 68\%]{CO\ 1\ atm,\ Fe(CO)_5,\ MeOH} PhCH_2COOMe \qquad (68)$$

In analogy with the sequences described in the preceding section (^-COOH), trihalomethanes[1596,1597] (Scheme 69) and tris(alkylthio)methyl carbanions[535,1598–1601] can be used for the synthesis of esters.

$$p\text{-}FC_6H_4CHO + HCBr_3 \xrightarrow[0\text{–}5^\circ\ to\ rt,\ 14\ h,\ 69\%]{MeOH,\ KOH} p\text{-}FC_6H_4\text{—}CHOH\text{—}COOMe \qquad (69)^{1597}$$

The procedure given for the sulfur-stabilized carbanionic species in Scheme (70)[1598] is particularly useful in the preparation of γ-keto esters.

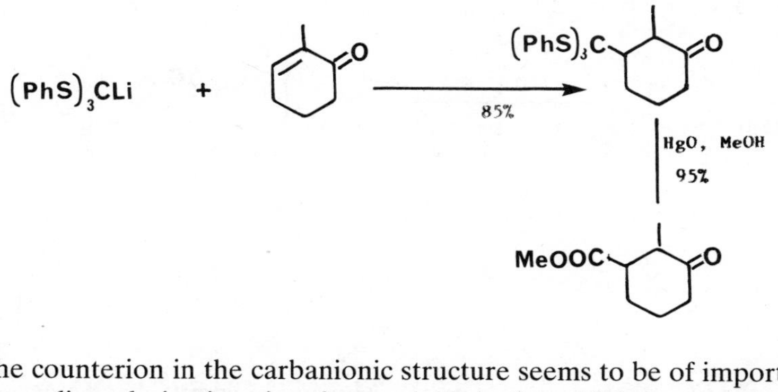

$(PhS)_3CLi$ + (cyclohexenone) → 85% → $(PhS)_3C$–(cyclohexanone) —HgO, MeOH, 95%→ MeOOC–(cyclohexanone) (70)

The counterion in the carbanionic structure seems to be of importance, since the sodium derivative gives lower yields in the addition reaction. α,β-Unsaturated aldehydes react in the 1,2-addition mode. Lactones give with

tris(methylthio)methyllithium, after hydrolytic work-up, variously functionalized esters; this sequence was used in the conversion of aldonolactones to the corresponding methyl 2-aldulosonates.[1602] Procedures for the hydrolysis of the *ortho* thioester intermediates are known.[576]

A useful solvolysis procedure for the readily available ketene thioacetals,[461,491,494] to obtain α-halo esters, is described in Scheme (71).[494,715]

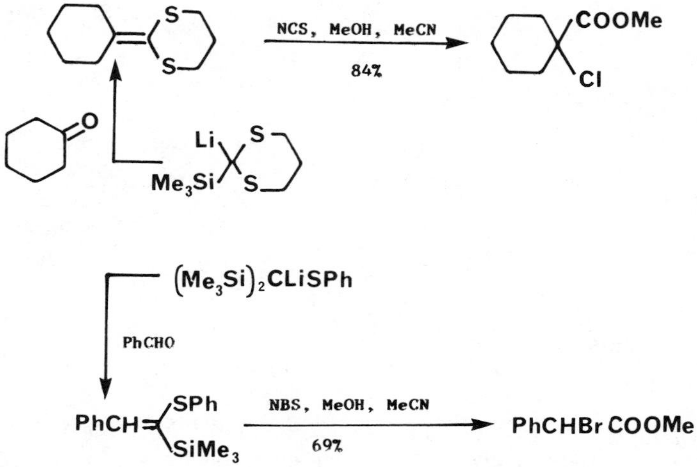

NCS, MeOH, MeCN
84%
COOMe
Cl
O
Li S
Me_3Si S

(71)

$(Me_3Si)_2CLiSPh$

PhCHO

PhCH=C(SPh)($SiMe_3$) —NBS, MeOH, MeCN / 69%→ PhCHBr COOMe

The routes reported therein complement each other, and make feasible the homologation of both aldehydes and ketones. Ketene thioacetals are also used in the procedure described in Scheme (72),[1603] serving a twofold purpose. They not only contain a latent carboxyl function, but also are excellent terminators of cationic acyliminium-initiated cyclizations, yielding interestingly substituted pyrrolizidines.

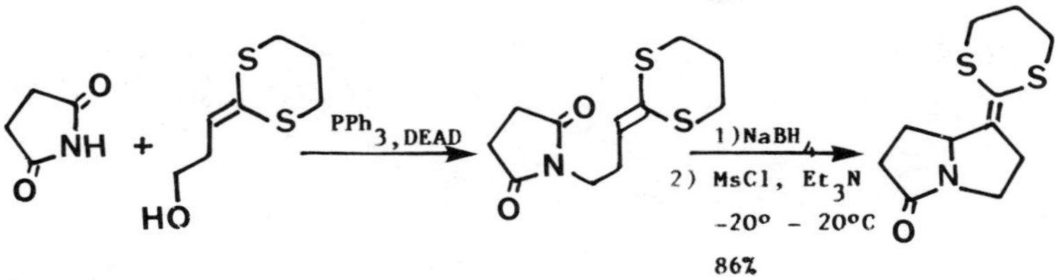

O NH O + S S HO —PPh_3, DEAD→ O N O S S —1) $NaBH_4$ 2) MsCl, Et_3N −20° − 20°C 86%→ S S N O (72)

α-Sulfinyl carbanions can be reacted with esters to afford carbonyl derivatives, which when elaborated further by oxidative procedures afford new esters, comprising a functionalized group in the α-position[1559,1604] (Scheme 73).

$$\text{PhCOOR} + \text{MeSOMe} \xrightarrow[88\%]{t\text{-BuOK}} \text{PhCO—CH}_2\text{SOMe} \xrightarrow[75\%]{\text{Br}_2}$$

$$\text{PhCO—CHBr—SOMe} \xrightarrow[60\%]{\text{MeOH}} \text{PhCO—COOMe} \qquad (73)^{1559}$$

Stabilized α-sulfinyl carbanions, such as those obtained from methyl methylthiomethyl sulfoxide (or sulfone), on reaction with aldehydes[713,1605–1607] (Scheme 74) or esters[708, 1606] (Scheme 75) give directly α-alkoxy or α-keto esters.

$$\mathrm{PhCHO + MeSCH_2{-}SO_2{-}Me \xrightarrow[70\%]{K_2CO_3,\ iPrOH} PhCH{=}C(SMe)(SO_2{-}Me)} \qquad (74)^{713}$$

$$\xrightarrow[\text{2) HCl, reflux, MeOH}]{\text{1) } H_2O_2\text{, AcOH, 2 d}} \mathrm{MeO{-}CHPh{-}COOMe}\ \ 76\%$$

$$\mathrm{PhCOOEt + MeSCH_2{-}SO{-}Me \xrightarrow[\text{(2) AcOH 84\%}]{\text{(1) NaH, THF}} PhCO{-}CH(SMe)(SOMe)}$$

$$\xrightarrow[\text{2) } Ac_2O/py;\ Et_3N\ 76\%]{\text{1) } NaBH_4\ 98\%} \mathrm{PhCH{=}C(SMe)(SOMe)} \xrightarrow[78\%]{} \mathrm{PhCH_2COOEt} \qquad (75)^{708}$$

The last step in sequence (75) proceeds equally well for intermediates derived from aryl[1607] and simple alkyl esters. The reaction with nitriles (in the presence of $CuCl_2$) (Scheme 76)[1608] or alkyl or aryl halides[1606,1609] (involving phase-transfer catalysis and a subsequent oxidation step) (Scheme 77) gives access to a variety of substituted esters.

$$\mathrm{PhCH_2CH_2CN + MeSCH_2SOMe \xrightarrow[74\%]{NaH} (PhCH_2CH_2)(H_2N)C{=}C(SMe)(SOMe)} \qquad (76)$$

$$\xrightarrow[70\%]{CuCl_2,\ H_2O/EtOH} \mathrm{PhCH_2CH_2{-}COCOOEt}$$

$$\mathrm{C_7H_{15}CH{=}CHCH_2Br + MeSCH_2{-}Ts \xrightarrow[93\%]{(C_8H_{17})_3MeNCl,\ TolH,\ NaOH}}$$

$$\mathrm{C_7H_{15}CH{=}CHCH_2CH(SMe){-}Ts \xrightarrow[\text{2) } H^+\text{, MeOH 54\%}]{\text{1) } H_2O_2\text{, AcOH}}} \qquad (77)^{1606}$$

$$\mathrm{C_7H_{15}CH{=}CHCH_2COOMe}$$

A sequence relying on the Wittig olefination, using $Ph_3P{=}CHOMe$ as the ylide reagent, has been applied to a variety of ketones and aldehydes; the yields were good in general. Some compounds, such as the one given in Scheme (78)[1610] are not accessible by other methods, since an abstraction of a cyclopropyl proton is involved; singlet oxygenation is the only viable route.[1611]

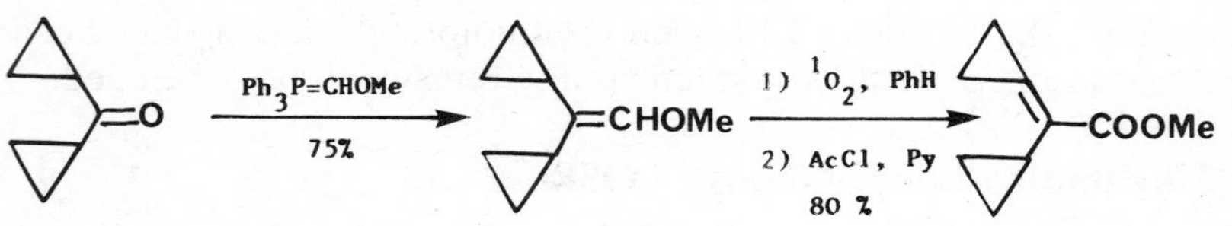

=O $\xrightarrow[75\%]{Ph_3P{=}CHOMe}$ =CHOMe $\xrightarrow[2)\ AcCl,\ Py\quad 80\ \%]{1)\ ^1O_2,\ PhH}$ –COOMe (78)

The use of carbanions obtained from 2-oxazolines[1528] or from enol ethers (Scheme 79)[1054] has also been reported for the synthesis of esters through umpoled hydroxycarbonyl or related synthons. In this case, the reported copper–lithium derivative gives access to γ-ketoesters after ozonolysis of the enol ether intermediate.

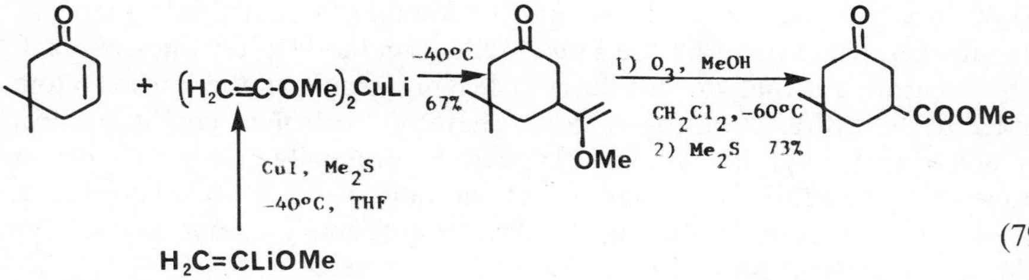

O + $(H_2C{=}C\text{-}OMe)_2CuLi$ $\xrightarrow[67\%]{-40°C}$ O, OMe $\xrightarrow[CH_2Cl_2,\ -60°C\quad 2)\ Me_2S\quad 73\%]{1)\ O_3,\ MeOH}$ O, COOMe

CuI, Me_2S, −40°C, THF

$H_2C{=}CLiOMe$

(79)

Dichloroacetic acid or ester enolates react with aldehydes and ketones and with a variety of work-up procedures to give α-keto esters in moderate yields (Scheme 80).[1612] Overall, oxidative $^-$COOMe synthon equivalency is involved here.

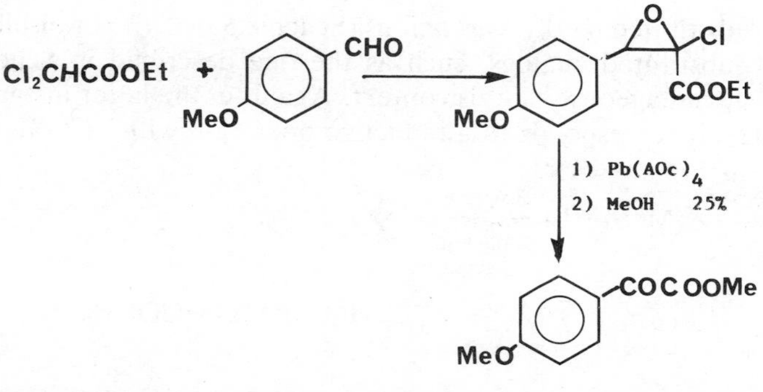

$Cl_2CHCOOEt$ + MeO, CHO ⟶ MeO, O, Cl, COOEt

1) $Pb(AOc)_4$
2) MeOH 25%

MeO, COCOOMe (80)

Also reported are sequences for the synthesis of ring fused butyrolactones from isonitriles[1613] by means of an 1,4-addition to α,β-unsaturated aldehydes and ketones (Scheme 81).

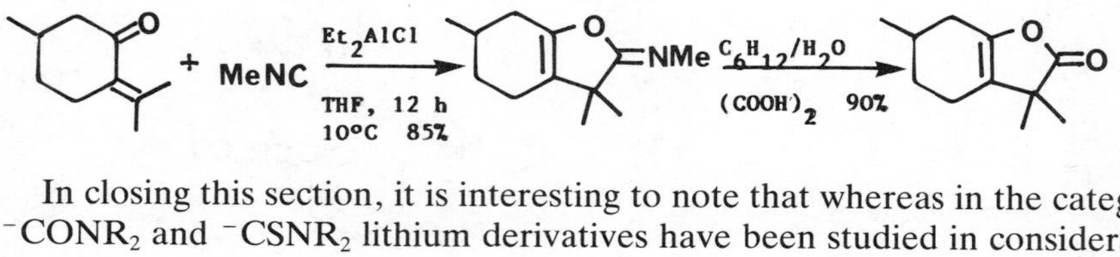

O + MeNC $\xrightarrow[THF,\ 12\ h\quad 10°C\quad 85\%]{Et_2AlCl}$ O, =NMe $\xrightarrow[(COOH)_2\quad 90\%]{C_6H_{12}/H_2O}$ O, =O (81)

In closing this section, it is interesting to note that whereas in the category $^-CONR_2$ and $^-CSNR_2$ lithium derivatives have been studied in considerable

detail (see Sect. 3.3.5. and 3.3.10.), ethyl lithioformate has in fact been generated and trapped[411] but its synthetic utility remains to be developed.

3.3.8. (Alkylthio)carbonyl Anions $^{-}$COSR

Thiol esters are synthetically useful in acylation where they often show enhanced or more specific reactivity as compared to their oxygen analogs.[1614,1615] Their synthesis through umpoled thioalkyl- or thioarylcarbonyl synthons is reported by a variety of methods, generally involving suitable thioalkyl- or -aryl-substituted precursors.

The reaction of HCN with *N*-tosylhydrazones[1616–1619] gives *N*-(1-cyanoalkyl)-*N'*-tosylhydrazines in high yields.[1441] Addition of ethanethiol to the carbon–nitrogen triple bond by the Pinner–Klein method[1620] furnishes *N*-alkyl-*N'*-tosylhydrazines that do not carry a leaving group at the carbon atom bonded to the hydrazine nitrogen atom. Therefore hydrolytic conditions now give access to *S*-alkyl thiocarboxylates, derived originally from aldehydes or ketones by a one-carbon elongation but without true $^{-}$COSR involvement, however, as the products are not α-hydroxy thiolesters (Scheme 82).[1441] The yields are in general good, and the procedure convenient.

N-NHTs —[1) HCN; 2) EtSH; 3) H_2O, steam dist.]→ COSEt (55%) + COOH (20–25%) (82)

Tris-alkylthiosubstituted methyl carbanions (Scheme 83),[1621,1622] or bisalkylthio trialkylsilyl-substituted analogs, such as the one described in Scheme (84)[1623] are used advantageously in this context. Again, in the latter instance, the overall net result corresponds to a reductive operation with $^{-}$COSR.

$$(MeS)_3CH \xrightarrow[-78^\circ]{BuLi,\ THF} (MeS)_3CLi \xrightarrow[2)\ BF_3\cdot Et_2O,\ HgO\ 80-95\%]{1)\ Me_2CO} HO{-}CMe_2{-}COSMe \qquad (83)^{1622}$$

and

MeO, MeO-benzaldehyde (CHO) + $LiC(SMe)_2SiMe_3$ → MeO, MeO-C$_6$H$_3$-CH=C(SMe)SMe —[CF_3COOH, 86%]→ MeO, MeO-C$_6$H$_3$-CH_2COSMe (84)

A recently described variant to this procedure employs methoxyphenylthiotrimethylsilylmethyllithium as a convenient reagent for the homologation of carbonyl compounds to the corresponding phenylthioesters.[756]

Sequences starting from dithiosubstituted precursors involve oxidative elaboration of the carbonyl carbon atom;[825] those using methyl methylthiomethyl sulfoxide precursors[715,1455,1624] as starting material rely on the formation of intermediate ketenedithioacetals (Scheme 85), which can then be transformed into thiolesters.[1624]

$$i\text{PrCN} + \text{MeSCH}_2\text{SOMe} \xrightarrow{\text{NaH}} (\text{H}_2\text{N})(i\text{Pr})\text{C}{=}\text{C}(\text{SMe})(\text{SOMe})$$

$$\xrightarrow[68\%]{(\text{C}_{11}\text{H}_{23}\text{CO})_2\text{O, py}} \text{C}_{11}\text{H}_{23}\text{CO—NH—C(SMe)(Pr-}i\text{)—COSMe} \qquad (85)$$

α-Sulfinyl carbanions were also shown to be useful in this respect,[823,1559] as illustrated in Scheme (86).

$$\text{PrCHO} \xrightarrow{\text{LiCH}_2\text{SOTol}} \text{HO—CHPr—CH}_2\text{SOTol} \xrightarrow[\text{CF}_3\text{COOH 70\%}]{\text{DCC, DMSO, py}}$$

$$\text{PrCOCH}_2\text{SOTol} \xrightarrow[\text{2 h, reflux 70\%}]{\text{Ac}_2\text{O, NaOAc}} \text{AcO—CHPr—COS—Tol} \qquad (86)^{823}$$

Analogous structures with additional α-substituents, such as α-chloro (Scheme 87),[1625] involving a silicon Pummerer rearrangement,[1626–1628] or α-nitro derivatives (Scheme 88)[1629] have also found application in the preparation of thiol esters.

$$\text{PhSOCH}_2\text{Cl} \xrightarrow[\text{2) PhCH}_2\text{Br 87\%}]{\text{1) LDA, }-78^\circ}$$

$$\text{PhSO—CHCl—CH}_2\text{Ph} \xrightarrow[\text{2) Me}_3\text{SiCl, }-78\text{–}60^\circ\text{, 2 h 60\%}]{\text{1) LDA, }-78^\circ} \text{PhS—COCH}_2\text{Ph} \qquad (87)$$

The spectrum of nucleophiles used in Scheme (88) spans from alcohols over amides (shown), imides, and sulfonic acids to malonates, and leads to thiol-

esters with a wide range of α substituents. Direct oxidation of the thioenol ether furnishes unsubstituted thiolesters.

$$PhS{-}CH_2NO_2 \xrightarrow[\text{2) MsCl, Et}_3\text{N, }-78^\circ\text{ to }0^\circ\text{, 60\%}]{\text{1) MeCHO, KOH, MeOH, }0^\circ} MeCH{=}C(SPh){-}NO_2 \xrightarrow[\text{2) O}_3\text{, MeOH, }-78^\circ\text{, 62\%}]{\text{1) DMF, CH}_2\text{FCONH}_2\text{, }-30^\circ} FCH_2CONH{-}CHMe{-}COSPh \qquad (88)$$[1629]

3.3.9. (Alkylamino- and Dialkylamino)thiocarbonyl Anions $^-CS{-}NHR$ and $^-CS{-}NR_2$

The synthesis of thioamides from isonitriles and aldehydes, using $Na_2S_2O_3$ as the source of sulfur, has been reported (Scheme 89):[1468]

$$CH_2O + Me_2NH + CX{-}NC \xrightarrow[\text{86\%}]{\text{Na}_2\text{S}_2\text{O}_3\text{, H}_2\text{O}} Me_2NCH_2{-}CS{-}NH{-}CX \qquad (89)$$

The generation of $LiCSNR_2$ and reaction with a large variety of electrophiles, such as aldehydes, ketones, α,β-unsaturated ketones (1,2-addition), esters (to give α-ketothioamides), and primary alkyl iodides, yields variously functionalized thioamides[2308,2309] (Scheme 90). The substitution pattern at the amide nitrogen atom was also investigated.

$$Me_2N{-}CSH \xrightarrow[-100^\circ\text{, 85\%}]{\text{LDA, THF}} Me_2N{-}CSLi$$

$$Me_2N{-}CSLi \xrightarrow[\text{75\%}]{\text{PhCHO}} HO{-}CHPh{-}CS{-}NMe_2$$

$$Me_2N{-}CSLi \xrightarrow[\text{85\%}]{\text{PhCOOMe}} PhCO{-}CS{-}NMe_2 \qquad (90)$$[2308]

$$Me_2N{-}CSLi \xrightarrow[\text{70\%}]{\text{EtI}} Et{-}CS{-}NMe_2$$

3.3.10. (Alkylamino)selenocarbonyl Anions $^-CSe{-}NHR$

Isonitrile chemistry in the presence of selenides has yielded the only synthesis of selenoamides by means of an umpoled hydroxycarbonyl synthon (Scheme 91).[1468]

$$R_2NH + i\text{-}PrCHO + C_6H_{11}{-}NC \xrightarrow{\text{Na}_2\text{Se}} R_2N{-}CH(Pr\text{-}i){-}CSe{-}NHC_6H_{11} \qquad (91)$$

4

HETEROATOM-SUBSTITUTED sp^3 CARBANIONIC SYNTHONS

Joseph E. Saavedra

CONTENTS

4.1. INTRODUCTION

The formation of new carbon–carbon bonds is central to organic synthesis. Since many target molecules contain heteroatoms, α-heterosubstituted carbanions are valuable intermediates that react with electrophiles to form new bonds. This allows further transformation, or branching of the molecule when a reactive group is introduced on the newly formed linkage.[374] This chapter surveys the preparation and subsequent reactions of carbanionic synthons substituted by a single heteroatom in the α-position. To limit the scope of this discussion, only synthons in which the carbanionic center is an sp^3 carbon atom will be covered. The α-heterosubstituent will be derived from elements of the fourth through the seventh main groups. Species from groups Va and VIa bearing a partial positive charge, such as phosphinyl, sulfinyl, sulfonyl, seleninyl, etc., will not be covered. Also omitted will be ylides, that is, heteroatoms from these two groups that bear a complete positive charge and their corresponding carbanions. The only digression will be in the case of nitrogen. Here, masked primary, secondary, and tertiary amino carbanions often bear a partial positive charge (*i.e.*, nitro, nitrosamines, amine oxides). The metal Met is generally lithium, but other alkali metals, magnesium, and copper are used on occasion.

C—C bonds are constructed, in general, by polar processes achieved through the interaction of a nucleophilic and an electrophilic center. The "philicity" of the carbon is determined by the electronic environment conferred by the activating group attached to it.[370] Heteroatoms Y in Scheme (1), such as sulfur,

phosphorus, and silicon, convey nucleophilic character to the α-carbon, thus these atoms are able to stabilize the adjacent carbanion. On the other hand, carbon atoms adjacent to a heteroatom Y′ (Scheme 2), such as nitrogen, oxygen, or halogen, are electrophilic centers. It is convenient now to adopt Seebach's[6] nomenclature of subscripts d (donor), and a (acceptor) to indicate the polarity of the molecules. For a heteroatom Y′ (d) to stabilize

$$\underset{\mathbf{d\ a}}{\mathrm{CH{-}Y}} \xrightarrow{\mathrm{R{-}Met}} \mathrm{Met^{+}}\ \underset{\mathbf{d\ a}}{\mathrm{^{-}C{-}Y}} \xrightarrow{\mathrm{El^{+}}} \underset{\mathbf{d\ a}}{\mathrm{El{-}C{-}Y}} \quad (1)$$

an adjacent carbanion, its normal reactivity mode must be reversed. The synthetic value of this reversal cannot be overstated. It has been the subject of numerous review articles.[1,6,370,374,1640,1641] This process, referred to by Corey[1] as "symmetrization of reactivity," by Evans[1640] as "charge affinity inversion operations," and by Lever[370] as "polarity inversions," is now universally called "umpolung."[1641] The complete process is outlined in Scheme (2). The donor heteroatom Y′ is converted into an acceptor site Z that is able to stabilize an α-carbanion. This newly formed donor carbon can undergo reactions with electrophiles followed by the restoration of the normal reactivity mode induced by Y′.

$$\underset{\mathbf{a\ d}}{\mathrm{CH{-}Y'}} \xrightarrow{\mathrm{Z}} \underset{\mathbf{d\ a}}{\mathrm{CH{-}Z}} \xrightarrow{\mathrm{R{-}Met}} \mathrm{^{+}Met}\ \underset{\mathbf{d\ a}}{\mathrm{^{-}C{-}Z}} \xrightarrow{\mathrm{El^{+}}}$$

$$\underset{\mathbf{d\ a}}{\mathrm{El{-}C{-}Z}} \xrightarrow{\mathrm{Y'}} \underset{\mathbf{a\ d}}{\mathrm{El{-}C{-}X'}} \quad (2)$$

Heteroatom exchange and modification are two of the general headings reviewed by Seebach[6] on methods used for reactivity umpolung. These two are of major importance in this chapter. Since our purpose is to obtain a donor site attached to a heteroatom, it will be necessary in some cases to use either of the above techniques to reverse its polarity transiently. This indicates that a heteroatom exchange is a temporary operation with sufficient flexibility to allow return to the original element. The modification of the heteroatom must be such that easy access to the original form is possible. To illustrate briefly the principles behind heteroatom exchange and modification, let us consider α-hydroxy carbanion synthons $^{-}$C—OH.

A simple esterification of an alcohol to its 2,4,6-triisopropylbenzoate derivative is a good example of heteroatom modification. Metallation takes place adjacent to the oxygen. Reaction with an electrophile El^{+}, and subsequent

demasking gives the corresponding alcohol. The entire process (Scheme 3) constitutes an umpolung method for primary alcohols.[1642]

$$\underset{\mathbf{a}\ \ \mathbf{d}}{CH_3OH} \longrightarrow CH_3OOC\text{–}Ar \xrightarrow{R^-\ Met^+} {}^-CH_2OOC\text{–}Ar\ {}^+Met$$

$$El\text{–}CH_2OH \xleftarrow[2)\ \text{hydrolysis}]{1)\ El^+} \ \simeq\ {}^-CH_2OH \qquad (3)$$

Dimesitylboron is representative of a heteroatom replacement process. Metallation of an alkyldimesitylborane[1643] provides an α-hydroxy synthon $^-CHR—OH$ (Scheme 4).

$$\underset{\mathbf{d}\ \mathbf{a}}{RCH_2\text{—}BMes_2} \longrightarrow {}^+Met\ {}^-CHR\text{—}BMes_2 \xrightarrow{El^+}$$

$$El\text{—}CHR\text{—}BMes_2 \longrightarrow \underset{\mathbf{a}\ \mathbf{d}}{El\text{—}CHR\text{—}OH} \qquad (4)$$

Carbanions can be obtained by direct metallation when the heteroatom confers a donor character to the adjacent carbon. Good candidates for direct metallation are alkyl phenyl thioethers RCH_2SAr,[1644] and some ethers RCH_2OR'.[1645,1646] Whether α-heteroatom carbanion formation is direct or umpoled, it only constitutes a type of substitution reaction, and is subject to limitations. There are many other well-established methods for the synthesis of organometallics apart from the hydrogen–metal exchange reaction. Some of these methods can be applied to heterosubstituted organometallics.[374] Halogen–metal exchange is an efficient way to prepare lithio halocarbenoids (Scheme 5).[1647,1648]

$$RCHBr_2 \xrightarrow[-110^\circ]{n\text{-BuLi}} R\text{—}CHBrLi \simeq {}^-CHBr\text{—}R \qquad (5)^{1648}$$

Other substitution reactions include metal–metal and metalloid–metal exchanges.[374] A metal–metal exchange is also referred to as a transmetallation reaction. The element being replaced, usually by lithium, belongs to group IVa. This is an excellent way of preparing certain α-heterosubstituted organolithium reagents. For example, tin–lithium exchange has become a general

procedure for obtaining alkylsubstituted α-alkoxyorganolithium synthons (Scheme 6).[1649,1650]

$$RCH_2{-}OR' \xrightarrow{n\text{-BuLi}} Li^+ \ {}^-CHR{-}OR' \tag{6}$$

A similar transmetallation reaction features in a pathway to α-tertiary amino carbanions.[1651] Carbon–metal cleavages for silicon, lead, or germanium have also been reported.[1652]

Similarly, through transmetallation and metal–halogen exchange, lithium can replace an element of groups Va and VIa to produce a new carbanion. Selenium-stabilized carbanions are obtained from C—Se bond cleavage by *n*-butyllithium (Scheme 7).[1556,1653] These metal–metalloid exchanges will be discussed in detail in Sections 4.6.2 and 4.7.

$$RSeCH_2SeR \xrightarrow{n\text{-BuLi}} Li^+ \ {}^-CH_2SeR \tag{7}$$

Addition of alkyllithium or Grignard reagents to heterosubstituted ethylenes represents another way of preparing α-heterosubstituted carbanions (Scheme 8). This has been observed with vinyl sulfides[1071,1072] and vinyl silanes.[1654] The R substituent on the heteroatom (Scheme 8) is limited to an aryl group or a group with electron-withdrawing substituents.[1651] Additions of alkyllithiums to vinyl phosphines,[1655] arsines,[1656] selenides,[374] and germanes[1657] have also been reported.

$$CH_2{=}CH{-}YR \xrightarrow{R'Met} R'CH_2CH^-YR \ {}^+Met \tag{8}$$

Factors contributing to the ease of formation and stability of α-heterosubstituted carbanions have been examined in several review articles.[374,1651,1652,1658,1659] There is no general agreement about the mechanisms by which heteroatoms stabilize a carbanionic center. Streitwieser and coworkers[1660,1661] determined by equilibrium acidity measurements that stabilization of carbanion by sulfur through polarization is the principal mechanism. This is in agreement with molecular orbital calculations in which stabilization by carbon, oxygen, and sulfur is compared,[1662,1663] and they concluded that *d*-orbital conjugation is irrelevant for $^-CH_2SH$ anion stabilization. The work of J.-M. Lehn and coworkers[1664] also supports the theory that polarization of electron distribution through the negative charge and not (*d-p*) bonding is responsible for carbanion stabilization. Bordwell and colleagues,[1665] on the other hand, concluded that the stabilizing ability of elements with low-lying *d*-orbitals comes about by conjugative effects utilizing those orbitals.

Regardless of these views, the stability of the α-heterosubstituted carbanion

is of major importance when considering a synthesis. An α-heterosubstituted carbanionic synthon should be easily generated, it must be reasonably stable to undergo reaction with electrophiles, and the heteroatom must be retained under normal conditions. The system used to generate the carbanion (*i.e.*, the alkylmetal, solvent, chelating agents) also plays an important role in the reaction. Lithium diisopropylamide in tetrahydrofuran and HMPA at −80° are conditions used generally for nitrosamine metallation.[1641] These conditions avoid nucleophilic addition by the alkyllithium.[1666] The use of LDA/potassium-*t*-butoxide was introduced by Seebach and coworkers[1667] as an improved base system for α-nitrosamino carbanion formation. Isocyanides are generally metallated with *n*-butyllithium,[1668] but the same organolithium fails to form the anion of isobutyl phenyl sulfide in good yield.[1644] In the latter case, the more reactive *t*-butyllithium is used to achieve quantitative metallation. Preparations of α-silylcarbanions have been carried out in THF with lithium naphthalenide.[253] α-Lithiated amides are best obtained with *s*-butyllithium/(TMEDA).[1669,1828] This same system can be used for the lithiation of the α-methylene group of an ester.[1642] Metallation of saturated ethers with lithium salt-free butyl potassium was recently reported by Schlosser and coworkers.[1646]

Groups attached directly to the heteroatom greatly influence the formation and stability of the carbanion. Electron-withdrawing groups on the heteroatom, such as phenyl and trifluoromethyl, enhance the acidity of the α-carbon. The stability of the carbanion has been shown to increase by charge dipole interaction,[1641,1659] which can be extended to chiral dipole-stabilized anions.[1670] Beak and colleagues,[1671] and Meyers and colleagues[1672] determined simultaneously, by infrared spectroscopy, the initial coordination between the lithium base and the heteroatom, a necessary step leading to dipole-stabilized carbanions. Once the carbanion is formed, there is an intramolecular chelation of the heteroatoms with the lithium and solvent molecules.[1673]

In the following sections, synthons will be divided according to individual heteroatoms. Nitrogen-substituted synthons comprise α-primary, α-secondary, and α-tertiary amino carbanions. Two major groups are included as oxygen-substituted carbanions. These are the α-hydroxy and α-alkoxy carbanionic synthons. α-Thioether, α-thiol, α-thiocarbamate, and α-thioester carbanions form the family of sulfur-substituted carbanionic synthons. Carbanions substituted with halogen are referred to as halocarbenoids.[1647] These are important synthons not only in their reaction with electrophiles, but also in the nucleophilic displacement of the halogen to form a new α-heteroatom-substituted carbanion. The chemistry of α-selenoorganometallics has been extensively reviewed, mainly by Krief[374] and Clive.[1674] This survey will include only reactions of α-seleno carbanions with electrophiles to form compounds in which the selenium heteroatom remains in the molecule. Subsequent reactions of selenium compounds to olefins and other systems is beyond our scope here. Section 4.7 will deal with the remaining heteroatoms from groups IVa, Va, and VIa. Compounds containing a heteroatom Y as well as an

activating group W on the α-carbon (Scheme 9) will be discussed in Section 4.8, as long as the anions do not behave as acyl anion equivalents.

$$Y{-}\overset{\ominus}{C}{-}W \xrightarrow{El^+} Y{-}C(El){-}W \qquad (9)$$

4.2 NITROGEN-SUBSTITUTED CARBANIONIC SYNTHONS

The lone pair of electrons on a nitrogen atom makes amines powerful nucleophiles. The α-carbon is relatively unreactive. However, via iminium ions, an amine becomes an efficient α-heterosubstituted cationic synthon (Scheme 10). This synthon undergoes reactions with nucleophilic reagents,[1675] such as alkyllithiums, Grignard reagents (Scheme 11), nitroalkanes (Scheme 12), and hydrides (Scheme 13).

$$C{=}N^+ \longleftrightarrow {}^+C{-}N{:} \qquad (10)$$

$$^+C{-}NR_2 \xrightarrow{R'Met} R{-}C{-}NR_2 \quad (Met = MgX, Li) \qquad (11)$$

$$^+C{-}NR_2 \xrightarrow{^-CH_2NO_2} O_2NCH_2{-}C{-}NR_2 \qquad (12)$$

$$^+C{-}NR_2 \xrightarrow{H^-} CH{-}NR_2 \qquad (13)$$

The electrophilic reactivity at the α-carbon atom can be seen in the Pictet–Spengler isoquinoline synthesis (Scheme 14),[1676] as well as in the Mannich reaction (Scheme 15).[1677]

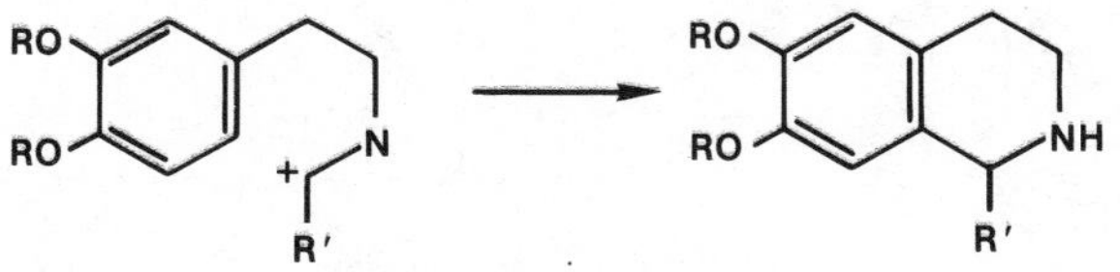

(14)

$$^+CH_2{-}NR_2 \xrightarrow{R'COCH_2^-} R'COCH_2CH_2NR_2 \qquad (15)$$

An electrophilic carbon adjacent to a nitrogen is present in oxazolidines,[1678] and in their acyclic congeners (alkylamino)alkoxymethanes.[1679] These compounds can react with Grignard reagents to form a new C—C bond (Scheme 16), or be subjected to hydride attack with a nucleophilic reducing agent

(Scheme 17)[1680] to form alkylalkanolamines. Recently *N,N*-bis-(trimethylsilyl)methoxymethylamine was introduced[214] as a convenient synthetic equivalent for $^{+}CH_2NH_2$.

$$ROCH_2NR_2 \simeq {}^{+}CH_2NR_2 \xrightarrow{R'MgX} R'CH_2NR_2 \quad (16)$$

$$\text{RCH}\begin{matrix}\diagup O \\ \diagdown NH\end{matrix}\Big] \simeq {}^{+}CHR{-}NHCH_2CH_2OH \xrightarrow{H^-} RCH_2NHCH_2CH_2OH \quad (17)$$

Because of the donor character of the nitrogen atom, direct metallation of the carbon adjacent to an amine cannot be carried through. Fortunately, the multiplicity of oxidation states inherent to nitrogen gives the chemist great flexibility in heteroatom modification to produce an umpoled synthon. Generally, all this requires is the introduction of an electron-withdrawing group on the amine nitrogen. A partial positive charge on the nitrogen will promote the formation of an α-carbanion, where dipole polarization factors are a major contribution to its stabilization.[1659] Primary and secondary amines are well suited to the umpoled approach. α-Tertiary-amino carbanions ($^{-}C{-}NR_2$) are best obtained by metal interchange.[1681]

4.2.1. α-Primary-Amino Carbanions $^{-}C{-}NH_2$

Of the limited number of methods that have been developed to form α-primary amino carbanions, all involve heteroatom modification (Scheme 18).

$$\underset{a\quad d}{CH{-}NH_2} \longrightarrow \underset{d\quad a}{CH{-}NY_2} \longrightarrow {}^{-}C{-}NY_2 \xrightarrow{El^+} El{-}C{-}NY_2 \longrightarrow \underset{a\quad d}{El{-}C{-}NH_2} \quad (18)$$

4.2.1.1. α-NITROSAMINOALKYL ETHERS

Lithiation of nitrosamines is a well-established method of considerable synthetic value for preparing masked α-secondary amino carbanions ($^{-}CH_2NHR$).[1641] The α-carbon polarity of a primary amine can be reversed by converting it to a chemical species similar to a secondary amine that can then be nitrosated. The nitrosamine can undergo electrophilic attack at the

α-carbon, and a demasking step will then regenerate the primary amine (Scheme 19).[1682]

$$\underset{\mathbf{a}\quad\mathbf{d}}{\text{CH—NH}_2} \xrightarrow[\text{(2) HONO}]{\text{(1) RY}} \underset{\mathbf{d}}{\text{CH—NR—NO}} \xrightarrow{\text{LDA}}$$

$$\underset{\mathbf{d}}{\text{Li}^+ \ {}^-\text{C—NR—NO}} \xrightarrow{\text{El}^+} \text{El—C—NR—NO} \xrightarrow[\text{—NO}]{\text{—R}} \underset{\mathbf{a}}{\text{El—C—NH}_2} \qquad (19)$$

To carry out the reactions outlined in Scheme (19), the primary amine, such as methylamine, is converted to an α-nitrosaminoalkyl ether **(1)** Scheme 20. The preparation of **1** can be carried out by condensation of the amine with acetaldehyde–methanol in the presence of nitrous acid,[1682,1683] or by acid-catalyzed addition of methanol to *N*-nitroso-*N*-methylvinylamine (**2,** Scheme 20).[1684]

$$\text{MeNH}_2 + \text{MeCHO} + \text{MeOH} \xrightarrow{\text{HONO}} \underset{\mathbf{1}}{\text{MeN(NO)—CHMe—OMe}} \qquad (20)$$

$$\underset{\mathbf{2}}{\text{Me—N(NO)—CH=CH}_2} \longrightarrow \mathbf{1}$$

N-Nitroso-*N*-methyl-1-methoxyethylamine **(1)** was metallated at −80° with LDA, followed by addition of methyl iodide to give *N*-nitroso-*N*-ethyl-1-methoxyethylamine **(3)** in 80% yield. Treatment of **3** with ethyl chloroformate in moist acetone, followed by acid hydrolysis gave the corresponding primary amine in 80% yield (Scheme 21).[1682]

$$\mathbf{1} \xrightarrow[\text{THF, }-80^\circ]{\text{LDA}} {}^-\text{CH}_2\text{N(NO)—CHMe—OMe} \simeq {}^-\text{CH}_2\text{NH}_2 \xrightarrow{\text{MeI}}$$

$$\underset{\mathbf{3}}{\text{EtN(NO)—CHMe—OMe}} \xrightarrow[\text{2) 6 }N\text{ HCl}]{\text{1) ClCOOEt}} \text{EtNH}_2\text{HCl} \qquad (21)$$

Acetone as the electrophile reacted with the lithioanion of **1** to give *N*-nitroso-*N*-(2-hydroxypropyl)-1-methoxyethylamine **(4).** Upon hydrolytic demasking 1-amino-3-methyl-2-propanol hydrochloride was obtained (Scheme 22).

$$\mathbf{1} \xrightarrow[\text{2) Me}_2\text{CO}]{\text{1) LDA}} \underset{\mathbf{4}}{\text{HO—CMe}_2\text{—CH}_2\text{N(NO)—CHMe—OMe}} \longrightarrow$$

$$\text{HO—CMe}_2\text{—CH}_2\text{NH}_2 \qquad (22)$$

Alkanolamines form *N*-nitrosooxazolidines in the presence of aldehydes and nitrous acid.[1683,1685] These compounds, which are cyclic congeners of α-nitrosaminoalkyl ethers, have acidic protons at the C-4 position, and can serve as umpoled synthons of β-alkanolamines.[1685,1686] These synthons are of limited utility since no regioselectivity of alkylation is observed in the reaction, and multiple alkylation occurs to a large extent. This can be seen in the metallation and subsequent methylation of *N*-nitroso-2-methyloxazolidine (**5**, Scheme 23), where isomeric *N*-nitroso-2,4-dimethyloxazolidines **(6)** were formed in 32% yield and *N*-nitroso-2,4,4-trimethyloxazolidine **(7)** was obtained in 11% yield.

O NNO CH3 5 —LDA, -80°→ $^{-}$Li^{+} O NNO CH3 ≡ $HOCH_2CH^{-}NH_2$

—MeI→ CH3 O NNO CH3 6 + CH3 CH3 O NNO CH3 7 (23)

Denitrosation of **6** with HCl in benzene followed by hydrolysis gave 2-aminopropanol hydrochloride (Scheme 24), whereas **7** gave 2-amino-2-methylpropanol hydrochloride (Scheme 25).

$$\mathbf{6} \longrightarrow HOCH_2\text{—}CHMe\text{—}NH_2HCl \qquad (24)$$

$$\mathbf{7} \longrightarrow HOCH_2\text{—}CMe_2\text{—}NH_2HCl \qquad (25)$$

4.2.1.2. Isocyanides

Alkyl isocyanides were first metallated by Schöllkopf and coworkers.[1687] Addition of aldehydes or ketones to the α-lithioisocyanide gave olefins together with oxazolines. Consequently, at the outset isocyanides were considered olefination reagents rather than sources for α-primary amino carbanion synthons. However, further work revealed that neutralization of the carbonyl adduct with glacial acetic acid at −70° gave 2-isocyano-1-alkanols[1688,1689] from which, upon hydrolysis, β-alkanolamines were obtained (Scheme 26).

$$CH_2N{=}C \xrightarrow{R'Li} Li^+ \ {}^{-}CHN{=}C \simeq {}^{-}CH_2NH_2$$

$$\downarrow R_2CO \qquad (26)$$

$$OH\text{—}CR_2\text{—}CH_2NC \xrightarrow{H^+} HO\text{—}CR_2\text{—}CH_2NH_2$$

Chain-lengthening of a primary amine is accomplished through α-lithiated isocyanides and alkyl halides (Scheme 27).[1690] Similarly, the α-metallated synthon opens epoxides and oxetane[1668] to give 3- and 4-aminoalcohols (Scheme 28). A limitation encountered with isocyanides is that secondary alkylisocyanides without other activating substituents cannot be metallated.[1691]

$$^{-}CH_2N{=}C \xrightarrow{RX} RCH_2N{=}C \xrightarrow{H^+} RCH_2NH_2 \qquad (27)$$

$$^{-}CH_2N{=}C \xrightarrow{A} HO{-}CHR'{-}CH_2CH_2N{=}C \longrightarrow HO{-}CHR'CH_2CH_2NH_2$$

or

$$HO(CH_2)_4N{=}C \longrightarrow HO(CH_2)_4NH_2 \qquad (28)$$

where A = alkyloxirane or oxetane.

4.2.1.3. Imines

Two independent groups, Kauffmann and colleagues[1692] and Hullot and colleagues,[1693] were responsible for the development of α-primary aminocarbanion synthons by way of diphenylmethyleneaminomethyllithium **(8a).** This is formed[1692] on metallation of *N*-(diphenylmethylene)methylamine **(8)** with lithium diisopropylamide in THF/ether at −45°. Metallation was also carried out[1693] using lithium diethylamide in HMPA/benzene at −70° or with *n*-butyllithium in THF at −78°. Chain-lengthening of methylamine takes place on addition of alkyl halides to **8a** (Scheme 29).[1693] Nucleophilic aminomethylation of ketones gives 2-aminoalkanols (Scheme 30),[1692] and epoxide opening yields 3-aminoalkanols (Scheme 31).[1693]

$$\underset{\mathbf{8}}{MeN{=}CPh_2} \longrightarrow \underset{\mathbf{8a}}{Li^{+}\,^{-}CHN{=}CPh_2}$$

$$\xrightarrow[2)\ H^+]{1)\ RX} RCH_2NH_2 \qquad (29)$$

$$\xrightarrow[2)\ H^+]{1)\ R_2CO} HO{-}CR_2CH_2NH_2 \qquad (30)$$

$$\xrightarrow[2)\ H^+]{1)\ Epoxide} HO{-}CHR{-}CH_2CH_2NH_2 \qquad (31)$$

Dithiocarboxylation of amines introduces the bis-(alkylthio)methylenamino group $(RS)_2C{=}N$. This masked primary amino function is capable of stabi-

lizing α-carbanions by allylic resonance.[1694] *N*-Alkyliminocarbonates are both masked α-amino and α-thio carbanions. However, this type of compound is an effective α-amino carbanion synthon only when the α-methylene bears an electron-withdrawing group. Potassium *t*-butoxide in THF at −70° is used to form the anions (Scheme 32).

$$\begin{aligned} &RCH_2N{=}C(SMe)_2 \longrightarrow {}^-CHR{-}N{=}C(SMe)_2 \longleftrightarrow \\ &\qquad RCH{=}NC^-(SMe)_2K^+ \xrightarrow{R'X} \\ &\qquad RR'CHN{=}C(SMe)_2 + RCH{=}N{-}CR'(SMe)_2 \end{aligned} \tag{32}$$

4.2.1.4. Nitroalkanes

The strong electron-withdrawing nitro group can activate a neighboring C—H group. Thus, the Henry reaction[1695] is a classical method for C—C bond formation by nitroaldol addition to give vicinal nitroalcohols. These compounds, to a limited extent, appear suitable for use as masked primary-amino carbanion synthons (Scheme 33).

$$\underset{\mathbf{a}\quad\mathbf{d}}{CH_3NH_2} \xrightarrow{(O),\ base} \underset{\mathbf{d}\quad\mathbf{a}}{{}^-CH_2NO_2} \xrightarrow{RX} RCH_2NO_2 \xrightarrow{(H)} RCH_2NH_2 \tag{33}$$

There are two drastic steps that limit their application. The amine must first be oxidized to the nitro group to reverse the polarity, and to complete the cycle, the nitro group has to be reduced back to a primary amine. Further limitations result from the tendency of nitro compounds to undergo oxygen, in preference to carbon, alkylation. Even though this problem rules out the use of alkyl halides, several techniques have been developed to improve yields in the nitroaldol addition. Rosini and coworkers[1696] prepared 2-nitroalkanols on alumina surfaces without solvent (Scheme 34).

$$RCH_2NO_2 + R'CHO \xrightarrow{Al_2O_3} HO{-}CHR'{-}CHR{-}NO_2 \tag{34}$$

Matsumoto[1697] carried out the reaction under pressure with n-Bu_4NF as catalyst. Seebach and team[1698] converted the alkyl nitrate into a silyl nitronate. Condensation with an aldehyde (ketones do not react) gave the protected vicinal nitroalcohol that was then reduced to the 2-aminoalkanol with lithiumaluminum hydride (Scheme 35).

$$\begin{aligned} &RC{=}N^+\begin{matrix} \diagup O^- \\ \diagdown OSiR'_3 \end{matrix} + R''CHO \longrightarrow R'_3SiO{-}CHR''{-}CHR{-}NO_2 \\ &\qquad \xrightarrow{LAH} HO{-}CHR''{-}CHR{-}NH_2 \end{aligned} \tag{35}$$

Because of the trialkylsilyl protective group, retroaldol reaction was minimized in the reduction step. To carry out the condensation, the dianion from the primary nitroalkane is formed first. The resulting intermediate is silylated *in situ* to the protected nitroalcohol prior to reduction (Scheme 36).

$$RCH_2NO_2 \xrightarrow[\text{THF/HMPA}]{\text{2BuLi}} RCNO_2Li_2 \xrightarrow[\text{2) } R'_3SiCl]{\text{1) } R''_2CO} R'_3SiO{-}CR''_2{-}CHR{-}NO_2 \quad (36)$$

Nitroaldol reactions in acyclic systems lack stereoselectivity. Methods were developed by Seebach and coworkers[1699] to overcome this problem and prepare diastereomerically enriched nitroaldols. Reduction of these nitroaldols lead to diastereomerically enriched 1,2-aminoalcohols.

4.2.1.5. *N*-Sulfinylamides

The electron-withdrawing properties of the *N*-sulfinylamide group provides a synthetic equivalent of an α-primary amino carbanion ($RCH_2NSO \simeq {}^-CHRNH_2$).[1700] The primary amine is converted on treatment with thionyl chloride to the corresponding *N*-sulfinylamide. Anion formation is carried out at −70° with lithium triphenylmethide or with potassium *t*-butoxide at 0° in glyme. The only type of electrophile used in this study was allyl halides. Regeneration of the primary amine is done hydrolytically in acid medium.

4.2.2. α-Secondary Amino Carbanions $^-C{-}NHR$

The introduction of an electron-withdrawing group on the nitrogen atom is generally sufficient to reverse the polarity on the α-carbon of a secondary amine. The nitrosamine function is able to provide the α-anion with an orbital system for delocalization of the electron pair (Scheme 37).[1701]

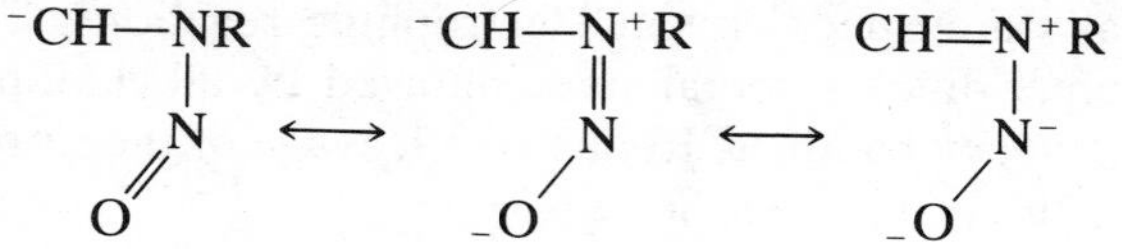

$$^-CH{-}NR(N{=}O) \longleftrightarrow {}^-CH{-}N^+R({=}N{-}O^-) \longleftrightarrow CH{=}N^+R(N^-{-}O^-) \quad (37)$$

Metallation of formamidine[1672] and amide[1671] derivatives of secondary amines gives dipole-stabilized carbanions (Scheme 38).

(38)

Nitrosamines, formamidines and amides are readily converted into the original amine by hydrolytic and reductive methods.

4.2.2.1. *N*-Nitrosamines

N-Nitrosamines have been known[1702] since 1863. However, the first hint of their usefulness as α-secondary amino carbanion synthons was reported by Keefer and Fodor[1703] in 1970. *N*-Methylnitrosamine was alkylated with methyl iodide in refluxing THF and sodium hydride as the base to give methylethylnitrosamine in 15% yield (Scheme 39). Later, Seebach and coworkers[1641] reported the quantitative metallation of dimethylnitrosamine with LDA in THF at −80°. Alkylation of the lithionitrosamine with methyl iodide gave methylethylnitrosamine in 75% yield (Scheme 39).[1704] Since that time, *N*-nitrosamines have become useful synthetic equivalents of α-secondary amino carbanions.[1641,1659]

$$\mathrm{MeN(NO)Me} \xrightarrow[15\%]{\mathrm{NaH,\ MeI}} \mathrm{EtN(NO)Me}$$

$$\mathrm{MeN(NO)Me} \xrightarrow[100\%]{\mathrm{LDA},\ -80^\circ} {}^-\mathrm{CH_2N(NO)Me} \xrightarrow[75\%]{\mathrm{MeI}} \mathrm{EtN(NO)Me} \qquad (39)$$

In addition to reactions with alkyl halides giving elongated or branched secondary amines (Scheme 40), metallated nitrosamines react with heteroatom-containing electrophiles (Scheme 41),[1705] acylating agents (Scheme 42),[1667] aldehydes, or ketones (Scheme 43),[1706] and undergo coupling reactions in the presence of iodine (Scheme 44).[1707] There is considerable regio- and stereospecificity in electrophilic reactions with cyclic nitrosamine anions.[1708–1710] This can be seen in the reaction of an electrophile and the anion of *N*-nitroso-4-phenylpiperidine[1711] to give only the axial substitution product. A second metallation followed by alkylation gave the 2,6-diaxial derivative. The synthesis of the *cis*-isomer of pseudo-conhydrin[1710] is a good illustration of the conformational stereospecificity in electrophilic reactions. The nitroso derivative of 3-piperidinol is metallated followed by alkylation with *n*-propyl iodide to give the axial adduct. Reductive cleavage of the nitroso group gave a 52% yield of the product (Scheme 45).

$${}^-\mathrm{CHR{-}NR'{-}NO} \xrightarrow{\mathrm{R''X}} \mathrm{RR''CH{-}NHR'} \qquad (40)$$

$${}^-\mathrm{CHR{-}NR'{-}NO} \xrightarrow{\mathrm{R''{-}YX}} \mathrm{R''Y{-}CHR{-}NHR'} \qquad (41)$$

$${}^-\mathrm{CHR{-}NR'{-}NO} \xrightarrow{\mathrm{R''COCN}} \mathrm{R''CO{-}CHR{-}NHR'} \qquad (42)$$

$${}^-\mathrm{CHR{-}NR'{-}NO} \xrightarrow{\mathrm{R''_2CO}} \mathrm{HOCR''_2{-}CHR{-}NHR'} \qquad (43)$$

$${}^-\mathrm{CHR{-}NR'{-}NO} \xrightarrow{\mathrm{I_2}} \mathrm{(R'NH{-}CHR)_2} \qquad (44)$$

LDA, PrI → [H] → (45)

Several procedures for demasking of nitrosamines have been developed. Hydrogenation over Raney-nickel catalyst is an effective method for cleaving the nitroso group.[1710] It has been reported recently that aluminum–nickel alloy in alkali rapidly reduces nitrosamines to the corresponding secondary amine.[1712] Nonreductive cleavages can be accomplished by passing hydrochloric acid gas through a warm solution of the nitrosamine in an organic solvent.[1706,1710]

An important property of the nitrosamine function is its ability to stabilize a positive charge, as well as the anionic species discussed above. Oxidative decarboxylation of *N*-nitrosoamino acids with lead tetraacetate is believed to go through *N*-nitrosoimminium ions.[1713] The reactivity of electronegative α-substituents (*i.e.*, acetoxy) has also been explained by an electronic stabilization of a carbonium ion by the nitrosamine function (Scheme 46).[1714,1715] Some stable *N*-nitrosoimminium ion salts have actually been isolated and characterized.[1715]

$$\mathrm{AcO{-}CHR{-}NR'{-}NO} \longrightarrow \mathrm{{}^{+}CHR{-}NR'{-}NO} \longleftrightarrow \mathrm{RCH{=}N^{+}R'{-}NO} \xrightarrow{\mathrm{Nu^-}} \mathrm{Nu{-}CHR{-}NR'{-}NO} \qquad (46)$$

The value of *N*-nitrosamines as anionic or cationic synthons has been overshadowed by their carcinogenic and toxicological properties. However, it is possible to carry out the nitrosation, alkylation, and final demasking step with minimum exposure to the chemist or the environment.[1641]

4.2.2.2. Formamidines

The substitution of amines on the α-position via dipole-stabilized carbanions from formamidines was first reported by Meyers and coworkers.[1716] These compounds are prepared by treating the *N*-formyl derivative of a secondary amine with dimethyl sulfate or triethyloxonium tetrafluoroborate to give the corresponding formamidinium salt, followed by reaction with an equivalent of a primary amine (Scheme 47).

$$\mathrm{R_2NH} \longrightarrow \mathrm{R_2N{-}CHO} \xrightarrow{\mathrm{Me_2SO_4}} \mathrm{R_2N^{+}{=}CHOMe} \xrightarrow[\text{2) HO}^-]{\text{1) R'NH}_2} \mathrm{R_2NCH{=}NR'} \qquad (47)$$

Metallation of the formamidine was carried out with *t*-butyllithium at $-78°$. After reaction of the dipole-stabilized carbanion with electrophiles, the elaborated amine could be obtained by hydrolytic or reductive[1717] cleavage (Scheme 48).

$$Me_2NCH=NR \xrightarrow{t\text{-BuLi}} \text{[HC(=NR→Li)N(Me)CH}_2\text{]} \xrightarrow{El^+} MeN(CH_2El)CH=NR \quad (48)$$

$$\xrightarrow{H^+} MeNHCH_2El$$

Hydrolytic removal of the formamidine function can be accomplished[1717] in aqueous ethanol with hydrazine and acetic acid at 53°. Basic hydrolysis in refluxing methanolic potassium hydroxide is also effective. Lithium aluminum hydride in refluxing tetrahydrofuran is used in the reductive demasking to the secondary amine.[1717]

The methods developed for metallation of formamidines and alkylation with electrophiles have been applied to the alkylation of tetrahydroisoquinolines at the 1-position (Scheme 49).[1717]

$$\text{tetrahydroisoquinoline (NH)} \xrightarrow{Me_2NCH=NtBu} \text{N-CH=NtBu derivative} \quad (49)$$

$$\xrightarrow[2.\ El^+]{1.\ s\text{-BuLi}} \text{1-El, N-CH=NtBu derivative} \xrightarrow{NH_2NH_2 \text{ or } KOH \text{ or } LAH} \text{1-El-tetrahydroisoquinoline (NH)}$$

This method allows the synthesis of optically active 1-alkyl-1,2,3,4-tetrahydroisoquinolines,[1670] accomplished by the use of chiral formamides to induce asymmetric alkylation of the α-carbon. Further elaboration of the ring provides an asymmetric approach to isoquinoline alkaloids.[1718]

Metallation of the 2-position in indolines, tetrahydroquinolines, and tetrahydrocarbolines has also been accomplished through their formamide derivatives.[1719,1720] The amino carbanion generated from the formamidine derivative of β-carboline has offered a novel entry into the indole alkaloids.[1719,1721]

4.2.2.3. Amides

Fraser and colleagues[1722] were the first to use an amide to enhance stabilization of the lithioanion to the nitrogen in dibenzylamine (Scheme 50).

$$\mathrm{Bzl_2NH} \longrightarrow \mathrm{Bzl_2NCOPh} \xrightarrow{\mathrm{LDA}} \mathrm{PhCH(Li\cdots{}^-O)\!-\!C(Ph)\!=\!N^+(Bzl)} \xrightarrow{\mathrm{El^+}} \mathrm{PhCHEl\!-\!N(Bzl)COPh} \quad 66\text{–}98\% \tag{50}$$

This finding came about during a parallel study of the metallation and alkylation of *N*-nitrosodibenzylamine. In both cases, the yields of the alkylation/acylation products were high, but the former did not involve a possibly carcinogenic intermediate. In spite of success in the electrophilic substitution of *N,N*-dibenzylbenzamide, these studies did not establish a general method for an α-secondary amino carbanionic synthon. This is because the stabilization of the anion can be attributed largely to delocalization through the aromatic ring, and not entirely to dipole stabilization by the amide function. In fact, attempts to form the carbanion of *N*-benzoylpiperidine with lithium diisopropylamide failed.[1722]

Beak and coworkers[1723] and Seebach and coworkers[1724] established amides as synthetic equivalents to α-lithio secondary amines. Conversion of a secondary amine to an amide is generally easy. The α-carbanion stabilization can be provided only by the amide function, and following the electrophilic reaction a straightforward hydrolysis to the elaborated secondary amine can be accomplished. Proton removal usually requires *s*- or *t*-butyllithium and tetramethylethylenediamine at −45° to −78°. Although benzamides have been alkylated successfully,[1725] it is best to use amides where the carbonyl group is sterically protected. Beak and coworkers have used 2,4,6-triisopropylbenzamides[1725] and 2,2-diethylbutyramides[1669,1828] to provide α-secondary amino carbanion equivalents. Derivatization of piperidines with these amides overcame the earlier failure to form the α-anion of *N*-benzoylpiperidine[1722] with LDA. However, the electrophilic substitution products of the 2,4,6-triisopropylbenzamides cannot be hydrolyzed to the corresponding amine. On the other hand, 2,2-diethylbutyramides undergo α-metallation and reaction with electrophiles to give products that can be cleaved to the substituted amine (Scheme 51).[1669]

$$\mathrm{RCH_2NHR'} \xrightarrow{\mathrm{R''COCl}} \mathrm{RCH_2N(R')C(=O)R''} \xrightarrow[\mathrm{TMEDA}]{s\text{-}\mathrm{BuLi}} \mathrm{RCH(Li\cdots O)\!-\!N^+(R')\!=\!CR''} \xrightarrow[2.\ \mathrm{H_3O^+}]{1.\ \mathrm{El^+}} \mathrm{RCHElNHR'} \quad \mathrm{R''=MeCH_2C(Et)_2\!-} \tag{51}$$

A noteworthy feature is observed in the hydrolysis of an α-hydroxyalkylated amide. There is an acid-driven *N*-to-*O* migration of the acyl group, forming an amino ester. Basic hydrolysis of this compound gives the corresponding alkanolamine (Scheme 52).

$$\begin{aligned} &\text{HO—CR}_2\text{—CHR}'\text{—NR}'''\text{—COR}'' \xrightarrow{\text{H}_3\text{O}^+} \\ &\quad \text{R}''\text{COO—CR}_2\text{—CHR}'\text{—NHR}''' \xrightarrow{^-\text{OH}} \text{HO—CR}_2\text{—CHR}'\text{—NHR}''' \end{aligned} \tag{52}$$

Seebach and coworkers have demonstrated that triphenylacetamides,[1724,1726] hindered urethanes,[1727] ureas,[1728] and pivaloylamides[1729] undergo lithiation and electrophilic substitution. Removal of the activating functionality was carried out hydrolytically in acid, or reductively with lithium aluminum hydride. In the case of tris(*t*-butyl)phenylurethane derivatives, a Friedel–Crafts transalkylation with subsequent basic hydrolysis gave the corresponding substituted amine.[1728] *N*-Pivaloyl derivatives of 1,2,3,4-tetrahydroisoquinolines were used to generate a highly nucleophilic site on the 1-position.[1729] This lithio anion reacts with aldehydes to give high yields of the 1-hydroxyalkyl product, but there is little or no stereoselectivity. Exchange of the lithium with magnesium bromide generated a diastereoselective reagent.[1730] Thus, addition of benzaldehyde to the 1-magnesium derivative of *N*-pivaloyl-1,2,3,4-tetrahydroisoquinoline gave only the *u*-isomer in 83% yield (Scheme 53).

NCOtBu, Li $\xrightarrow{\text{MgBr}_2}$ NCOtBu, MgBr

$\xrightarrow{\text{PhCHO}}$ H, NCOtBu, Ph, OH

(53)

Like amides, thioamides are synthetically useful sources of α-secondary amino carbanions.[1731] Thus, *N,N*-dimethylpivalothioamide **(9)** undergoes α-lithiation with *s*-butyllithium/TMEDA at −78° to give the dipole-stabilized intermediate **(9a)**. The anion can be trapped with a variety of electrophiles, and the products hydrolyzed to secondary amines (Scheme 54).

$$(Me)_2NCSt\text{-}Bu \ (\mathbf{9}) \longrightarrow \begin{matrix} Li\text{----}S^- \\ | \qquad | \\ CH_2\text{—}\overset{+}{N}\text{=}Ct\text{-}Bu \\ | \\ Me \end{matrix} \ (\mathbf{9b}) \xrightarrow{El^+} \tag{54}$$

$$ElCH_2N(Me)CS\text{-}tBu \longrightarrow ElCH_2NHMe$$

4.2.2.4. Phosphoramides

Dipole-stabilized carbanions can be formed on a carbon adjacent to the nitrogen of phosphoramide.[1732,1733] Benzylpentamethylphosphotriamide **(10)** was metallated with *n*-butyllithium, and the anion **10a** treated with isopropyl iodide to give the branched-chain phosphotriamide **11** in 80% yield (Scheme 55).[1734]

$$\underset{\mathbf{10}}{(Me_2N)_2PO\text{—}NMe\text{—}Bzl} \xrightarrow[-70^\circ]{BuLi} \underset{\mathbf{10a}}{(Me_2N)_2PO\text{—}NMe\text{—}CH^-Ph} \xrightarrow[80\%]{Me_2CHI} \underset{\mathbf{11}}{(Me_2N)_2PO\text{—}NMe\text{—}CHPh\text{—}CHMe_2} \tag{55}$$

When the anion **10a** was quenched with acetophenone, the diastereomers **12** and **13** were formed (Scheme 56).[1732]

$$\mathbf{10a} \xrightarrow{PhCOMe} \underset{\mathbf{12}}{(Me_2N)_2PO\text{—}NMe\text{—}\underset{H}{\overset{Ph}{C}}\text{—}\underset{Me}{\overset{Ph}{C}}\text{—}OH} + \underset{\mathbf{13}}{(Me_2N)_2PO\text{—}NMe\text{—}\underset{H}{\overset{Ph}{C}}\text{—}\underset{Ph}{\overset{Me}{C}}\text{—}OH} \tag{56}$$

Treatment of the widely used solvent hexamethylphosphoramide with *s*-butyllithium in dimethoxyethane forms solutions of α-lithiohexamethylphosphoramides.[1733] These lithioanions are quenched with carbonyl compounds to form

the β-hydroxy adducts in 40–83% yield. Acid hydrolysis of the alkylated phosphoramides results in the formation of the corresponding secondary amine.[1262] In the case of hydroxyalkylated phosphoramides, there does not appear to be a clear-cut demasking procedure to a secondary alkanolamine.[1733]

4.2.3. α-Tertiary Amino Carbanions $^{-}C—NR_2$

There are relatively few methods leading to the formation of α-tertiary amino carbanions. As for umpoled synthons for this species, none have been thoroughly developed. Quaternary ammonium compounds can be metallated on the α-position to form trialkylammonium methylides,[1735] which can undergo alkylation at this position. A selective dealkylation of the quaternary compound must be carried out to regenerate the tertiary amine. Umpolung is also observed in some amine oxides.[1735,1736] A more efficient way to generate an α-tertiary amino carbanion does not involve an umpolung route at all but a transmetallation by means of aminomethyltrialkylstannanes.[1737] Here, the α-heterosubstituted carbanion will undergo electrophilic substitution to give a new tertiary amine directly.

4.2.3.1. Ammonium Ylides

Tetramethylammonium chloride reacts with phenyllithium to form $Me_3N^+CH_2^-$. This stable ylide reacts with benzophenone to form $HO—CPh_2—CH_2N^+Me_3$, which on pyrolysis is demethylated to the corresponding tertiary amine[1738a] $HO—CPh_2—CH_2NMe_2$. Ylides of the *N,N*-dimethylpyrrolidinium system can also be trapped with benzophenone to the corresponding β-hydroxy derivative in 52% yield; however, demethylation has not been reported. Ammonium ylides seem to have very limited value as anionic synthons for tertiary amines. They are subject to Hofmann elimination if a labile β-proton is present. In the absence of a β-proton, ammonium ylides may undergo Stevens or Sommelet rearrangements.

4.2.3.2. Amine Oxides

The only case where an amine oxide has been successfully used to reverse the polarity of the α-carbon was reported by Barton and coworkers.[1736] Quinuclidine oxide **(14)** was metallated with *t*-butyllithium at −78° and the anion trapped with various electrophiles. The products were converted to the tertiary amine with hexachlorodisilane. This established a one-step synthesis of the ruban skeleton (Scheme 57).

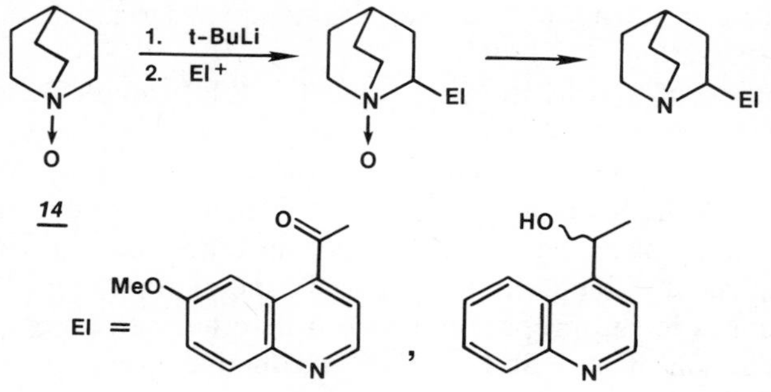

1. t-BuLi
2. El+

14

El = (57)

MeO

N

HO

N

4.2.3.3. Transmetallation Reactions

Transmetallation of *N,N*-disubstituted aminomethyltributyltin with *n*-butyllithium furnished the corresponding α-tertiary amino carbanion[1737] whose treatment with benzaldehyde gave the corresponding tertiary alkanolamine in 70–80% yield (Scheme 58).

$$Bu_3SnCH_2NR_2 \xrightarrow[0°]{BuLi} {}^-CH_2NR_2 \xrightarrow[70-80\%]{PhCHO} HO—CHPh—CH_2NR_2$$

R = Me

3,4-$(MeO)_2C_6H_3CHO$ 95% ; 83% 1) EtCOMe 2) PhCOCl (58)

$3,4\text{-}(MeO)_2C_6H_3—CHOH—CH_2NMe_2$ **15** $PhCO—CMe(Et)—CH_2NMe_2$ **16**

Quintard and coworkers[1738b] reported the synthesis of the alkaloids macromerine **(15)** and stovaine **(16)** from the lithioanion of trimethylamine and the appropriate carbonyl electrophile. Only primary lithioanions prepared via transmetallation have been reported.

4.3. OXYGEN-SUBSTITUTED CARBANIONIC SYNTHONS

A carbon atom doubly bonded to an oxygen, that is, a carbonyl carbon, is an effective α-heterosubstituted cationic synthon ($R_2C{=}O \simeq {}^+CR_2—OH$). The most common examples of addition to the positive carbon pertain to the nucleophilic attack by organometallic compounds. Grignard reagents add to the aldehyde and ketone to form primary and secondary alcohols, respectively.[1739] Similarly, trialkylaluminum[1739] and alkyl and aryllithiums add to the acceptor site of the carbonyl function. In the Reformatsky reaction,[1740] also of significant synthetic value, an alkylzinc is added to the carbon adjacent to an oxygen to form β-hydroxy esters. Hydride reductions of carbonyl compounds, cyanohydrin formation, ketalization, attack by halogen nucleophiles,

and the acid-catalyzed addition of olefins and alkynes to aldehydes all involve a type of α-oxocation, $^{+}C{-}OH$.

Anions adjacent to a hydroxyl group, $^{-}C{-}OH$, cannot be generated directly. Heteroatom modification[1642] or replacement[1643] is necessary to reverse the polarity of the oxygen atom. Transmetallation of tri-*n*-butylstannylmethanol with butyllithium provides a nucleophilic hydroxymethylating agent.[179] Direct generation of an anion adjacent to an ether oxygen, $^{-}CH_2OR$, has been accomplished with butylpotassium as the metallating agent.[1646,1649] Metal–metal exchanges in the preparation of α-lithioether carbanions were developed by Peterson[1652] and Still *et al.*[1650] Reductive lithiation of α-(phenylthio)ethers has been introduced as a general preparative method for α-lithioethers.[1741]

4.3.1. α-Hydroxy Carbanions $^{-}C{-}OH$

Donor sites adjacent to an hydroxyl function can be established, by means of umpoled synthons, by conversion to an ester,[1642] and by a temporary replacement with triarylboranes,[1643] or the diisopropoxymethylsilyl group.[1742] A nonumpoled approach to the anion was achieved by Meyer and colleagues,[179] by transmetallation of stannylated methanol.

4.3.1.1. Esters

Primary alcohols are easily converted to 2,4,6-triisopropylbenzoate **(17)** or 2,6-bis(dimethylamino)-3,5-diisopropylbenzoate **(18)** esters. The esters can be metallated with *s*-butyllithium/TMEDA at −78°; the resulting carbanion can then undergo electrophilic substitution with a variety of acceptor molecules. Cleavage was carried out reductively with lithium aluminum hydride for the esters of **17,** and hydrolytically for those of **18** (Scheme 59).

$$\begin{array}{l} RCH_2OCOAr \xrightarrow[\text{TMEDA}]{s\text{-BuLi}} {}^{-}CHR{-}OCOAr \simeq {}^{-}CHR{-}OH \xrightarrow{El^+} \\ \qquad\qquad El{-}CHR{-}OCOAr \xrightarrow[\text{or } H^+]{\text{LAH}} El{-}CHR{-}OH \end{array} \qquad (59)$$

where Ar = 2,4,6-tri-i-propylphenyl **(17)** or 2,6-bis(dimethylamino)-3,5-di-i-propylphenyl **(18).**

A variation of the arylester approach to heteroatom modification involves the formation of dialkylcarbamates[1743] followed by metallation with *n*-butyllithium/TMEDA (Scheme 60). This method has been confirmed only with benzyl alcohols.

$$\begin{array}{l} ArCH_2{-}OCONR_2 \xrightarrow[\text{TMEDA}]{\text{BuLi}} {}^{-}CHAr{-}OCONR_2 \xrightarrow{El^+} \\ \qquad\qquad El{-}CHAr{-}OCONR_2 \end{array} \qquad (60)$$

4.3.1.2. ALKYLDIMESITYLBORANES

It has been demonstrated by Rathke and Kow[1744] that hindered bases such as lithio-2,2,6,6-tetramethylpiperidine metallate the α-carbon of certain boranes. Boron-stabilized carbanions were obtained from the mesitylboranes of primary and secondary alcohols.[1745] Anion formation is accomplished with lithiumdicyclohexylamide at 0°, and alkylation takes place at room temperature. An oxidation step with hydrogen peroxide is required to generate the corresponding alcohol (Scheme 61).

$$RCH_2BMes_2 \xrightarrow[0^\circ]{CX_2NLi} {}^-CHR{-}BMes_2 \simeq {}^-CHR{-}OH \xrightarrow[rt]{El^+} El{-}CHR{-}BMes_2 \xrightarrow{H_2O_2} El{-}CHR{-}OH \quad (61)$$

4.3.1.3. TRANSMETALLATION

Treatment of tri-*n*-butylstannylmethanol (**19**) with *n*-butyllithium yields a pentane-soluble hydroxymethylating agent.[179] It was determined by nuclear magnetic resonance (NMR) measurements that the expected dilithiated methanol (**20**) was not formed, but, instead, a tin derivative (**21**) is produced. The complex reacts well with electrophiles at −80° (Scheme 62).

$$\underset{\mathbf{19}}{Bu_3SnCH_2OH} \xrightarrow{BuLi} \underset{\mathbf{21}}{\left[BuSn{-}CH_2 \text{ (bridged by O)}\right]^{2-}} 2Li^+ \xrightarrow{El^+} El{-}CH_2OH \quad (62)$$

$$Bu_3SnCH_2OH \not\rightarrow \underset{\mathbf{20}}{LiCH_2OLi}$$

A nucleophilic hydroxymethylating agent of organic halides was developed by Tamao and colleagues.[1742] This reagent is the diisopropoxymethylsilylmethyl Grignard $(i\text{-}PrO)_2MeSiCH_2MgCl$. The reaction proceeds by means of metal-catalyzed cross-coupling and subsequent cleavage of the silicon–carbon bond with hydrogen peroxide.

4.3.2. α-Alkoxy Carbanions ⁻C—OR

Formation of α-alkoxy carbanions can be carried out through metal–metal exchange,[1649,1652] metal–metalloid exchange,[1741] and by direct metallation.[179,1646]

4.3.2.1. TRANSMETALLATION

α-Lithioether carbanions R′O—CHRLi are readily prepared[1652] from α-alkoxystannates R′O—CHR—$SnBu_3$ and butyllithium at −78°. It has been

shown that some α-alkoxyorganolithium reagents are configurationally stable. They are prepared by a fast, low-temperature exchange of a stereochemically defined organostannate.[1746] Quenching of the resulting α-alkoxyorganolithium reagent with electrophiles proved the reaction to be totally stereospecific. Alkoxymethyllithiums ($LiCH_2OR \simeq {}^-CH_2OR$) are also prepared by treatment[179] of trihalostannylmethyl ethers (X_3SnCH_2OR) with four equivalents of *n*-butyllithium at $-78°$ for 1 h. On addition of benzaldehyde, these ethers give the β-hydroxy adducts ($PhCHOHCH_2OR$) in 65–95% yields. The unstable trihalostannylmethyl ethers are generated from halomethylalkyl, or aryl, ethers with stannous chloride–lithium bromide and are reacted with butyllithium *in situ*.

A lithium–sulfur exchange reaction to prepare α-alkoxy anions involves the reductive lithiation of α-(phenylthio)ethers ($PhSCH_2OR$) with LDMAN or LN.[1741] The stability and ready availability of α-(phenylthio)ethers makes this method one of great value in many reactions.

4.3.2.2. Direct Metallation

Direct lithiation of benzylmethyl ether on the benzyl side was accomplished with *s*-butyllithium-tetramethylenediamine. No *ortho*-lithiation or Wittig rearrangements were observed, and electrophilic substitution was possible on the benzyl methylene site.[1747] When metallating ethers without activating groups, *s*-butyllithium/potassium *t*-butoxide must be used to form the α-potassioalkyl ether.[1645,1646] To avoid a mixture of lithium and potassium cations, dibutylmercury is used to prepare butylpotassium,[1646] which readily metallates ethers such as tetrahydrofuran or diethyl ether at $-75°$.

4.4 SULFUR-SUBSTITUTED CARBANIONIC SYNTHONS

This section includes the preparation of α-thiol carbanions by means of their α-thiocarbamate and thioester derivatives, and α-thioether carbanions. The carbanionic center we are dealing with is derived from an sp^3 hybridized carbon adjacent to a single sulfur atom in the neutral oxidation state. These α-sulfur anions are excellent nucleophiles in carbon–carbon bond formation. However, it is important to note that α-carbanions of sulfoxides, sulfones, and sulfonium systems are useful synthetic intermediates that can be trapped with electrophilic agents.[374,1748] Moreover, thioformyl, dithiane and other dithioacetal groups are effective masked carbonyl anions as is discussed in Chapter 2.

Sulfur confers a nucleophilic character to the α-carbon, and can stabilize the adjacent anion. Stabilization of an adjacent carbonium ion ($^+C{-}SR$) occurs mechanistically during the solvolysis of thioacetals, or vinyl sulfides, to give an *O,S*-acetal (Scheme 63).[373] Although the carbon adjacent to the sulfur in thioketones may undergo nucleophilic addition, the sulfur side of the molecule is also prone to attack by nucleophiles.[1749]

$$\text{CH—C(SR)}_2 \text{ or C=C—SR} \longrightarrow \text{CH—C}^+\text{SR} \xrightarrow{\text{R}'\text{O}^-} \text{CH—C(SR)—OR}' \quad (63)$$

4.4.1. α-Thiol Carbanions $^-$C—SH

Conversion of a mercaptan to an ester or a carbamate[1750] gives derivatives that can be metallated on the carbon adjacent to sulfur. These dipole-stabilized carbanions provide the α-lithioalkanethiol synthons. Ideally, sterically hindered 2,4,6-triethyl or -triisopropylbenzoate esters of the thiol are used in the reaction sequence,[1750] but unsubstituted benzoate esters can also be effective.[1751] Ethanethiol 2,4,6-triethylbenzoate ester **22** gives a formally dipole-stabilized carbanion **23** on metallation with *s*-butyllithium/TMEDA at −78°. Trapping of the anion with electrophiles such as benzyl bromide gives compound **24** in 72% yield. The demasking of **24** to 2-phenylethanethiol proceeds in 77% yield, and involves a reductive cleavage of the ester with lithium aluminum hydride (Scheme 64).

$$\underset{\mathbf{22}}{\text{MeCH}_2\text{SC(=O)Ar}} \xrightarrow[-78°]{n\text{-BuLi}} \underset{\mathbf{23}}{\text{MeCH(Li}\cdots{}^-\text{O)}\text{—S}^+\text{=CAr}} \xrightarrow{\text{BzlBr}} \underset{\mathbf{24}}{\text{MeCH(Bzl)SC(=O)Ar}} \xrightarrow{\text{LAH}} \underset{\mathbf{25}}{\text{MeCH(Bzl)SH}} \quad (64)$$

Ar = 2,4,6-triethylphenyl

Lithiation of thiocarbamates,[1750] such as ethyl-*N,N*-dimethylthiocarbamate **(25)**, is accomplished with either *n*-butyllithium or *s*-butyllithium/TMEDA at −78°. The dipole-stabilized carbanion **26** undergoes reaction with a variety of electrophiles to give the alkylated or hydroxyalkylated product **27** in yields of 50–85%. Hydrolysis in a basic medium gives the substituted thiol **28** (Scheme 65).

$$\underset{\mathbf{25}}{\text{EtS—CONMe}_2} \xrightarrow[\text{TMEDA}]{s\text{-BuLi}} \underset{\mathbf{26}}{\text{Me—CHLi—SCONMe}_2} \xrightarrow{\text{El}^+} \underset{\mathbf{27}}{\text{El—CHMe—SCONMe}_2} \xrightarrow{\text{HO}^-} \underset{\mathbf{28}}{\text{El—CHMe—SH}} \quad (65)$$

Thiocarbamates and thioesters are not limited to serving as synthetic equivalents of α-thiol carbanions. They are also effective olefin, thiirane, and β-diketone precursors.[1750,1751]

4.4.2. α-Thioether Carbanions $^{-}C—SR$

Carbanions on the carbon adjacent to a thioether group can be formed by many methods. These include direct metallation of a thioether,[580] metal–metalloid exchange,[1752] metal–halogen exchange,[1753] and addition of alkyllithiums to vinyl sulfides.[1071] Thioimidates form dipole-stabilized carbanions adjacent to sulfur on treatment with alkyllithium.[1754] These compounds can be trapped with electrophiles to give an elaborated chain, but they do not constitute synthetic equivalents for either α-thiol or α-thioether carbanions. However, they have been widely used and are useful precursors of thiiranes.[1659]

4.4.2.1. Direct Metallation

Metallation at the methyl group of thioanisole carried out with *n*-butyllithium resulted in a low yield.[764] Nearly quantitative yields of phenylthiomethyllithium **(29)** were produced[580] from thioanisole using *n*-butyllithium complexed with DABCO at 0° in THF. Alkylation of **29** with isopropyl iodide, for example, gave phenyl isobutyl sulfide **30** in 55% yield (Scheme 66).

$$\mathrm{MeSPh} \xrightarrow[\text{DABCO, }0^\circ]{n\text{-BuLi}} \underset{\mathbf{29}}{\mathrm{LiCH_2SPh}} \xrightarrow{i\text{-PrI}} \underset{\mathbf{30}}{\mathrm{Me_2CHCH_2SPh}} \qquad (66)$$

Butyllithium/TMEDA is an effective metallating agent in the conversion of dimethylsulfide to methylthiomethyllithium.[1755] Metallation of higher homologs of phenylmethyl sulfides was carried out[1644] using *t*-butyllithium/hexamethylphosphoramide at −78°. Addition of *n*-butyllithium to the double bond of phenyl vinyl sulfide **(31)** at 0° in ether gave the α-lithiothioether **32.**[1071] The lithioanion gave phenyl hexyl sulfide **(33)** in 55% yield on protonation, and α-phenylthioheptanoic acid **(34),** in 52% yield, on addition of carbon dioxide (Scheme 67).

$$\underset{\mathbf{31}}{\mathrm{CH_2{=}CHSPh}} \xrightarrow{\mathrm{BuLi}} \underset{\mathbf{32}}{{}^{-}\mathrm{CH(C_5H_{11}){—}SPh}} \xrightarrow{\mathrm{H_2O}} \underset{\mathbf{33}}{\mathrm{Me(CH_2)_5{—}SPh}}$$

$$\mathbf{32} \xrightarrow{\mathrm{CO_2}} \underset{\mathbf{34}}{\mathrm{Me(CH_2)_4{—}S{—}CHPh{—}COOH}} \qquad (67)$$

4.4.2.2. Metal–Metalloid and Metal–Halogen Exchange

The action of *n*-butyllithium at −78° on 1-thiophenyl-1-selenophenyl-2-methylpropane **(35)** gave the α-thiolithio anion **36** by displacement of butyl phenyl selenide.[1752] The lithio anion **36** gave the 1,2-adduct **37** with cyclohexenone in THF at −78°. When the solvent system was changed to THF/HMPA, the α-thioether anion **36** added 1,4 to cyclohexenone to form **38** (Scheme 68).

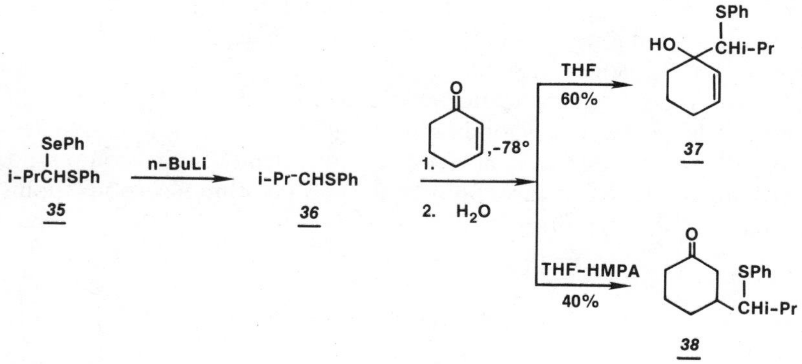

SePh
i-PrCHSPh (35) —n-BuLi→ i-Pr⁻CHSPh (36) —1. cyclohexenone, −78°; 2. H_2O→ THF, 60%: 37 (1-(1-phenylthio-2-methylpropyl)cyclohex-2-en-1-ol; HO, SPh, CHi-Pr); THF-HMPA, 40%: 38 (3-substituted cyclohexanone; SPh, CHi-Pr)

(68)

The formation of an arylthiomethyl Grignard reagent ($ArSCH_2MgCl$) has been reported.[1753] This reagent was formed from the corresponding arylchloromethyl sulfide ($ArSCH_2Cl$) and magnesium at 10–20° in THF. The reaction of the Grignard reagent occurs at 50° with alkyl halides, at room temperature with trimethylsilyl chloride, and at −78° with benzaldehyde.

4.4.2.3. Thioimidates

2-Alkylthio-2-oxazolines **(39)** or -2-thiazolines **(39a)** are useful precursors of thiiranes.[1754,1756,1757] The latter compounds are, in turn, important intermediates in olefin synthesis.[1756] Thus, elaboration of the alkylthio side chain of **39** is central to the preparation of alkenes. This is simplified by the ability of thioamidates to form dipole-stabilized lithioanions **(40)** that react with electrophilic reagents such as aldehydes and ketones. The alcohols that are formed can yield thiiranes **(41)** on acid, base, or heat treatment (Scheme 69).[1757]

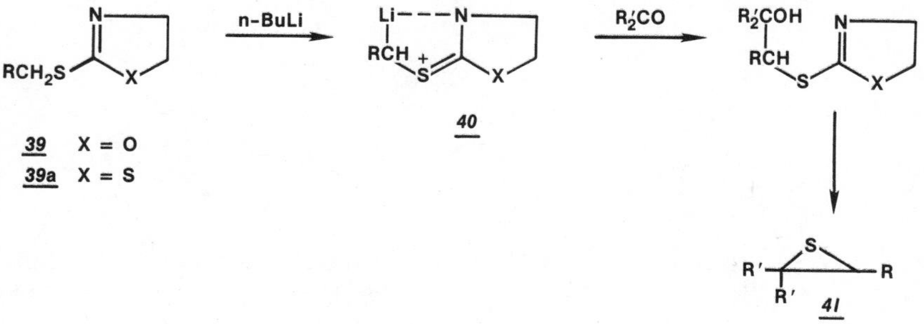

RCH_2S–(N, X ring) (39 X = O; 39a X = S) —n-BuLi→ Li---N, RCH, S+ (40) —R'_2CO→ R'_2COH–RCH–S–(N, X ring) → thiirane 41 (S, R', R', R)

(69)

Similar synthetic transformations can be effected with 2-alkylthiopyridines,[1757] dithiocarbamates,[1758] and 2-thiomethyl-Δ^1-pyrroline.[1759]

4.5. HALOGEN-SUBSTITUTED CARBANIONIC SYNTHONS $^-$C—X

Lithioanions on a carbon adjacent to one or more halogens are stable species at temperatures below −70°, and are known as lithium halocarbanoids. This section will survey only halocarbenoids containing a single halogen atom adjacent to the carbanionic center. Because the negative halogen (X) is a leaving group, and the geminal lithium is a positive center, the adjacent carbon atom exhibits both donor and acceptor properties. As donors, they react with electrophiles to form the corresponding halogen-containing adduct. Nucleophilic displacement of the halogen gives a new organolithium compound. Another reaction common to carbenoids is the carbon atom insertion leading to cyclopropanes (Scheme 70).[1207,1647,1760]

$$R_2\overset{a}{\underset{d}{C}}(Li)Hal \xrightarrow{El^+} R_2\overset{a}{C}(El)Hal$$

$$R_2\overset{a}{\underset{d}{C}}(Li)Hal \xrightarrow{\text{alkene}} \text{cyclopropane (R, R)}$$

$$R_2\overset{a}{\underset{d}{C}}(Li)Hal \xrightarrow{Nu^-} R_2\underset{d}{C}(Nu)Hal \qquad (70)$$

4.5.1. Lithium–Halogen Exchange

The existence of halocarbenoids and their donor properties were first demonstrated by Köbrich and Trapp[1761] via trapping experiments with electrophiles. Lithiation of an alkyl bromide leads to a halogen-free organolithium compound at the expense of the lithium halocarbenoid.[1760] α-Bromo- or α-chloroalkyllithiums were efficiently prepared by Villieras and coworkers[1648] from 1-bromo-1-chloroalkanes or 1,1-dibromoalkanes, respectively, by bromolithium exchange with *n*-butyllithium at −110°. Regiospecific syntheses of α-haloketones using these reagents are feasible in good yields (Scheme 71).[1762]

$$\text{X—CHR—Br} \xrightarrow[-115^\circ]{\text{BuLi}} \text{X—CHR—Li} \xrightarrow{\text{R'COOEt}} \text{X—CHR—COR'} \qquad (71)$$

where X = Cl, Br

This method failed to give reasonable yields of the monohalomethyllithiums ($LiCH_2X$) as shown by trapping experiments with carbon dioxide.[1763] How-

ever, stabilization of the carbenoid species was accomplished by addition of an equivalent of lithium bromide.[1764] α-Halomethyllithium compounds have thus become versatile intermediates for the preparation of halohydrins, α-halomethylketones, and epoxides (Scheme 72).

$$BrCH_2X \xrightarrow[-115°,\ LiBr]{s\text{-}BuLi} LiCH_2X \longrightarrow LiBr \begin{cases} \xrightarrow{R_2CO\ (X\,=\,Cl)} HO\text{—}CR_2\text{—}CH_2Cl \\ \xrightarrow{RCOOEt} RCOCH_2X \\ \xrightarrow{R_2CO\ (X\,=\,Br)} R_2C\text{—}CH_2 \text{ (epoxide, bridged by O)} \end{cases} \quad (72)$$

4.5.2. Heteroatom Exchange and Modification

Diphenylarsenylmethyllithium (Ph_2AsCH_2Li) undergoes reaction with alkyl halides, aldehydes, and ketones to form the corresponding alkyl or hydroxyalkyldiphenylarsenic compound.[1765,1766] Treatment of the diphenylarsenyl derivatives with bromine gives a nearly quantitative yield of the corresponding alkyl bromide.[1765] Thus, diphenylarsinylmethyllithium (Ph_2AsCH_2Li) is a synthetic equivalent of α-halomethyllithium (XCH_2Li).

Lithiochloromethyl phenyl sulfoxide **(42)** is used as a synthetic equivalent of α-chloromethyllithium ($ClCH_2Li$) in the preparation of chloromethylketones.[1767] The reaction of **42** with an aldehyde at −78° in THF gave the β-hydroxy-α-chlorosulfoxide **43** which, on heating in xylene, produced the chloromethylketone **44** in excellent yield (Scheme 73). Alkylation and pyrolysis of **42** leads to vinyl chlorides.[1768] Base treatment of the β-hydroxy-α-chlorosulfoxide **43** produces α-epoxysulfoxides which, in turn, are pyrolyzed to α,β-unsaturated aldehydes.[1769]

$$\underset{\mathbf{42}}{PhSO\text{—}CHCl\text{—}Li} \xrightarrow[THF\ -78°]{RCHO} \underset{\mathbf{43}}{HO\text{—}CHR\text{—}CHCl\text{—}SOPh} \xrightarrow[Xylene]{\Delta} \underset{\mathbf{44}}{RCOCH_2Cl} \quad (73)$$

4.6. α-SELENOALKYL CARBANIONS $^{-}$C—SeR

The chemistry of organoselenium compounds, including α-selenoalkyl carbanions,[374,1674,1770,1773] has been extensively reviewed in recent times.[374,1674,1770–1773] Selenoxides, which are oxidation products of selenides, undergo *syn* elimination[1774] under mild conditions to olefins.[1775,1776] Thus, se-

lenoxides are synthetic equivalents of olefins, hydroxyalkyl selenides are equivalents of allylic alcohols,[1777] and β-ketoselenides are converted to α,β-unsaturated ketones. The need for efficient methods of introducing C=C unsaturation into organic molecules has been responsible for the rapid development of organoselenium chemistry. The use of selenides as precursors of alkylhalides, carbonyl compounds, epoxides, alcohols, and oxetanes is also an important aspect of their chemistry.[374]

These compounds have been prepared by various routes.[1674] The classical process involves the opening of an epoxide with phenylselenide ion, a soft nucleophile.[1777] On the other hand, phenylselenenyl chloride or bromide acts as soft electrophilic selenium.[1771] Our scope limits us to the preparation and reactions of α-selenoalkyl anions. These species are highly nucleophilic, producing a stable selenide following the formation of a new carbon–carbon bond. Phenyl- and alkylselenoalkyllithium are stable for several hours at −78°. Phenylselenomethyllithium is actually stable at 20° for 0.5 h in THF.[374] Lithioanions adjacent to a selenium atom can be prepared directly by proton removal, provided that additional stabilization is present in the molecule, such as a benzyl[1778,1779] or an allyl group.[1780] Selenoacetals and ketals are the most versatile sources of α-selenoalkyl anions[374,1556,1674] involving a nucleophilic attack on selenium by an alkyllithium. In other words, a metal–metalloid exchange process takes place. A metal–halogen exchange to form an α-selenoalkyl anion occurs when butyllithium is added to α-bromoalkylselenides.[1778,1781] Addition of alkyllithiums across the double bond of vinylphenylselenides is another way to obtain α-lithioalkylphenylselenides.[1656,1781]

4.6.1. Direct Metallation

Phenylmethylselenide was metallated with *n*-butyllithium/TMEDA in only 38% yield[1556] at a temperature above 0°. The low production of the organometallic is the result of two competing reactions that take place when a selenide is treated with an alkyllithium. There is proton removal, which depends on proton acidity, and nucleophilic attack on selenium resulting in cleavage (Scheme 74).

$$\mathrm{PhSeBu} + \mathrm{MeLi} \xrightarrow{n\text{-BuLi}} \mathrm{PhSeMe} \xrightarrow[\text{TMEDA}]{n\text{BuLi}} \underset{\mathbf{38\%}}{\mathrm{PhSeCH_2^-}} \tag{74}$$

However, hydrogen–metal exchange on the carbon adjacent to selenium is possible in good yields if an additional activating group is present in the molecule. The lithioanion of benzyl phenyl selenide ($^-$CHPh—SePh) was formed with LDA in THF at room temperature, and on reaction with a benzyl or alkyl halide generated new selenides Ph—CHR—SePh.[1778,1779] Deprotonation of allylic selenides is carried out with LDA or lithium diethylamide at

temperatures ranging from 0° to −100°, depending on the acidifying contribution of the gamma substituent.[1780] The phenylallylselenide **45** is a representative example of this family of compounds. Anion formation takes place at −78° within 10 min in THF. The allyl-stabilized anion **46** undergoes electrophilic attack predominantly at the α-position; however, there is competition between α- **(47)** and γ- **(48)** attack (Scheme 75).[1674,1780] The allylic selenide products are subsequently oxidized to give allylic alcohols and enones.[1780]

$$\underset{\mathbf{45}}{PhSeCH_2CH{=}CH_2} \longrightarrow \underset{\mathbf{46}}{{}^{-}\overset{\alpha}{C}H(SePh){-}\overset{\beta}{C}H{=}\overset{\gamma}{C}H_2} \xrightarrow{El^+} \underset{\mathbf{47}}{El{-}CH(SePh){-}CH{=}CH_2} + \underset{\mathbf{48}}{PhSeCH{=}CHCH_2{-}El} \qquad (75)$$

4.6.2. Metal–Metalloid Exchange

The low-lying energy 4*d* orbitals of the selenium atom in selenides may be responsible for their electrophilic properties, and result in the cleavage of the C–Se bond in preference to proton removal when selenides are treated with alkyllithiums.[1770] The lability of the Se–C bond has proven to be a most versatile feature in the preparation of α-selenocarbanions.[1653] Bis(phenylseleno)methane **(49)** gives phenylselenomethyllithium **(50)** (and butyl phenyl selenide) when treated with *n*-butyllithium in THF at −78°. The carbanion **50** reacted with benzophenone to give the β-hydroxy adduct **51** in 85% yield (Scheme 76).[1653] This reaction sequence established that selenoketals and acetals are synthetic equivalents of α-selenoalkyl carbanions.[1653,1782] The reaction of **49** with lithium diisopropylamide results in proton removal to give the selenoacetal anion **(52)**; no C–Se bond cleavage takes place (Scheme 76).[1653]

$$\underset{\mathbf{49}}{(PhSe)_2CH_2} \xrightarrow[-78^\circ]{n\text{-BuLi}} \underset{\mathbf{50}}{{}^{-}CH_2SePh} \xrightarrow{Ph_2CO} \underset{\mathbf{51}}{HO{-}CPh_2{-}CH_2SePh}$$

$$\underset{\mathbf{49}}{(PhSe)_2CH_2} \xrightarrow{LDA} \underset{\mathbf{52}}{(PhSe)_2CH^{-}} \qquad (76)$$

The discovery of this convenient route to α-selenoalkyl carbanions and their reactions with carbonyl compounds has prompted the development of new synthetic applications based on the C–Se bond cleavage. The β-hydroxy adducts of type **53,** which result from the addition of aldehydes and ketones to the α-phenylselenoalkyl carbanion, are regiospecific allyl alcohol precur-

sors.[1782] Oxidation of **52** with hydrogen peroxide[1777] gives the corresponding allyl alcohols (Scheme 77).

$$\mathrm{RCH_2CH^-SePh} \xrightarrow{\mathrm{R'CHO}} \underset{\mathbf{53}}{\mathrm{RCH_2CH(SePh)\text{—}CHOH\text{—}R'}} \longrightarrow \mathrm{RCH{=}CH\text{—}CHOH\text{—}R'} \qquad (77)$$

Methylselenoacetals,[1783] as well as ethyl- and heptylselenoacetals,[374] can also be quantitatively cleaved by *n*-butyllithium. The β-hydroxy adducts of the α-methylselenoalkyl carbanions offer a route to epoxides.[1783] The synthesis of alkylidene cyclopropanes **(56)** from aldehydes and ketones was accomplished with the β-hydroxy adduct **(55)** of methyl- or phenylseleno-1-lithiocyclopropane (**54,** Scheme 78).[1784] The β-hydroxyselenide **(55),** on treatment with *p*-toluenesulfonic acid in benzene, gives the cyclobutanone (**57,** Scheme 78). A similar type of ring expansion with methyl- or phenylselenocyclobutanes allowed the regioselective synthesis of polyalkylated cyclopentanones.[1785] The lithio anion **(54)** will undergo an electrophilic reaction with primary alkyl halides, *n*-decyl bromide, for example, to form the corresponding selenide **(58)** in 75% yield (Scheme 78).[1786]

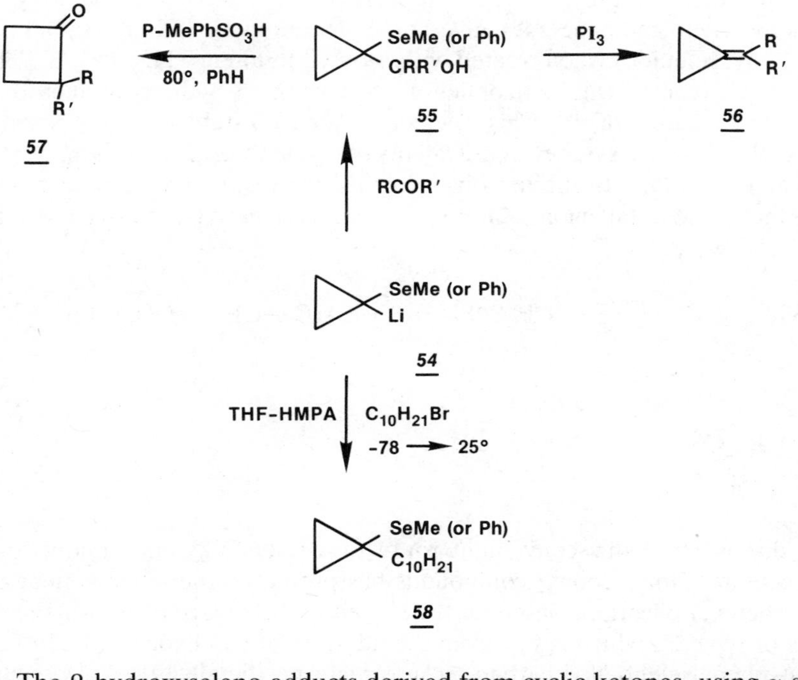

O
R
R′
57
P-MePhSO$_3$H
80°, PhH
SeMe (or Ph)
CRR′OH
55
PI$_3$
R
R′
56
RCOR′
SeMe (or Ph)
Li
54
THF-HMPA
$C_{10}H_{21}Br$
-78 → 25°
SeMe (or Ph)
$C_{10}H_{21}$
58

(78)

The β-hydroxyseleno adducts derived from cyclic ketones, using α-selenocarbanions bearing two alkyl groups, are valuable reagents for the ring ex-

pansion of ketones,[1787] as well as for the synthesis of hindered epoxides and olefins.[1788] The use of acylating and formylating electrophiles to trap α-selenoalkyllithiums established synthetic routes to α-seleno carbonyl compounds ($PhSeCR_2COR'$).[1789] These compounds, upon oxidation, are excellent sources of α,β-unsaturated aldehydes, ketones, carboxylic esters, and acids.

4.6.3. Metal–Halogen Exchange

α-Bromoalkyl phenyl selenides (PhSeCHRBr) react with *t*-butyllithium[1778] or *n*-butyllithium[1790] at −78° in THF to form the corresponding lithio anion (PhSeCHRLi) and the alkyl bromide. The alkyl bromide side product will react with the lithioanion to form a new selenide. *n*-Butyl bromide, for example, will form *n*-pentyl phenyl selenide with α-phenylselenomethyllithium ($PhSeCH_2Li$) in 90% yield after 1 h. However, since aldehydes and ketones react much faster with the lithioanion, it is possible to form β-hydroxyselenides in yields of 62–75%. α-Chloroalkyl selenides cannot be used in lithium–halogen exchange reactions, since nucleophilic attack of the organolithium occurs on the selenium atom.[1790]

4.6.4. Addition of Alkyllithiums to α-Selenosubstituted Ethylenes

Alkyllithiums, including *t*-butyllithium, add to vinyl phenyl selenide **(59)** to give the corresponding α-lithioalkyl phenyl selenide **(60)**.[1656,1781] The resulting anion **(60)** reacts with various electrophiles to give adducts **(61)**. Thus, we can consider vinyl phenyl selenide to be a source of the $^{-}CH(SePh){-}CH_2^{+}$ synthon (Scheme 79). The reaction with *n*-butyllithium is run at 0° in dimethoxyethane, and the reactions involving isopropyllithium or *t*-butyllithium are carried out in ethyl ether at 0°.

$$\underset{\mathbf{59}}{CH_2{=}CHSePh} \xrightarrow{RLi} \underset{\mathbf{60}}{^{-}CH(SePh){-}CH_2R} \xrightarrow{El^+} \underset{\mathbf{61}}{El{-}CH(SePh){-}CH_2R} \qquad (79)$$

4.7. OTHER CARBANIONIC SYNTHONS WHERE THE HETEROATOM IS AN ELEMENT FROM GROUP IVa, Va, OR VIa

Reagents for nucleophilic lithioalkylation where the heteroatoms belong to group IVa are best prepared *via* halogen–lithium exchange, as in the case of germanium, tin, and lead.[1652,1765,1766] Tin- and lead-substituted carbanions can be formed by metal–metal exchange, whereas silicon-substituted ones are

best obtained from α-(phenylthio)silanes through a sulfur–lithium exchange.[253,257] Metal–metalloid and metal–halogen exchanges are efficient methods for the synthesis of α-antimony[1765,1766] and α-phenyltelluro-substituted carbanions.[1766,1791] Synthetically useful carbanions adjacent to phosphorus are either phosphoryl- (R—PO—C^-) or phosphorane- ($R_3P^+C^-$) stabilized anions.[320,324,1651,1652] However, α-trivalent phosphorus carbanions have also been studied.[1655]

4.7.1. Carbanions Adjacent to Silicon $^-C—SiR_3$

In addition to alkene synthesis via the Wittig reaction,[320] sulfinamide,[1792] or selenoxide[1774–1776] eliminations, β-functionalized organosilicon compounds can also be used for carbon–carbon double bond formation.[1792] The addition of α-silylalkyllithiums to carbonyl compound[773,1793] provides the desired β-substituted silane intermediate involved in the elimination reaction to the alkene (Scheme 80). Thus, the formation of carbanions adjacent to silicon is an important step in olefin synthesis at the outset.

$$^-CHR—SiR''_3 + R'_2C{=}O \longrightarrow HO—CR'_2—CHR—SiR''_3 \longrightarrow R'_2C{=}CHR \qquad (80)$$

Trimethyl- and triarylsilyl-substituted organolithium ($^-C—SiMe_3$ and $^-C—SiAr_3$) reagents have been prepared by direct metallation of the corresponding trimethyl or triarylsilyl compound.[773,1793,1794] However, an additional activating group on the α-carbon is necessary for anion formation. A versatile method for preparing α-silyl carbanions involves halogen–metal exchange in α-halosilanes (Scheme 81) with magnesium[1795] or alkyllithiums.[1794] The only limitation to this method is the accessibility of the starting halide.

$$R_3SiCH_2X \begin{cases} \xrightarrow{Mg^\circ} R_3SiCH_2MgX \\ \xrightarrow[R'Li]{} R_3SiCH_2Li + R'X \end{cases} \simeq {}^-CH_2SiR_3 \qquad (81)$$

The lithium–sulfur exchange in α-phenylthioalkylsilanes ($PhSCR_2SiMe_3$) with LDMAN provides a general method[257] for the production of α-silyl carbanions ($^-CR_2SiMe_3$) in yields ranging from 71% to 96%. Metal–silicon exchange studies in bis(trisubstituted silyl)phenylmethane $(R_3Si)_2CHPh$ indicate high yields of α-silyl carbanions.[1796] Metal alkoxides were used in the exchange reaction, and the yield of the anion was based on the final olefinic product (stilbenes) and not on the β-hydroxysilane intermediate.

Trialkyl- or triarylvinylsilanes ($CH_2{=}CHSiR_3$) offer an alternative method for the preparation of α-silyl carbanions accompanied with chain lengthening.

The generation of the α-silylalkyllithium takes place on addition of aryl and alkyllithiums to the double bond (Scheme 82).[1654,1797]

$$CH_2{=}CHSiR_3 \simeq {}^+CH_2CH^-SiR_3 \xrightarrow{R'Li} R'CH_2CH^-SiR_3 \qquad (82)$$

The yield of the α-silyl carbanion depends on the relative reactivity of the entering organolithium reagent, and its steric interaction with the substituents on the phenyl group. A further limitation of this reaction is the polymerization of the unsaturated silane which competes with addition of the organometallic.[1797]

As mentioned above, the addition of α-silyl carbanions to aldehydes and ketones is an important reaction leading to β-hydroxysilanes which are precursors of olefins. However, α-silyl carbanions have been trapped with other electrophiles as well. With acyl chlorides, the corresponding β-silylketones are formed.[1798] The α-position of the silane can be deuterated, methylated, carboxylated, or silylated by trapping the anion with D_2O, methyl iodide, carbon dioxide, or trimethylsilyl chloride, respectively.[1794]

4.7.2. Carbanions Adjacent to Tin, Lead, and Germanium: $^-C{-}SnR_3$, $^-C{-}PbR_3$, and $^-C{-}GeR_3$

Triphenylstannylmethyllithium ($^-CH_2SnPh_3$) and triphenylplumbylmethyllithium ($^-CH_2PbPh_3$) are generated by lithium–metal exchange from bis(triphenylstannyl)methane[1799] in 36% yield and from bis(triphenylplumbyl)methane[1765] in quantitative yield, respectively. Lithium–halogen exchange provides a better synthesis of triphenylstannylmethyllithium (98%). This α-stannyl anion is formed on treatment of triphenylstannylmethyl iodide (ICH_2SnPh_3) with *n*-butyllithium. Similar exchange with lead as the heteroatom only gives a 58% yield of the α-plumbyl anion. Triphenylgermaniummethyllithium ($^-CH_2GePh_3$) is generated from triphenylgermanium methyl bromide ($BrCH_2GePh_3$) by exchange with *n*-butyllithium.[1795] Bis(triphenylgermanium)methane does not undergo metal–metal exchange when treated with *n*-butyllithium. The α-germanium anion is much more reactive than the stannyl and plumbyl counterparts. All three react with carbonyl compounds, but only triphenylgermaniummethyllithium reacts with alkyl halides.[1652]

4.7.3. Carbanions Adjacent to Phosphorus, Antimony, and Tellurium: $^-C{-}PR_2$, $^-C{-}SbR_2$, and $^-C{-}TeR$

The strong affinity of phosphorus for oxygen, and the availability of 3*d* orbitals for bonding are characteristics that make phosphorus-containing compounds synthetically useful. Reactions of ylides or alkylideenephosphoranes are ex-

amples of the versatility and value of phosphorus-containing reagents. These reagents are prepared by proton abstraction at the carbon adjacent to a quaternary phosphorus with *n*-butyllithium or some other strong base. On the other hand, metallation at a carbon adjacent to trivalent phosphorus is not as well known, and has not been developed as a useful entry to α-phosphorus-substituted carbanionic synthons. However, work carried out in this field has suggested 3*d* orbital stabilization of the incipient α-phosphino carbanion by trivalent phosphorus.[1651,1800]

Benzyldiphenylphosphine ($BzlPPh_2$) was shown to be metallated at the benzylic position with phenyllithium at reflux.[1801] Methyldiphenylphosphine ($MePPh_2$), dimethylphenylphosphine (Me_2PPh), and di-*n*-hexylmethylphosphine are metallated at the methyl carbon with *t*-butyllithium forming the corresponding α-phosphorus-substituted anion ($^-CH_2PRR'$) in yields ranging[1800,1802] from 14 to 54%. The phosphine carbanions were treated with carbon dioxide and then with elemental sulfur and the products isolated as the phosphine sulfides $RR'P(S)—CH_2COOH$. An explanation of the low yields of phosphinocarbanions was found using ^{31}P-NMR. The studies revealed that dimethyl-alkyl- and -aryl-phosphines also form bimetallated phosphines in up to 35% yield.[1803] With *n*-butyllithium at 25°, nucleophilic substitution at phosphorus in tertiary phosphines takes place as well as deprotonation to the carbanion.[1804] Direct α-metallation of phosphorus-bonded alkyl groups other than methyl is negligible.[1800] However, addition of *n*-butyl- or *t*-butyllithium to vinyldiphenylphosphine ($Ph_2PCH{=}CH_2$) gives the corresponding α-phosphinoalkyl carbanions $Ph_2PCH^-CH_2Bu$-*n* (or -*t*) in fair yields.[1655] The reaction is limited to vinyldiarylphosphines, since the use of di-*n*-butylvinylphosphine leads to polymeric products.

Bis-diphenylstibanylmethane ($Ph_2SbCH_2SbPh_2$) reacts with phenyllithium at $-70°$ in THF to give diphenylstibanylmethyllithium (Ph_2SbCH_2Li) quantitatively.[1765,1766] This lithio anion reacts with aldehydes and ketones to give the corresponding β-hydroxy derivatives in fair yields, but it reacts poorly with alkyl halides (12–14%). Exchange of the lithio cation with copper gave diphenylstibanylmethylcuprate (Ph_2SbCH_2Cu), a reactive species toward alkyl halides (Scheme 83).[1766]

$$\begin{array}{ccc} Ph_2SbCH_2Li & \xrightarrow{R_2CO} & Ph_2SbCH_2—CR_2—OH \\ \downarrow{\scriptstyle CuCl} & & \\ Ph_2SbCH_2Cu & \xrightarrow{RX} & Ph_2SbCH_2R \end{array} \qquad (83)$$

In group VIa, α-telluroorganometallics are not as well known as their seleno counterparts, but they do show some similarities.[1652] Thermally stable α-phenyltelluromethyllithium ($PhTeCH_2Li$) is synthesized by cleaving the C–

Te bond in bis(phenyltelluro)methane with an alkyllithium.[1791] The telluryl moiety is better able to stabilize the α-carbanion than is the selenyl congener.[1652] Bond cleavage in telluroacetals is easier than lithium–selenium exchange in selenoacetals.[1770] Treatment of bis(phenyltelluro)methane with lithiumdiisopropylamide did not result in C–Te bond cleavage but in the formation[1791] of bis- (phenyltelluro)methyllithium $(PhTe)_2CHLi$. This type of metallation had been previously observed with the selenium counterpart (Scheme 76).[1653] Very little has been reported on the electrophilic substitution of the α-phenyltelluromethyl anion. It can be trapped with D_2O to form the monodeuterated telluride ($PhTeCH_2D$), as well as with benzaldehyde or benzophenone to form the corresponding β-hydroxy adducts in 60 and 40% yields, respectively.[1791]

4.8. α-HETEROATOM CARBANIONIC SYNTHONS WITH ADDITIONAL ACTIVATING GROUPS

Scheme (9) gives the general structure of compounds containing a heteroatom Y and an additional activating group W. The group W may be the principal or a partial contributor to the activation and stabilization of the α-anion. Typically, this type of structure encompasses formyl, acyl,[7] and R-functional acyl synthons.[8] These are synthons that provide carbonyl compounds by direct alkylation and hydroxylation, or upon a straightforward demasking step. It is our intention in this section to survey α-heterocarbanions with additional activating groups that do not fall into the category of masked carbonyls. Heteroatom-bearing molecules with allyl, propargyl, carbonyl, cyano, or any other electron-withdrawing group in general on the α-carbon provide a highly reactive methylene site for anion formation and electrophilic substitution. This is especially true if the heteroatom itself has been modified, or umpoled, to convey a nucleophilic character to the α-carbon. A relevant example relates to the synthesis of amino acids by phase-transfer reactions.[1805] Glycine ethyl ester is further activated through its Schiff base **(62)** with benzophenone. Electrophilic addition to **62** can be accomplished using LDA/THF or phase-transfer catalysts to form the adduct **63.** Acid hydrolysis of **63** provides the substituted amino acid **64** (Scheme 84). Benzamide,[1806] isocyanate,[1807] formamide,[1808] or bis(alkylthio)methyleneamino[1809] groups have also been used instead of the Schiff base.

$$\underset{\mathbf{62}}{Ph_2C{=}NCH_2COOEt} \xrightarrow[\text{LDA or } R_4N^+X^-]{El^+} \underset{\mathbf{63}}{PhC{=}N{-}CHEl{-}COOEt} \longrightarrow \underset{\mathbf{64}}{H_2N{-}CHEl{-}COOEt} \qquad (84)$$

Allylic carbanions adjacent to amines and other heteroatom functionalities have been described.[1810] In these cases we must consider that the heteroatom and the double bond may be part of a conjugated system (Scheme 85).

$$R'CH{=}CR{-}\overset{\alpha}{C}H^{-}Y \longleftrightarrow {}^{-}\overset{\gamma}{C}HR'{-}CR{=}CH{-}Y \quad (El^{+}) \tag{85}$$

Thus, reactions with electrophiles to give the desired substitution α- to the heteroatom may be encumbered by competing γ-substitution. In the case of allyl nitrosamines, it has been shown that under kinetic conditions, electrophilic addition to the α-carbon is the predominant reaction.[1811] Under thermodynamic reaction conditions double bond isomerization to the corresponding nitrosoenamine takes place.[1713,1811,1812] The introduction of a carbomethoxy group on the nitrogen of 3-pyrroline specifically directed lithiation and alkylation to the α-carbon. This reaction provided an entry to the synthesis of prostaglandin derivatives,[1713] as well as pyrrolizidine and indolizine carbon skeletons.[1813]

α-Aminonitriles are well-known sources for carbonyl anion synthons (see Chapt. 2) and reagents for enamine synthesis upon ^{-}CN elimination.[867] However, it has been found that α-cyanoamines can also be used as α-amino carbanion synthons.[862] Anion formation takes place at the α-position of the cyanoamine, which is then readily alkylated, and the cyano group in the resulting product can be reductively cleaved to the corresponding tertiary amine with sodium borohydride (Scheme 86).

$$NC{-}CHR{-}NR'_2 \xrightarrow[2)\ El^{+}]{1)\ LDA} NC{-}CR(El){-}NR'_2 \xrightarrow{NaBH_4} El{-}CHR{-}NR'_2 \tag{86}$$

A secondary amine substituted with a cyano group on the α-position is benzoylated prior to anion formation. A specific example is 2-cyanopiperidine, which is reacted with benzoyl chloride to give the cyanobenzamide **65.** Metallation with LDA followed by electrophilic substitution with benzaldehyde gave the adduct **66.** Reduction with sodium borohydride cleaved the benzoyl as well as the cyano group to give predominantly the *threo*-phenylcarbinol **67** in 70% overall yield (Scheme 87).[862]

(87)

Stereocontrolled syntheses of vicinal aminoalcohols were achieved from *N*-carbomethoxyamines containing a diphenylphosphine oxide (Ph_2PO—) group on the α-carbon.[1814] An illustrative example of this method is the preparation of (±)-conhydrine **(72)**. Anion formation with LDA took place on the carbon vicinal to phosphorus and nitrogen of compound **68,** and hydroxyalkylation with propionaldehyde formed the oxazolidone **69.** Demasking to the aminoalcohol required three operations. An elimination of the diphenylphosphinyl group took place thermally to give the 2-oxazolone derivative **70.** This was reduced stereoselectively over PtO_2 to the *erythro*-isomer **71,** which, on basic hydrolysis, yielded (±)-conhydrine **72** (Scheme 88). This method has also been employed for the preparation of racemic ephedrine and *N*-methylephedrine.[1814]

68 → (1. LDA; 2. EtCHO; 62%) → 69 → (1. Δ; 80%) → 70 → (H_2, PtO_2; 95%) → 71 → (^-OH; 91%) → 72

(88)

The preparation of 2-substituted propargylamines as potential central nervous system enzyme inhibitors required the development of an equivalent for the propargylamine anion synthon $^-CH(NH_2)—C≡CH$.[1815] This was accomplished by the initial Schiff base formation with benzaldehyde followed by introduction of a trimethylsilyl group on the terminal carbon to give TMSC≡CCH_2N=CHPh. Metallation of the α-carbon was readily effected with *n*-butyllithium at −78°. Reaction with electrophiles followed by acid hydrolysis gave the 2-substituted propargylamine El—$CH(NH_2)$—C≡CH. α-Alkylation of (*S*)-(+)serine was carried out *via* the lithium enolate **(74)** of methyl (2*R*,4*S*)-2-*t*-butyl-3-formyl-oxazolidine-4-carboxylate **(73).** Reaction with electrophiles occurs mainly on the Re-face of the nucleophilic center to give **75,** which upon hydrolysis gave the alkylated or hydroxyalkylated serine **76** (Scheme 89).[1808]

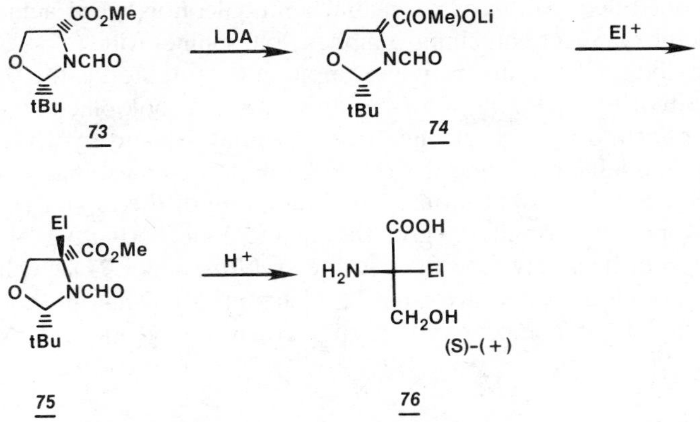

(89)

Alkoxycarbanions having a thioester group on the α-carbon ($ArCOS—CH_2OR$) were investigated as possible Wittig rearrangement intermediates,[1816] thus providing a procedure for homologation of an alcohol.[1750] However, it was found that carbanions of methoxymethyl- and butoxymethyl-2,4,6-triisopropylthiobenzoates ($ArCOS—CH^-OMe$ or —OBu-*n*) are stable at −98°. The anions are trapped with electrophiles to give the corresponding adducts in 72–90% yield. Methoxymethyltrimethylsilane (Me_3SiCH_2OMe) provides not only a formyl carbanion (^-CHO) equivalent but also of source of the α-methoxymethylene carbanion $^-CH_2OMe$.[1220] The lithiocarbanion ($Me_3Si—CH^-OMe$) is formed with *s*-butyllithium in THF at −78°, and it is trapped with aldehydes or ketones. With cyclohexanone, the adduct **77** is formed. Treatment of **77** with potassium hydride at room temperature results in $^-OSiMe_3$ elimination to the enol ether **(78).** Cesium fluoride in DMSO, on the other hand, converts **77** into the β-hydroxy ether **(79)** (Scheme 90).

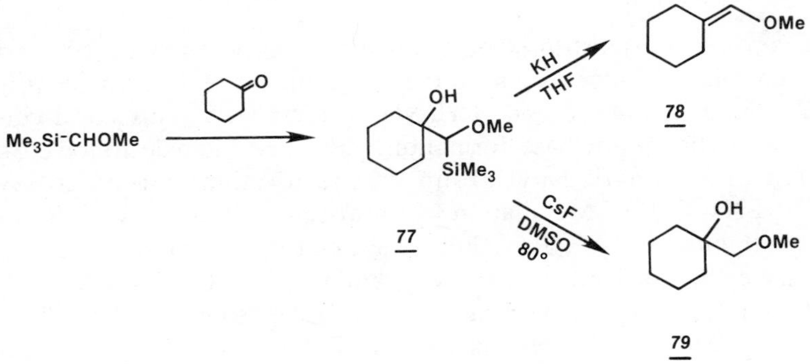

(90)

A camphor-based oxazoline **(80)** was developed as a reagent for the chiral glycolic acid carbanion synthon $^-CHOH—COOH$.[1817] The anion **81** is generated with *n*-butyllithium and trapped with a variety of alkylating agents to

form the substituted oxazoline **82.** Acid hydrolysis then gives the corresponding α-hydroxy acid **83** in high enantiomeric excess (Scheme 91).

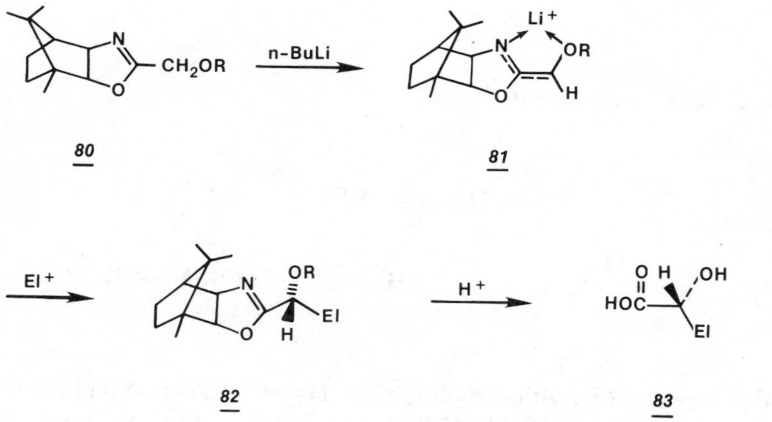

80 81

(91)

82 83

Lithio carbanions derived from allylic ethers or protected allylic alcohols react with electrophiles mainly at the γ-position.[1810,1818,1819] However, it was found that the regioselectivity of addition to allylic ether carbanions, with carbonyl electrophiles, is counterion dependent.[1818] It is possible to direct the reaction to the α-position exclusively by replacing lithium with zinc. The lithioanion of methyl allyl ether, when treated with one equivalent of zinc chloride, forms the reagent **84,** which on trapping with cyclohexanone adds exclusively in the position α to the oxygen, forming the adduct **85** in 92% yield (Scheme 92).

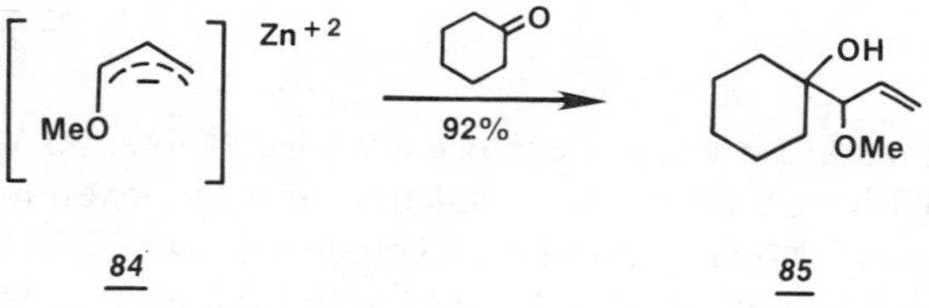

(92)

84 85

The carbanions from allylic thioesters, and protected allylic thiols, tend to react with alkyl halides at the α-position.[1750,1820–1822] However, aldehydes and ketones exhibit opposite regioselectivity.[1750,1823] Allyl 2,4,6-triisopropylthiobenzoate ($ArCO{-}SCH_2CH{=}CH_2$) gave the α′-lithiothioallyl anion on treatment with LDA or *n*-butyllithium at −98°. Addition of methyl iodide or *n*-butyl iodide gave the corresponding α-adducts in 46 and 45% yields, respectively. Acetaldehyde, on the other hand, added exclusively to the γ-position.[1750] It is possible, with allylic thioethers, to reverse the regioselectivity of carbonyl addition from the normal γ-addition to obtain reaction at the

α-center. This is accomplished through a borate intermediate **(87).** The lithiocarbanion derived from an allyl thioether **(86)** with borane leads to the complex **87.** This complex reacts exclusively at the α-position to form **88** (Scheme 93).[1823]

$$Li^+ \; CH_2\bar{C}HCH{-}SR' + BR''_3 \longrightarrow$$

86

$$Li^+ \; {}^-BR''_3{-}CH_2CH{=}CHSR' \xrightarrow{R_2C=O} \qquad (93)$$

87

$$CH_2{=}CH{-}CH(SR){-}CR_2{-}OH$$

88

Methylthio- and *t*-butylthioallyllithium ($CH_2{=}CHCHLi{-}SR$) in the presence of one equivalent of HMPA react with 2-cyclopentenone almost exclusively by α-1,4-addition to give **89,** and only small amounts (<5%) of the γ-1,4-adduct **90** was detected (Scheme 94).[1820] In the absence of HMPA the reaction proceeds by way of α- and γ-1,2-addition.

$CH_2 = CH^-CHSR$ —(2-cyclopentenone)→ **89** (3-substituted cyclopentanone bearing CH(SR)CH=CH$_2$) + **90** (3-substituted cyclopentanone bearing CH$_2$CH=CH~SR) (94)

Michael additions are observed exclusively with 1-potassio-1-selenopropionates ($MeSe{-}CMe^-COOMe$) and chalcone or cyclohexenone.[1824] The lithio analog undergoes a Michael addition to chalcone, but gives a mixture of the 1,2- and 1,4 products in a ratio of 70:30 with cyclohexenone.

Reagents with two activating groups of the type $Ph_3Met{-}CH^-I$, where Met is silicon, germanium, or tin, were recently reported by Kauffmann *et al.*[1825,1826] These organometallics react with benzaldehyde at −70°, with subsequent protonation at −65°, to give the corresponding iodohydrin $Ph_3Met{-}CHI{-}CHOH{-}Ph$. The resulting iodohydrins are important precursors to Ph_3Met-substituted oxiranes and olefins. Phenylthio(triphenylstannyl)-methyllithium (Ph_3SnCH^-SPh) and phenylthio(triphenylplumbyl)methyllithium (Ph_3PbCH^-SPh) can be used for stereospecific carbonyl olefination, as shown in Scheme (95) for the α-triphenylstannyl anion

91.[1827] The reaction of the lithioanion **91** with benzaldehyde gives a 1:1 ratio of the *threo* **92a** and *erythro* **92b** adducts, which upon heating yield the (*E*) **93** and (*Z*) **94** olefins, respectively.

Ph_3SnCH^-SPh (**91**) $\xrightarrow{PhCHO}$ 44% **92a** (Ph₃Sn, H, SPh / H, Ph, OH; *threo*) $\xrightarrow[93\%]{\Delta}$ **93** (*E*)-PhS–CH=CH–Ph; 44% **92b** (PH₃Sn, H, SPh / Ph, H, OH; *erythro*) $\xrightarrow[90\%]{\Delta}$ **94** (*Z*)-PhS–CH=CH–Ph (95)

5

CARBONYL α-CATIONS: $^{+}C—C=O$

Tapio A. Hase and Jorma K. Koskimies

CONTENTS

5.1. INTRODUCTION

In contrast to the other chapters in this book, this one is concerned with umpoled *cationic* synthons. Thus, it would seem that overall, umpoled carbanionic heteroatom-containing synthons are much more important than their cationic counterparts. This is, of course, a result of the electronegative nature of the usual heteroatoms, which confer a^1 character [using Seebach's[6] acceptor/donor (a/d) nomenclature] to the heteroatom-carrying carbon atom, and

d^2 character to the vicinal carbon atom. If then looked at in umpoled form, it is only the latter that have acceptor, that is, cationic character.

Among the classical synthons, there are only two really important anionic classes: the various organometallics (R^-, Ar^-, etc.) and carbonyl α-anions, that is, enolate anions with their vast number of synthetically useful reactions.[5,9,1633,1634] On the other hand, classical, nonumpoled synthon chemistry is rife with cationic a^n systems: a^1 synthons are involved in S_N substitution, Friedel–Crafts, and Grignard reactions, for example, whereas a^3 synthons are seen in Michael additions.

Quite unlike most of the anionic d^n synthons discussed in this book, the a^2 ketone α-cations are actually capable of existence. Thus, for example, the α-cations $^{+}CPh_2—CO—Ph$ and $^{+}CPh_2—COOMe$ have been observed directly by ^{13}C NMR.[1831] There are various means of setting up an a^2 situation in the form of a positive charge next to a carbonyl group, usually at the synthon level but in α-halocarbonyl compounds as an actual reagent. In fact, placing a halogen atom to the α-position in ketones, esters, nitriles, and similar compounds results in enhanced electrophilic reactivity compared with the corresponding alkyl halides,[1832–1836] although it seems that the relative rate is very much dependent on the nucleophile. In the formation of ketone α-cationic intermediates, stabilization by means of the enolonium structure is significant.[1835] In a review of his own work on the reactions of α-keto mesylates and triflates, Creary[1837] concludes that these compounds can solvolyze by a variety of mechanistic pathways, including the intervention of carbonyl-substituted cations. Oxiranyl ions may be alternative intermediates,[1838] theoretical calculations[1839] suggesting them to be lower in energy than the open α-keto cations. α-Mesyloxy amides, which were designed to evaluate the propensity for carbonyl participation in solvolysis, however generally solvolyze giving α-keto cations.[1840] In a study of the reaction of various nucleophiles with α-chloroisobutyrophenones, it was concluded that substitution in α-haloketones by an electron-transfer mechanism ($S_{RN}1$) is unimportant even when the S_N2 substitution is sterically hindered.[1841]

In practice, α-halogenated carbonyl compounds are frequently quite unpredictable in their chemical behavior, a consequence of the fact that these compounds have several potentially electrophilic sites. For example, a nucleophile can attack an α-haloketone at the C=O group, at the halogen-carrying carbon atom or at the halogen atom itself, or cause proton abstraction at the α, β, or α′ site.

Besides halogen atoms, other electrophilic functionalities can be present at a carbon atom α to a carbonyl. No umpolung operations on the reagent are then required. In this way, a formal $^{+}C(OH)—COOR$ synthon situation can be established simply by running a selective Grignard addition at the carbonyl group of an α-aldoester or α-ketoester.

Other trivial solutions to the general problem may easily be found by

suitable oxidative procedures. An allylic halide $X—CH_2—CR{=}CH_2$ will furnish the $^+CH_2—CO—R$ synthon if reacted with a nucleophile (this also takes advantage of the enhanced reactivity of the allylic halide) followed by ozonolysis. In this way allyl bromide has been used as an acetaldehyde cation equivalent for reaction with a nitronate anion.[2307] The synthons $^+C—C—OH$ (an a^2 synthon) are encountered in nucleophilic reactions on epoxides. If the product of such a reaction is oxidized to a carbonyl compound, the overall process becomes equivalent to carbonyl α-cation (also an a^2 synthon) intermediacy. It is in fact something of a weakness of the a/d method of classification of synthons that varying states of oxidation in otherwise similar systems are not readily discerned.

In any case, we will concentrate in this chapter on some of the more elaborate solutions, such as the reliable carbonyl α-cation synthon systems that are available in the form of various Michael-type additions to vinyl sulfones, nitro olefins, and ketene cyanohydrin derivatives.

Although there is a separate table (Table 5.4.) for vinylogous systems, in the text these are taken up in connection with the parent functionality. Thus, the vinylogous aldehyde synthons $^+CH_2CH{=}CH—CHO$ are discussed along with other aldehydes in Section 5.2. Finally, although any route that involves a $C{=}C^+—COR$-type synthon can be utilized for $CH—CH^+—COR$ purposes by a simple $C{=}C$ reduction, we have left the former aside because they arise mostly by routes that are entirely different from those used for the saturated α-cationic ketone synthons.

5.2. ALDEHYDE α-CATIONS $^+C—CHO$

Direct substitution reactions at the α-site of suitably constructed aldehydes are seldom feasible. Many divergent reaction paths will be available for such substrates, such as the attack of the nucleophile on the carbonyl, or the elimination of an α-proton by a basic nucleophile. One of the rare examples where α-haloaldehydes perform well as $^+C—CHO$ and $^{2+}C—CHO$ synthons is the reaction (Scheme 1) of α-chloro- and α,α-dichloroaldehydes with a full range of thiolate anions,[1842] mostly in good yield.

$$\begin{aligned} &Cl—CRR'—CHO + R''S^- \longrightarrow R''S—CRR'—CHO \\ &Cl_2—CR—CHO + R''S^- \longrightarrow (R''S)_2—CR—CHO \end{aligned} \quad (1)$$

where R = alkyl, R′ = H or alkyl, and R″ = alkyl or aryl.

For carbon–carbon bond formation, acetal-protected haloaldehydes have been used, as shown in Scheme (2) for 1,1-diethoxy-2-bromoethane.[427,428]

$$BrCH_2CH(OEt)_2 + \text{2-lithio-1,3-dithiane} \longrightarrow \text{2-(2,2-diethoxyethyl)-1,3-dithiane} \xrightarrow[H_2O,\ 68\%]{8\%\ HClO_4} \text{(1,3-dithian-2-yl)acetaldehyde (CHO)} \tag{2}$$

The addition of alkyl- or aryllithiums to acetals of 2-enals works in the required sense for cinnamaldehyde acetals only (Scheme 3); the key here is benzylic anion stabilization.[1843]

$$PhCH{=}CH{-}CH(OEt)_2 \xrightarrow{RLi} {}^{-}CHPh{-}CHR{-}CH(OEt)_2 \longrightarrow \longrightarrow PhCH_2{-}CHR{-}CHO \tag{3}$$

Silyl enol ethers[1635] of α-bromoaldehydes react with alkyl and aryl Grignards under Ni/P catalysis to give the Br → alkyl (or aryl) exchange product that can be hydrolyzed to the free aldehyde[1844] (see Scheme 24).

The majority of $^{+}C—CHO$ synthon sources involve Michael-type additions to vinyl sulfoxides or sulfilimines, or to nitro- or cyanoalkenes. Two general operational principles are in use: Either the carbon atom that carries the electron-withdrawing group already is at the correct oxidation level (Scheme 4), or the product obtained from the addition of a nucleophile is oxidized to an aldehyde (Scheme 5). It should be noticed that the products of addition of a nucleophile [step (1) in Schemes (4) and (5)] can be quenched with electrophiles other than a proton. Therefore, $^{+}C—CO^{-}$-type synthon equivalents also become available, leading to ketones as discussed below (Scheme 8) and in Section 2.17.5.1. (Chapt. 2).

$$C{=}C\begin{smallmatrix}Y\\W\end{smallmatrix} \xrightarrow[2)\ H^+]{1)\ Nu^-} Nu{-}C{-}CH\begin{smallmatrix}Y\\W\end{smallmatrix} \xrightarrow{\text{Hydrolysis}} Nu{-}C{-}CH{=}O \tag{4}$$

where Y = OR, SR, or NR_2 and W = SO—R, SO—Ar, or CN.

$$\text{C=CH—W} \xrightarrow[\text{2) H}^+]{\text{1) Nu}^-} \text{Nu—C—CH}_2\text{—W} \xrightarrow{\text{(O)}} \text{Nu—C—CH=O} \quad (5)$$

where W = SO—R, SO—Ar, or NO_2.

The starting materials in Scheme (4) are derivatives or analogs of either ketene cyanohydrins or acetals. Along with 2-alkylidene-1,3-dithianes,[534–536] probably most thoroughly studied are the ketene dithioacetal monoxides **(1)**. The simplest members are commercially available, whereas the alkylketene derivatives can be made from formaldehyde dithioacetal (50% overall)[717] or from methyl alkanedithioates (50–70%)[719] (Scheme 6).

$$\text{(MeS)}_2\text{CH}_2 \xrightarrow[\text{3) AcCl}]{\text{1) BuLi 2) MeCHO}} \text{(MeS)}_2\text{CH—CHMe—OAc} \xrightarrow[\text{MeOH}]{\text{NaIO}_4}$$

$$\text{Me—SO—CH(SMe)—CHMe—OAc} \xrightarrow[\text{PhH}]{\text{KOH}} \text{(Me—SO)(MeS)C=CH—Me} \;\; \mathbf{1} \quad (6)$$

$$\text{MeS—CS—CH}_2\text{R} \xrightarrow[\text{2) MeI}]{\text{1) HO}^-} \text{(MeS)}_2\text{C=CHR} \xrightarrow{\text{MCPBA}} \text{(Me—SO)(MeS)C=CH—R} \;\; \mathbf{1}$$

The ketene dithioacetal monoxides react cleanly with ketone and ester enolates[717–719,1845] (Scheme 7), enamines,[717] lithio imines,[719] and with malonate, ketoester, and β-diketone anions.[717,718] Hydrolysis can be brought about by using $HClO_4$ or HBF_4 in aqueous acetonitrile.

$$\text{MeCH}_2\text{CH}_2\text{CH}_2\text{COOMe} \xrightarrow{\text{1) LDA 2) CH}_2\text{=C(SMe)—SO—Me } (\mathbf{2})}$$

$$(7)\;\; \xrightarrow{\text{3) HClO}_4\text{/H}_2\text{O/MeCN}} \text{MeOCO—CHEt—CH}_2\text{—CHO}$$

$$(8)\;\; \xrightarrow{\text{3) MeI 4) HClO}_4\text{/H}_2\text{O/MeCN}} \text{MeOCO—CHEt—CO—CH}_3$$

$$\text{CH}_2\text{=CHCH}_2\text{COOEt} \xrightarrow[\text{3) MeI 4) hydrolysis}]{\text{1) LDA 2) } \mathbf{2}} \text{(EtOCO)(CH}_2\text{=CH)CR—CH}_2\text{—CHO} \quad (9)$$

A delicate interplay of relative acidities can be put to use for further reaction. Thus, while the sulfoxide **(2)** will provide, in reaction with alkanoic ester enolates, a $^{+}CH_2—CO^{-}$ synthon (Scheme 8), alkylation of the first-formed anion from **2** and malonates or β,γ-enoic esters (Scheme 9) proceeds differently, all this giving considerable latitude for synthetic design.

Vinyl disulfones such as $CH_2{=}C(SO_2—Ph)_2$ react very readily with enamines[1846] or enolizable ketones[1847] to give Michael-type products, $RCOCH_2CH_2CH(SO_2Ph)_2$. Although this approach has not been systematically exploited as a method for $^{+}CH_2—CHO$ expression, it would seem that there are possibilities along those lines. Geminal disulfones can be reduced to monosulfones $R_2CH—SO_2Ph$, and the latter oxidized to a ketone R_2CO;[830,832,1848] if this is feasible for the geminal disulfones $RCH(SO_2Ph)_2$, $^{+}CH_2CHO$ synthon equivalency is at hand. Alternatively, it might be possible to reduce[1849] the geminal disulfone to the corresponding disulfide; the latter are actually dithioacetals whose hydrolysis is quite straightforward.

Similar behavior is shown by the ketene cyanohydrin analog **3** (Scheme 10)[1630] which reacts with singly and doubly stabilized enolates and with lithio imines. In the example with pinacolone, omitting step (4) will lead to the ketoaldehyde *t*BuCO—CHMe—CH_2—CHO. If no second alkylation is performed [steps (4–6) not carried out], the product is *t*BuCO—CH_2CH_2CHO, showing $^{+}CH_2CHO$ synthon involvement in its simplest form.

$$tBu—COCH_3 \xrightarrow[\text{3) HMPA or KOBu-}t]{\text{1) LDA 2) }CH_2{=}C(CN)—NMe—Ph\ \mathbf{(3)}} tBuCOCH_2CH_2\overset{\overset{\large NMePh}{|}}{\underset{\underset{\large CN}{|}}{C^{-}}} \longrightarrow$$

$$tBuCO—CH^{-}CH_2CH(CN)—NMePh \xrightarrow[\text{6) }PhCH_2Br\ \ \text{7) }H_3O^{+}]{\text{4) LDA 5) MeI}} \qquad (10)$$

$$\begin{matrix} tBuCO \\ \quad\diagdown \\ \quad\quad CH—CH_2COCH_3 \\ \quad\diagup \\ PhCH_2 \end{matrix}$$

The remaining $^{+}CH_2CHO$ synthons require a final oxidation step for full expression of the synthon equivalency (Scheme 5). For carbon–carbon bond formation, the Michael-type addition to vinyl methyl or aryl sulfoxides works very well with cuprates,[1631] enolates,[1850] and nitronate, malonate, and keto-ester anions.[1851,1852] Again, the product from the initial addition may be quenched either with a proton source ($\simeq {}^{+}CH_2CHO$) or with other electrophiles ($\simeq {}^{+}CH_2CO^{-}$). In addition to the carbon nucleophiles, amines, alcoholates, and thiolates will also add to the vinyl sulfoxides.[1852] However, there appears to be no systematic study on the conversion of the addition products to aldehydes. This should be no problem with the cuprate addition products,

using the Pummerer reaction,[1853,1854] or a variation such as reaction with an *O*-silylated ketene acetal in the presence of zinc iodide,[1639] but it is less certain how well the more highly functional addition products respond to these conditions. There is also the interesting idea,[1855] used in connection with 1,4-diketone synthesis (see Scheme 35), of trapping the intermediate α-sulfinyl carbanion with ^{+}SMe (from Me_2S_2) to give a dithioacetal monoxide which is then hydrolyzed in the usual way (see Schemes 8 and 9 above, and Sect. 2.4. in Chapt. 2).

Diastereoselective Michael additions of malonate or β-ketoester anions to optically active tolyl styryl sulfoxide have also been reported.[1856] Although the authors were content just to desulfurize the major (4:1) diastereomer, it is possible that the conversion to an aldehyde might also be feasible.

It has been known for many years that nitroethene reacts smoothly with enamines, giving γ-nitroketones after hydrolysis.[1857–1859] Seebach and coworkers[474] have established that ketone, ester, and amide lithium enolates, 2-alkyl-, and 2-acyl-1,3-dithiane anions, lithio orthothioformates (a $^{-}COOR$ source, see Chapt. 3), and lithiated *N*-nitrosoamines (a $^{-}CH_2NH_2$ source, see Chapt. 4) add smoothly to nitroalkenes and nitrostyrenes (Scheme 11). The products were not converted to aldehydes (or ketones), but it is well known that there are several methods for that purpose, including various oxidation conditions and the Nef reaction (for literature references, see Sect. 2.12.2.2.).

$$\text{RCH}{=}\text{CR}'{-}\text{NO}_2 \xrightarrow[\text{2) AcOH}]{\text{1) Nu}^-} \text{Nu}{-}\text{CHR}{-}\text{CHR}'{-}\text{NO}_2 \qquad (11)$$

where R = H, Me or aryl and R′ = H (or Me).

Later, the Seebach group went on to study possibilities for stereochemical control in these Michael additions. It was shown[1860] that optically active β-hydroxyester enolates furnish the *erythro* products **4**; however, no stereocontrol operated in the formation of the new chiral centers α and β to the NO_2 group. In this case, the addition products were hydrogenated to aminoesters that cyclized to γ-lactams.

$$\begin{array}{c} \text{OH} \\ | \\ \text{Me}{-}\text{C}{-}\text{COOEt} \\ | \\ \text{RCH}{-}\text{CHR}'{-}\text{NO}_2 \end{array}$$

4

where R = H, alkyl, or aryl and R′ = H (or methyl).

The nitroalkene geometry has a clear-cut effect on product stereochemistry when certain aldehyde, ketone, or ester enolates are added.[1861] Thus, the two

1-nitropropenes react with cyclopentanone enolate to give the diastereomeric products **(5)** and **(6)**.

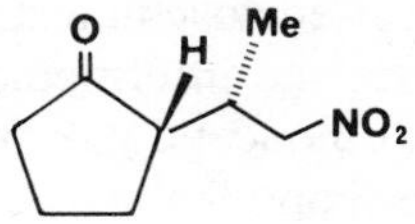

(5) 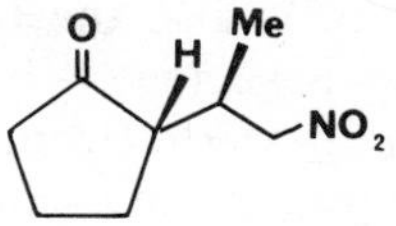 (6)

From (*E*)-1-nitropropene, the **(5)**:**(6)** ratio was 85:15 whereas the (*Z*)-isomer gave a 21:79 ratio. Enolates of cyclohexanone and of *t*-butyl propionate react in a comparable yield (60–80%) and with a similar diastereoselectivity; the chemical yield is poor with butanal enolate.

As far as vinylogous systems are concerned, the epoxy acetal shown in Scheme (12) is an efficient source for the (*E*) enal synthon $^{+}CH_2CH{=}CHCHO$.

$$H_2C\text{—}CH\text{—}CH_2CH(OEt)_2 \ (\text{epoxide, O bridging } H_2C \text{ and } CH) \xrightarrow{RMgX} \qquad (12)$$

$$RCH_2\text{—}CHOH\text{—}CH_2CH(OEt)_2 \xrightarrow{H_3O^+} RCH_2CH{=}CH\text{—}CHO$$

The reagent is made in 35% yield from allylmagnesiumchloride and triethyl orthoformate.[1862] A free choice of alkyl or aryl Grignards may be used to give the hydroxyacetals in 78–92% yield, with subsequent hydrolysis and dehydration taking place in 50–85% yield.

5.3. KETONE α-CATIONS $^{+}C—COR$ AND $^{+}C—COAr$

5.3.1. α-Haloketones

A thorough review on the reactivity (and synthesis) of α-halogenated ketones has appeared recently.[1832] Therefore, we will present only the salient features of nucleophilic substitution in these compounds.

As noted in the introduction to this chapter, α-halocarbonyl compounds are somewhat notorious as sources for carbonyl α-cation synthons, owing to the many opportunities for side reactions with nucleophiles, including halohydrin formation (possibly followed by epoxidation), HX elimination, reduction by loss of X^+, or the Favorskii rearrangement. Thus, *hydrolysis* to α-hydroxyketone is sometimes feasible,[1863–1872] but in many instances the Favorskii rearrangement will intervene. A more reliable method involves a substitution with a carboxylate (usually formic or acetic acid) salt[1873–1877] to

give the α-acyloxyketone, which can be readily hydrolyzed or alcoholyzed to the α-hydroxyketone (Scheme 13).[1878]

$$CH_3COCH_2Br \xrightarrow[54\text{–}58\%]{\text{HCOOK, MeOH}} CH_3COCH_2OH \qquad (13)$$

Phenacyl chloride reacts smoothly with pyruvate salts as well.[1879] Of course, any carboxylate can be reacted with the phenacyl halides, the reaction forming the basis of an important means of characterization of carboxylic acids. However, acetolysis of α-haloketones may sometimes lead to exceptional products.[1880]

In the words of Verhé and De Kimpe,[1832] "there is no doubt that the reaction of α-haloketones with *alkoxides* is the most profoundly investigated in the field of reactivity of α-halogenated carbonyl compounds." There are many examples[1871,1881–1886] of successful substitutions, but, again, several reaction pathways are available and the final outcome of a particular reaction depends on the degree of substitution in the haloketone, as well as on the nature of the halide, solvent, and the nucleophile and its concentration. It is interesting that the α-alkoxyketones, when formed, are not necessarily products of direct nucleophilic substitution. It has been in fact proved that these end products can arise through the epoxy ethers (Scheme 14).[1887]

$$Me_2CH—COCH_2Br \xrightarrow[\text{MeOH}]{\text{MeO}^-} Me_2CH—C(O^-)(OMe)—CH_2Br \longrightarrow$$

$$Me_2CH—\underset{\diagdown O \diagup}{C(OMe)—CH_2} \xrightarrow[54\%]{\text{MeOH}} Me_2CH—COCH_2OMe \qquad (14)$$

Alkoxyepoxides derived from suitably constituted haloketones can open up in the alternative fashion to give hydroxyketals (Scheme 15).[1881] When the latter are required, this one-step reaction is clearly the procedure of choice, as the intermediate methoxyepoxide (Scheme 13) could be isolated in a mere 33% yield, giving the hydroxyketal in 40% yield.

$$PhCO—CHCl—Ph \xrightarrow[73\%]{\text{MeO}^-\text{, MeOH}} Ph—C(OMe)_2—CHOH—Ph \qquad (15)$$

Instead of being subjected to a substitution by an alkoxide or phenoxide, α-haloketones sometimes respond well to simple alcoholysis conditions.[1882,1888]

Ammonia and primary and secondary amines react satisfactorily with

XCH_2COR- and X—CHR—COR-type haloketones, particularly chlorides.[1873,1889–1892] Halogen on a fully substituted α-carbon atom will lead to elimination, but this problem can be circumvented by using azide anion as the nucleophile.[1892–1894] Tertiary *amines* give the expected ammonium salts. Primary α-aminoketones are sometimes best prepared using the *K-phthalimide* method[1895] or by reduction of intermediary *N*-benzylaminoketones.[1832] Other amides have also been used occasionally.[1896,1897] *Thiols, thiophenols, and thiocarboxylates*[1871,1873,1896,1898–1907] or *selenocarboxylates*[1908] often give good yields of the α-substituted ketones. Thus 2-bromocyclobutanone furnishes the 2-phenylthio- and 2-(2-hydroxyethyl)thio-derivatives in 90–95% and 92% yield, respectively.[1906]

In a review on α-ketonitriles, the reaction of the *cyanide* anion with α-haloketones is discussed;[1909] the general review on α-haloketones has a chapter on a number of carbon nucleophiles.[1832] As for the cyanide ion, there are many examples[1910–1915] of successful nitrilations of α-haloketones (Scheme 16).

$$\mathrm{RCO{-}CH_2Br \xrightarrow{^{-}CN,\ H_2O/EtOH} RCO{-}CH_2CN} \qquad (16)$$

where R = tBu, 80% yield[1910] and R = aryl, 40–57% yield.[1911]

In some cases, particularly with aliphatic ketones, the reaction end products are cyanohydrins or cyanoepoxides (Scheme 17).[1909]

$$\mathrm{RCO{-}CH_2Cl \xrightarrow{^{-}CN} ClCH_2CR(OH)(CN) \longrightarrow H_2\overset{O}{C{-}CR}{-}CN} \qquad (17)$$

The latter may, however, be converted by dilute mineral acids or boron trifluoride to the same ketonitriles that would have resulted from a direct Cl → CN substitution in the α-haloketone.

In more recent times, tetraalkylammonium cyanides have been shown to be effective for the haloketone-to-cyanoketone conversion (Scheme 18).[1912]

$$\mathrm{RCO{-}CR'R''{-}X \xrightarrow[MeCN\ or\ DMSO]{Et_4N^{+}\ ^{-}CN} RCO{-}CR'R''{-}CN} \qquad (18)$$

where R = Ph, R′ = R″ = H, 70% yield; R = alkyl or phenyl, R′ = H, R″ = Me, 33% yield; R,R′,R″ = alkyl or phenyl, 82–85% yield.

Enamines[1890,1916] react smoothly with α-bromoketones, giving 1,4-diketones in a nice example of combination of the synthon counterparts $R'COCH_2^-$ and $RCOCH_2^+$ (Scheme 19).[1916]

$$\mathrm{RCOCH_2Br\ (or\ RCO{-}CHMe{-}Br) + CH_2{=}CR'{-}NR''_2 \longrightarrow \xrightarrow{H_2O} RCO{-}CH_2CH_2{-}COR'}\quad 51\text{–}87\%\ \text{yield} \qquad (19)$$

Doubly activated *enolate anions*[1917–1920] give the expected tricarbonyl compounds on reaction with α-haloketones (Scheme 20).[1921]

$$\mathrm{MeOCOCH_2CO(CH_2)_3CH{=}CH_2 + BrCH_2COMe \xrightarrow[86\%]{\text{NaH, dioxane}} MeOCO{-}\underset{\displaystyle CH_2COMe}{\underset{|}{C}H}{-}CO(CH_2)_3CH{=}CH_2} \quad (20)$$

There are no reports of successful alkylations under basic conditions using simple enolate anions. However, tin enolates couple with α-haloketones under Pd or Ru catalysis to give unsymmetrical 1,4-diketones,[1922] the one shown in Scheme (21) being then used for the synthesis of dihydrojasmone.

$$\mathrm{Bu_3SnCH_2COCH_3 + BrCH_2CO(CH_2)_5CH_3 \xrightarrow[66\%]{PdCl_2(PhCN)_2} CH_3CO{-}CH_2CH_2{-}CO(CH_2)_5CH_3} \quad (21)$$

Sterically hindered ketones on one hand and α-bromoacetone or α-bromoaldehydes on the other fail to react satisfactorily. Lewis acid catalysis ($TiCl_4$, pyridine) leads to aldol self-condensation of α-chloroketones, that is, both chlorine atoms are retained in the reaction [1923] and no $^{+}C—CO$ chemistry is involved. The Weiler-type β-ketoester dianions also do not behave well in attempted reactions with α-haloketones or their ketals.[1924]

Carbon–carbon bond formation using *organometallics* can be expected to be problematic in the present connection, because of the basicity of the reagents and because of their high affinity toward carbonyl groups. Thus it would seem surprising that various alkyl and aryl Grignards or lithiums can be used for the conversion of α-haloketones to the α-alkylated or arylated ketones. In truth, these reactions often actually proceed by a nucleophilic addition to the carbonyl group, giving a vicinal halohydrin, which then rearranges to the ketone (Scheme 22). In this manner, α-chlorocyclohexanone and phenylmagnesium bromide give α-phenylcyclohexanone in 68% yield;[1925] α-chlorocyclopentanone reacts similarly.[1926]

$$\mathrm{RCO{-}C{-}Cl \xrightarrow{R'Li} (HO)(R')CR{-}C{-}Cl \xrightarrow{rearr.} R{-}CO{-}C{-}R'} \quad (22)$$

Methyllithium will deliver stereoselectively an angular methyl group into 1-decalone or perhydroindanone by way of the corresponding α-chloro compound.[1927]

In contrast to the Grignards and alkyllithiums, cuprates actually react by nucleophilic substitution, although halogen–metal exchange, leading to the

same substitution products, can be a competing mechanism. This reaction is particularly useful for the synthesis of highly hindered ketones (Scheme 23).[1928,1929]

$$\text{Br—CMe}_2\text{—CO—CMe}_2\text{—Br} \xrightarrow[33\%]{\text{Me}_2\text{CuLi}} t\text{-Bu—CO—Bu-}t \qquad (23)$$

A penalty of decreasing yields must be accepted when going from primary to secondary to tertiary R groups in R_2CuLi. For this alkyl group at least, methylcyanocuprate is superior to lithium dimethylcuprate in reaction with various α-bromoketones.[1930]

Friedel–Crafts alkylations, where the aromatic substrate serves as an Ar^- synthon, can in some instances be performed with α-bromoketones.[1888] An Organic Syntheses[1931] procedure details the preparation in 53–57% yield of 1,1-diphenylacetone from 1-bromo-1-phenylacetone.

5.3.2. Other Sources

5.3.2.1. Grignard Reagents or Lithium Reagents As Nucleophiles

2-Bromo-1-trimethylsilyloxyalkenes, accessible[1635] from α-bromoketones, react with Grignard reagents in the presence of a Ni/phosphine catalyst to give, after hydrolysis, the α-branched ketones in yields ranging from 27 to 94% (Scheme 24).[1844]

$$\text{Br—CR=CR}'\text{—OSiMe}_3 \xrightarrow[\text{Ni[Ph}_2\text{P(CH}_2)_3\text{PPh}_2]\text{Cl}_2]{\text{R}''\text{MgX}}$$

$$\text{R}''\text{—CR=CR}'\text{—OSiMe}_3 \xrightarrow{\text{H}_3\text{O}^+} \text{RR}''\text{CH—CO—R}' \qquad (24)$$

where R = H or alkyl, R′ = H, Me or Ph, and R″ = alkyl or aryl.

Nitroalkenes bearing a suitably located leaving group (pivaloyloxy) add alkyl- or aryllithiums. Thus, 3-pivaloyloxy-2-nitro-1-propene **(7)** gives 2-nitro-1-alkenes, including 3-aryl-2-nitro-1-propenes, in 21–77% yields (for RLi) or 69–88% yields (for ArLi) (Scheme 25).[1932] The RLi reagent can also be a $C=C^-$ or $C\equiv C^-$ source.

$$\underset{\mathbf{7}}{\text{H}_2\text{C=C(NO}_2)\text{—CH}_2\text{O—COBu-}t} \xrightarrow{\text{RLi}} \text{RCH}_2\text{—C(NO}_2)\text{=CH}_2 \dashrightarrow \text{RCH}_2\text{—CO—CH}_3 \qquad (25)$$

In a similar manner 2-nitro-3-butyl-1-cyclohexene can be made in 92% yield, and the 3-phenyl analog in 89% yield.[573] Appropriately substituted styrenes can also be used (Scheme 26).

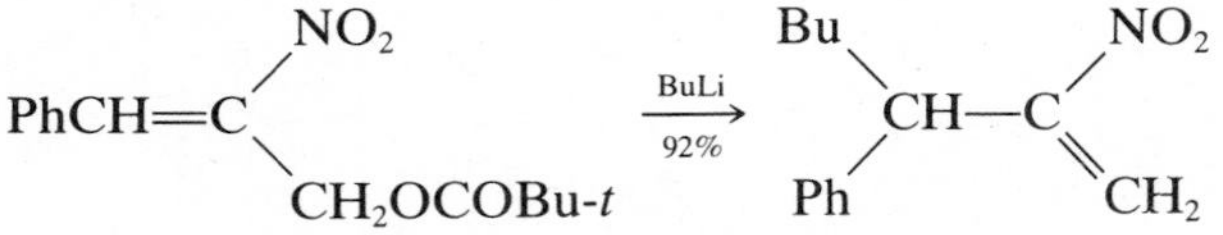

$$PhCH{=}C(NO_2)CH_2OCOBu\text{-}t \xrightarrow[92\%]{BuLi} (Bu)(Ph)CH{-}C(NO_2){=}CH_2 \qquad (26)$$

The above alkylation products (2-nitro-1-alkenes) were not carried over to the ketones (see Scheme 25), but there are numerous methods available for this conversion, some of the more recent ones involving Raney-Ni with sodium hypophosphite,[1933] a cathodic reduction,[1934] or chromyl chloride.[1935]

The addition of *t*-BuLi to vinylic acetals, shown above (Scheme 3) for cinnamaldehyde diethyl acetal, is very effective for the analogous ketone derivatives as well, such as benzalacetone diethyl ketal.[1843]

5.3.2.2. Cuprates As Nucleophiles

Various tosylhydrazones have been α-brominated and alkylated or arylated *in situ*[1936] (Scheme 27).

$$TsNH{-}N{=}CR{-}CH_3 \longrightarrow TsNH{-}N{=}CR{-}CH_2Br \xrightarrow{R'_2CuLi} TsNH{-}N{=}CR{-}CH_2R' \qquad (27)$$

where R = alkyl or aryl, R′ = Me 55–72%, and R′ = Ph 35–40%.

α,α′-Dialkylations can also be performed in 50–88% yield. The product tosylhydrazones were not hydrolyzed in this case, but that can be done with any of a rich variety of methods, for example, with clay-supported ferric nitrate[1937] or cupric chloride,[1938] to cite two of the most recent ones.

A related α-arylation route[1939] for ketones is shown in Scheme (28).

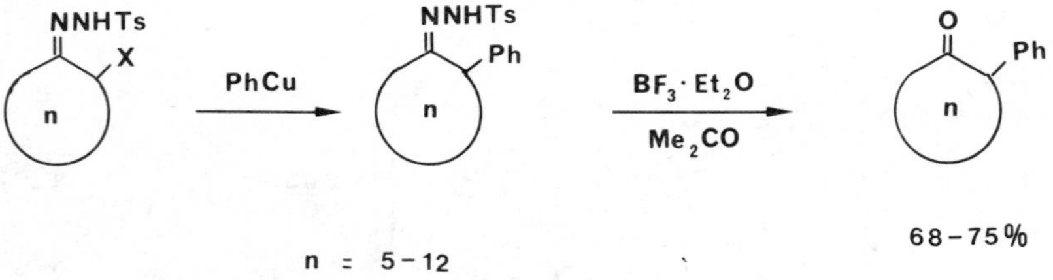

$$\text{cyclo-}(CH_2)_n\text{ C(=NNHTs)–CHX} \xrightarrow{PhCu} \text{cyclo-}(CH_2)_n\text{ C(=NNHTs)–CHPh} \xrightarrow[Me_2CO]{BF_3\cdot Et_2O} \text{cyclo-}(CH_2)_n\text{ C(=O)–CHPh} \qquad (28)$$

n = 5–12; 68–75%

In addition to the cycloalkanone derivatives, phenacyl bromide tosylhydrazone was α-phenylated in 69% yield. Furthermore, a functional arylcopper

reagent, *o*-dimethylaminomethyl-phenylcopper **(8)** can also be used. Rather than by direct substitution, these reactions apparently proceed by *in situ* formation of the azoene **(9)**, which then acts as a Michael acceptor.

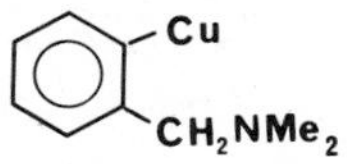

Cu
CH_2NMe_2
(8)

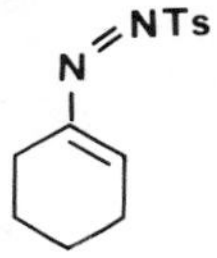

NTs
N
(9)

4-En-3-one *N*-methyltosylhydrazone derivatives in the steroid series react with lithium dimethylcuprate, followed by hydrolysis with 2 *N* HCl, to give the 2α-methyl-4-en-3-ones in 40–45% yield.[1940] Again, azoene intermediates are involved, this time formed by the elimination of TsH from the starting material.

Acetoxy oxiranes yield α-alkylated ketones in very moderate yields on treatment with lithium dialkylcuprates.[1941] The product mixture consists of the α-branched ketone and of the ketone without the α-branch, which will cause severe separation problems. In this manner 5-nonanone and 4-methyl-5-nonanone were obtained in 32 and 43% yields, respectively, from the reaction of lithium dimethylcuprate with 4-acetoxy-4-nonene epoxide. Other dialkylcuprates, or other substrates such as 1-acetoxycyclohexene epoxide, perform even less satisfactorily.

Acyl oxiranes can be made to serve in certain cases as highly efficient sources of $^+C—CO—C=C$ type synthons.[1942] The enol trimethylsilyl ether **(11)**, made from the cyclohexenone epoxide **(10)**, undergoes an S_N2' attack by lithium dimethylcuprate (Scheme 29, R = Me), to give 3,6-dimethyl-2-cyclohexenone in 97% yield. Other cuprates (Scheme 29, R = Ph, 95% yield, or R = vinyl, 93% yield) and enols in other guises (enol phosphate, or even the enolate anion itself) can be used.

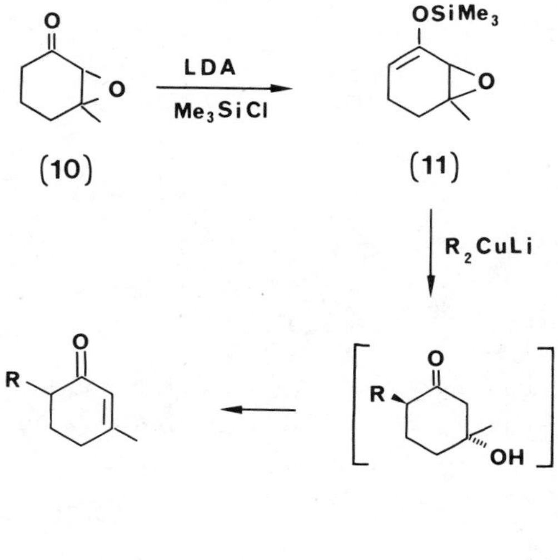

O
O
(10)
LDA
Me_3SiCl
$OSiMe_3$
O
(11)
R_2CuLi
(29)
R
O
R
O
OH

The cyclohexenone moiety may also be a steroid ring A; in the five-membered ring ketone series, the reaction is distinctly less effective with yields in the 40% range. Finally, other means for setting up a system for S_N2' attack were explored. Thus, α-chloro- or α-acetoxyketones, including acyclic compounds, gave moderate yields of the α'-alkylation products (Scheme 30).[1942]

$$CH_3CO—CMe_2—Cl \xrightarrow[44\%]{Ph_2CuLi} PhCH_2—CO—CHMe_2 \quad (30)$$

5.3.2.3. Enolate Anions and Related Nucleophiles

Under this heading, all nucleophiles that correspond to a $^{-}$C—CO—R unit will be discussed. Thus we are dealing with the combination of the synthon counterparts $^{-}$C—CO—R (normal) and $^{+}$C—CO—R' (umpoled), a process that will necessarily give 1,4-diketones R—CO—C—C—CO—R'. Enol silyl ethers and enoxyborates will be presented first, followed by enolates, doubly stabilized enolates, and lithio imines.

Symmetrical 1,4-diarylbutane-1,4-diones are accessible from enol trimethylsilyl ethers of aryl-substituted acetophenones (2 eq.) and PhI—OBF_3 (1 eq.); the latter apparently provides an equivalent of the intermediate ArCO—CH_2—IPh—OBF_2 which is then attacked by another equivalent of the fluoride-derived remaining enolate to attack. The yields are moderate (45–62%).[1943]

Conjugated nitro olefins **(12)** undergo attack by enol silyl ethers in the presence of a Lewis acid to give 1,4-diketones, with hydrolysis and the Nef reaction being carried out in the same flask (Scheme 31).[1944,1945]

OSiMe$_3$ + R(H)C=C(R')NO$_2$ —Lewis acid→ (31)

R = R' = H (SnCl$_4$) 85%
R = H, R' = Me (TiCl$_4$) 76%
R = Me, R' = H (AlCl$_3$) 63%

Methyl-substituted silyloxycyclohexenes and various silyloxycyclopentenes could also be used, as was a 2-silyloxy-1-alkene. Ketene methyl silyl acetals, made from the alkanoic acid methyl esters, react with nitroalkenes or nitrocyclohexenes in the presence of a Lewis acid and titanium tetraisopropoxide to give γ-ketoesters in good yield.[1944,1945] A stannous triflate-promoted rearrangement of β-ketosulfoxides leads to $^{+}$CH(SAr)—COR and thence to $^{+}CH_2$COR, which were used for combination with silyl enol ethers (see Scheme 47).

Potassium enoxyborates, prepared from cyclic ketones, can be alkylated with 2,3-dichloropropene in the presence of a palladium catalyst (Scheme 32) in 72–86% yield.[1946]

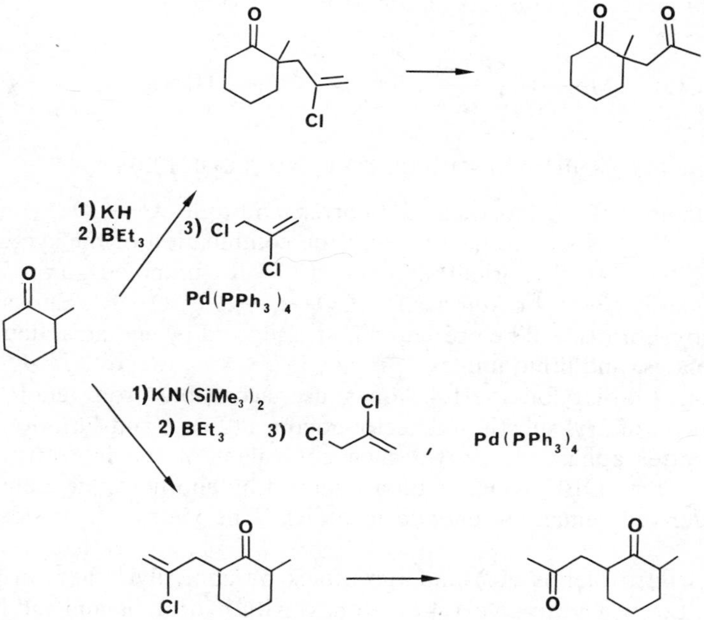

1) KH
2) BEt_3
3) Cl⁀C(Cl)=CH_2
$Pd(PPh_3)_4$

1) $KN(SiMe_3)_2$
2) BEt_3
3) Cl⁀C(Cl)=CH_2, $Pd(PPh_3)_4$

(32)

The resulting chloroallylketones are then hydrolyzed in 88% formic acid with mercuric acetate, giving the α-acetonyl ketones in 55–67% overall yield. It is quite likely that the reaction is applicable to ketones other than cyclopentanones and cyclohexanones. Incidentally, the homologous 1,5-diketones are available by an analogous reaction using 1,3-dichloro-2-butene; this reaction certainly works with acyclic starting ketones as well.

The versatile nitro olefins, already discussed above in connection with the $^{+}C—CHO$ synthon sources (Sect. 5.2.) and as electrophiles in a Mukaiyama-type reaction (Scheme 31), will also accept enolate, that is, carbonyl α-anions in $^{+}C—CO—R$ synthon situations. Thus, the nitro olefin chemistry of Scheme (11)[474] (see also **4**)[1860] involves ketone α-cation synthons if R′ = alkyl; R′ = methyl was specifically mentioned in both instances. An application of this route to γ-diketones is shown in Scheme (33), part of a sequence leading to the synthesis of (±)-cedrene.[1947]

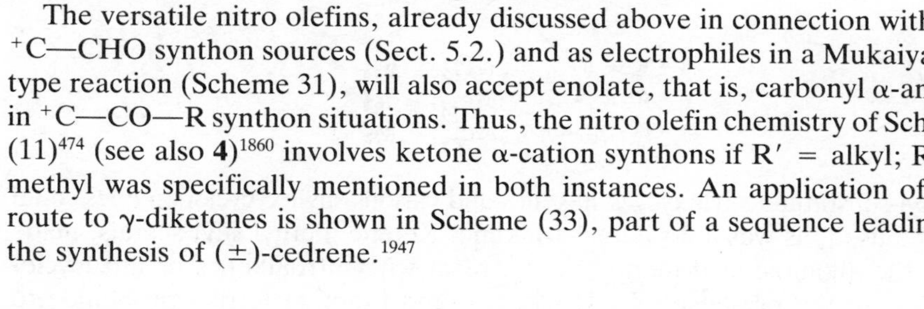

1) $MeCH=CMe-NO_2$, HMPA (90 %)
2) $NaNO_2$, PrONO, DMSO (44 %)

(33)

There is a limitation to these reactions in that nitrocyclohexene does not react[573] with lithium enolates or dithianes, or with enamines, which all add to open-chain $C{=}C{-}NO_2$.

The nitro olefins **(12)** react with dianions of carboxylic acids or α-anions of esters; both may also be α-branched. Isolated yields of the product γ-ketoacids or γ-esters, obtained after an acidic work-up, are in the 24–88% range for the acids and in the 41–81% range for esters.[1945,1948]

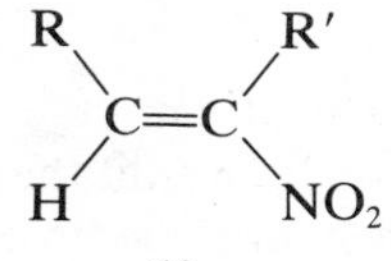

R, R′, H, NO_2 on C=C

12

where R = H or Me; R′ = Me or Et or R + R′ = $(CH_2)_4$.

3-Pivaloyloxy-2-nitro-1-propene **(7),** discussed above (Scheme 25) as an acceptor for alkyl- or aryllithiums, will also accept ketone or ester enolate nucleophiles in 44–87% yields. The products are 4-nitro-4-alken-1-ones (or -4-alkenoic esters) such as **(13)** from the corresponding cyclopropanecarboxylate, and need to be converted to the methyl ketones for full expression of the $^+CH_2COCH_3$ synthon.

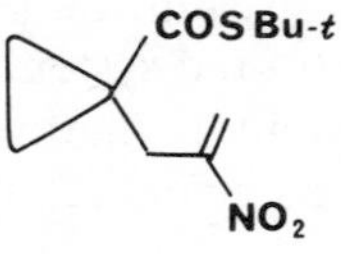

COSBu-*t*

NO_2

(13)

It was not lost on Seebach and Knochel[1932] that their products, being still nitro olefins, should be able to accept another nucleophile, thereby providing the bidentate synthon $^+CH_2{-}CO{-}CH_2^+$. This idea was exploited with many combinations of nucleophiles in good yield (66–75%); an example is given in Scheme (34).

(34)

(7) 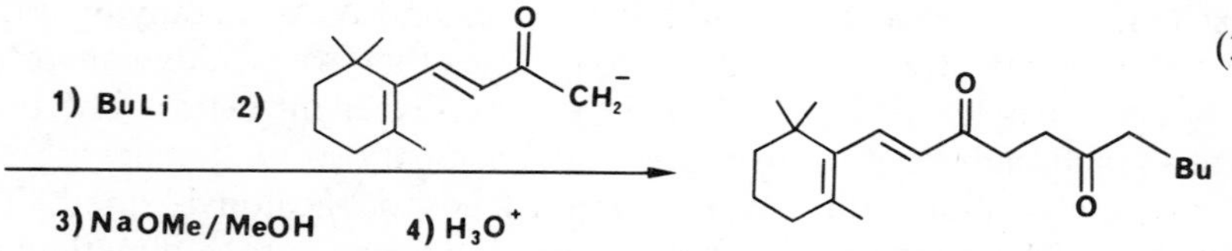 —— 1) BuLi 2) [enone]CH_2^- 3) NaOMe/MeOH 4) H_3O^+ ——→ [product] Bu

Just as enolates will add to unsubstituted vinyl phenyl sulfoxide (a $^+CH_2CHO$ synthon source; see Sect. 5.2.),[1850] 2-alkenyl phenyl sulfoxides react with car-

bonyl α-anions[1855] (Scheme 35). The product sulfoxide may be taken to the ketone (65%) using the Pummerer rearrangement.

$$Me_2CO \xrightarrow[2)\ CH_2=C(SO{-}Ph)(CH_2)_6CH_3\ \mathbf{14}]{1)\ LTMP} CH_3COCH_2CH_2CH(SO{-}Ph)(CH_2)_6CH_3 \quad (63\%) \qquad (35)$$

There was also an unsuccessful attempt to trap the intermediate α-sulphinyl carbanion with ^{+}SMe; the product would then have been a dithioketal monoxide with many precedented hydrolytic procedures. However, this strategy worked very nicely with the addition product of **14** and the dianion of ethyl acetoacetate, the latter having reacted at the $^{-}CH_2$ terminus.

Doubly stabilized enolate anions, such as those derived from malonic esters or β-ketoesters, have been reacted with various types of $^{+}C—COR$ synthon sources. 2-Nitropropene and 2-nitro-1-butene, as well as 2-nitro-2-butene, all combine with *neutral* β-diketones or ketoesters in the presence of potassium fluoride, but really good yields were obtained[1945,1949] with 2-methyl-1,3-cyclohexanedione only. The triketone products were obtained in these reactions directly, showing that a Nef-type reaction occurs *in situ*.

β-Dicarbonyl nucleophiles have been reacted with a number of acetonyl enolonium synthons $^{+}CH_2COCH_3$, but the principles of some of these may well be applicable to more substituted systems. 3-Chloro-2-trimethylsilyloxy-1-propene is alkylated in high yield by diethyl sodiomalonate or ethyl sodiocyanoacetate, giving acetonylated esters after hydrolysis (Scheme 36).[1950]

$$CH_2=C(OSiMe_3)(CH_2Cl) + Na^{+}\,{}^{-}CH(COOEt)_2 \xrightarrow{THF} \xrightarrow{H_2O} CH_3COCH_2CH(COOEt)_2 \qquad (36)$$

Dimethyl sodioethylmalonate will also react satisfactorily if sodium iodide is present, presumably a matter of overcoming the slow rate of reaction by Cl → I exchange in the reagent. α-Lithiated imines and *N,N*-dimethylhydrazones are also very effective, giving alkylated imines that are readily converted to the γ-diketones by mild hydrolysis in good-to-excellent yields. The two nitrogen functionalities give diverging regiochemistry so that, starting from 2-methylcyclohexanone, the cyclohexylimine leads to 2-acetonyl-2-methylcyclohexanone as the major product, whereas going by way of the dimethylhydrazone results in the 2,6-disubstituted ketone (in 97% yield but as a 3:7 diastereomeric mixture).

Another acetonyl enolonium synthon source is 3-acetoxy-2-chloro-1-propene, $CH_2{=}CCl{-}CH_2OAc$, which can be used for the alkylation of cyclic β-diketones or 2-methoxycarbonylketones in the presence of $Pd(OAc)_2$. Mercuric acetate-mediated hydrolysis then gives α-acetonylketones in 31–59% overall yield.[1951] Yet another $^{+}CH_2COCH_3$ source is the silylalkyne (**15,** Scheme 37),[1924] which was successfully used in a synthesis of dihydrojasmone.

$$C_5H_{11}C^{-}(COCH_2^{-})(COOMe) + \underset{\mathbf{15}}{BrCH_2C{\equiv}C{-}SiMe_3} \longrightarrow$$

$$C_5H_{11}CH(COCH_2CH_2C{\equiv}C{-}SiMe_3)(COOMe) \quad (>90\%) \xrightarrow[79\%]{Hg^{2+}/H_3O^{+}/THF} \tag{37}$$

$$C_5H_{11}CH(COCH_2CH_2COCH_3)(COOMe) \longrightarrow \text{Dihydrojasmone}$$

It is noteworthy that treatment of the dianion with 3-bromo-2-methoxy-1-propene only led to the γ-bromo-β-ketoester.

2-Alkanone-1-cations $RCH_2COCH_2^{+}$ are available for reaction with sulfonylacetates, as shown in Scheme 38.[1952]

$$RCHO + CH_2{=}CLi{-}OEt \xrightarrow[62\%]{Ac_2O}$$

$$AcO{-}CHR{-}C(OEt){=}CH_2 \xrightarrow[Pd(PPh_3)_4, DBU\ 65\%]{PhSO_2{-}CH_2{-}COOPr\text{-}i} \tag{38}$$

$$RCH{=}C(OEt){-}CH_2CH(SO_2{-}Ph)(COOPr\text{-}i) \xrightarrow[99\%]{H_3O^{+}} RCH_2{-}COCH_2CH(SO_2Ph)(COOPr\text{-}i)$$

The source reagents, 3-acetoxy-2-ethoxy-1-alkenes, are made using lithiated ethyl vinyl ether as a CH_3CO^{-} synthon. The product γ-keto-α-sulfonylacetates can be further elaborated, to give either the saturated ketoesters $RCH_2COCH_2CH_2COOPr$-*i* or the unsaturated analogues $RCH_2CO{-}CH{=}CH{-}COOPr$-*i*. Starting from α,β-unsaturated aldehydes in an otherwise identical sequence, an access to the vinylogous synthons $MeCOCH{=}CR{-}CH_2^{+}$ (R = H or Me) and $MeCOCH{=}CR{-}CHR'^{+}$ was also developed.[1952] Mesityl oxide, when converted to the π-allylpalladium

complex, reacts with the malonate, ethyl cyanoacetate, or ethyl methylsulfonylacetate anion in good yield[1953] as an alternative source of the $MeCOCH{=}CMe—CH_2^+$ synthon. A third form of vinylogous ketone cation synthons, for reaction with malonate or malonitrile anions, involves 2-cyclohexenone 4-cationic systems. Reviewed by Pearson,[1954] these equivalents are based on the methoxydiene-$Fe(CO)_3$ complexes.[1955,1956]

Finally, the synthon equivalency $^+CHPh—CO—R$ (R = Me, Ph, or $PhCH_2$) results when the corresponding ketone tosyl- or dinitrophenylhydrazones are α-brominated (using phenyltrimethylammonium perbromide) and treated *in situ* with the diethyl malonate anions.[1957] Overall yields are 35–70% for the malonated hydrazones, but these were not taken to the ketomalonates. Based on the same type of chemistry, there were also attempts at the α,α′-dicationic synthon $^+CHPh—CO—CHPh^+$ but yields were only in the 10–20% range. Azoene intermediates are involved in the above reactions; 1-tosylazocyclohexene **(9)** has in fact been prepared and reacted with β-diketones or ketoesters.[1958] The cyclic keto group was not demasked, as the products were utilized as precursors to tetrahydroindoles.

5.3.2.4. Other Carbon Nucleophiles

There are some examples of the coupling of carbonyl α-cation sources with phosphorus-stabilized carbanions, dithioacetal anions, and arenes, the latter behaving as Ar^- in Friedel–Crafts situations.

A number of β-diphenylphosphinoylketones were prepared for use as homoenolate equivalents (see Sect. 6.3., Chapt. 6) by the reaction of $^-CHR—POPh_2$ (R = H, Me, or Et) with 2,3-dichloropropene or with 3-methoxy-2-butanone (31–61%).[1959] 2,3-Dichloropropene also furnishes γ-ketophosphonate esters in an analogous reaction (Scheme 39; R = H or Me, yield 60–73%).[1960]

$$(EtO)_2PO—CH_2R \xrightarrow[\text{3) } CH_2{=}CCl—CH_2Cl]{\text{1) BuLi 2) CuI}} (EtO)_2PO—CHR—CH_2—CCl{=}CH_2 \xrightarrow{H_2SO_4} (EtO)_2PO—CHR—CH_2COCH_3 \tag{39}$$

Nitro olefins and nitrostyrenes furnish β-dithianylnitroalkanes when treated with 2-lithiodithiane; the latter may also carry a 2-alkyl or aryl group.[474] The products, formed by the combination of the two umpoled synthons RCO^- and $R'COCH_2^+$, can be viewed as potential β-diketones where the two carbonyls are differentiated by virtue of unlike maskings. Instead of acyl anions, a ^-COOR synthon source (see Chapt. 3) can also be used in the form of the trithioorthoformate anions.[474]

3-Pivaloyloxy-2-nitro-1-propene **(7)**[1932] and -cyclohexene[573] are able to exchange, overall, their acyloxy group for the dithiane group[573] (63%) or 2-

aryldithiane group[1932] (85%) (see Scheme 25). The products, 3-dithianyl-2-nitro-1-alkenes, are again potential β-dicarbonyl systems carrying unlike protecting groups. In addition, the nitroalkene moiety can be used for reaction with another nucleophile (compare Scheme 34) prior to unmasking of either carbonyl.

In contrast to acylation, acylmethylation of the aromatic ring under Friedel–Crafts conditions is difficult. Among the solutions to this problem, the one that relies on chloroalkyl sulfides (Scheme 40)[1961] only emerges as involving $RCOCH_2^+$ synthons by virtue of the reductive desulfurization step. Desoxybenzoins $ArCH_2COAr'$ are also available by this route.

$$C_6H_6 + MeS—CHCl—COMe \xrightarrow[88\%]{SnCl_4} C_6H_5—CH(SMe)—COMe \xrightarrow[94\%]{Zn/AcOH} C_6H_5CH_2COMe \qquad (40)$$

Another source of the acetonyl enolonium cation $^{+}CH_2COCH_3$ for use in Friedel–Crafts acylmethylation of aromatic substrates is the bromo-oxime ether (**16**, Scheme 41).[1962]

$$\underset{\mathbf{16}}{BrCH_2—CMe{=}N—OMe} \longrightarrow {}^{+}CH_2—CMe{=}N—OMe \longleftrightarrow CH_2{=}CMe—N{=}O^{+}Me \xrightarrow[3)\ H_3O^+]{1)\ ArH,\ AgBF_4\quad 2)\ KCN/H_2O} Ar—CH_2COCH_3 \qquad (41)$$

Both isomers (*E* and *Z*) can be prepared, and both react by way of the resonance-stabilized cation with various methoxy- or acetoxy-substituted aromatics, giving the corresponding arylacetones in 74–82% yields.

Electron-rich aromatics such as furan, thiophene, *N*-methylpyrrole, or 1,3,5-trimethoxybenzene react with 2-phenylselenoallyl cation precursors (Scheme 42)[1963] in the presence of silver perchlorate and a weak base.

Furan + $CH_2{=}C(SePh)CH_2Br$ —(62%)→ 2-furyl-$CH_2C({=}CH_2)SePh$ (42)

A number of simple side-chain-substituted analogs were also made. It was suggested by the authors that the actual reactive species is **17,** rather than the allylic cation. The product vinyl selenides, formed in 60–70% yield, unfortunately resisted hydrolysis to the ketones when sulfuric acid or mercuric chloride was tried; thus, full $^{+}CH_2COCH_3$ synthon equivalency is lacking.

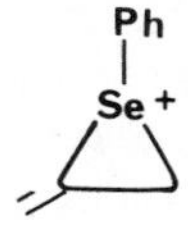

(17)

5.3.2.5. Heteroatom Nucleophiles

α-Aminoketones or their precursors are accessible using the following types of amines: R_2NH,[1964] ArNHR,[573,1932] $ArNH_2$,[1964] and Ar_2NH.[1932] α-Alkoxyketones[1943,1965] and α-alkyl- or α-arylthioketones[1932,1965,1966] are obtained similarly.

5.4. CARBOXYLIC ACID α-CATIONS $^{+}C—COOH$ AND DERIVATIVES

Very few carbon–carbon bond forming reactions based on the combination of $^{+}C—COOR$ synthons with Grignards, alkyl- or aryllithiums, or cuprates are on the record. An example is the reaction of methyl bromoacetate with the cyclopentadiene anion.[1967] Bromoacetonitrile gives 3-arylpropanenitriles (20–61%) when treated with benzylic halides in the presence of nickel metal.[1968] Since benzylnickel intermediates are presumably involved, it is legitimate to classify this reaction as one dealing with the synthons $ArCH_2^-$ and $^{+}CH_2CN$. The vinylogous synthon **18** has been recently reported.[1969] The lactol **19** reacts with alkyl and aryl Grignards in high yield, the products being then thermolyzed in a retro-Diels-Alder reaction (Scheme 43). No yields were given for the last step.

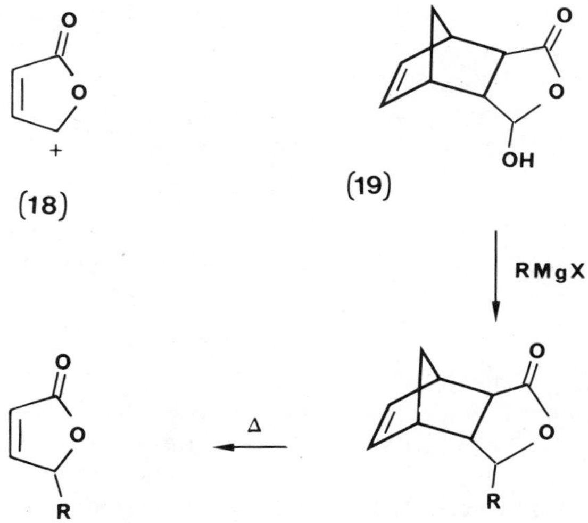

(18) (19)

RMgX

Δ

(43)

There are some examples of $^{+}C—COOR + {}^{-}C—COR$ matchings, leading to γ-ketoesters (or succinates). Ketoesters result from the reaction of enamines with α-bromoesters (*cf.* Scheme 19).[1916]

Chloro-thioalkyl-esters (Scheme 44), made by chlorination (NCS) of the corresponding α-phenylthioesters, react with enol silyl ethers in the presence of a Lewis acid ($TiCl_4$ or $ZnBr_2$) in good-to-excellent yield.[1970,1971]

$$Me_3SiO—CR{=}CHR' + PhS—CR''(Cl)—COOMe \longrightarrow$$

$$RCO—CHR'—CR''(SPh)—COOMe \xrightarrow{\text{Raney-Ni}} \qquad (44)$$

$$RCO—CHR'—CHR''—COOMe$$

where R″ = H or Me.

If the products are desulfurized, $^{+}CR''—COOMe$ synthon equivalency becomes apparent. Obviously, there are also other possibilities for further elaboration of the intermediate phenylthio-ketoesters.

A recent paper[1972] discusses the preparation of the Reformatsky reagent tBuOCO—CH_2ZnBr **(20)** and its reactions with α-bromoesters and their vinylogs (Scheme 45).

$$tBuOCO—CH_2CH_2COOR \xleftarrow{BrCH_2COOR} \mathbf{20} \xrightarrow{BrCH_2CH=CHCOOR} \qquad (45)$$

$$tBuOCOCH_2CH_2CH{=}CHCOOR$$

R = Me, 38%, R = tBu, 94%, R = Me, 56%, and R = tBu, 51%.

The ability to differentiate the two succinate ester groups is an added bonus; it would be interesting to know if the choice of electrophiles is strictly limited to bromoacetate and bromocrotonate.

γ-Bromocrotonate has also been reacted with a number of lithium ester enolates.[1973] Two reaction pathways are available, a direct S_N2 displacement and a Michael-initiated ring closure (MIRC), the latter leading in this case to 2-substituted cyclopropanecarboxylates. It was found that the less reactive, doubly stabilized anions, such as those derived from malonate, Meldrum's acid, or phenylacetate, mainly gave the substitution product, particularly in the presence of HMPA. Alkanoate α-anions lead to cyclopropanation. The above substitution on γ-bromocrotonate by the malonate anion involves the synthon pair $^{+}CH_2CH{=}CHCOOR$ and $^{-}CH(COOR)_2$; the former is also accessible in another guise, namely as the ethyl crotonate π-allylpalladium complex,[1953] and gives the malonate anion combination product in quantitative yield.

Aldehydes RCHO may be converted to 1-phenylthio-1-nitroalkenes, which will react with the malonate anion, in addition to certain *O*- and *N*-nucleo-

philes. The resulting nitronate salts (Scheme 46) were not isolated but were directly ozonolyzed at $-78°$, giving the α-substituted *S*-phenyl thioester [Nu = $CH(COOMe)_2$, R = Me, 60%].[1629]

$$RCHO + PhSCH_2NO_2 \longrightarrow \longrightarrow RCH{=}C(NO_2)SPh \qquad R = Me, 60\%;\ R = iPr, 31\% \qquad (46)$$

$$\xrightarrow{Nu^-} Nu—CHR—C(SPh){=}NO_2^- \xrightarrow{O_3} Nu—CHR—CO—SPh$$

α-Nitro nitriles **(21)** lacking α-protons undergo a substitution reaction by nitroalkane anions (Scheme 47),[1974] either $^-CH_2NO_2$, $^-CHR—NO_2$, or $^-CR_2—NO_2$. There was evidence that the reaction is an electron-transfer chain process.

$$^-CMe(Et)—NO_2 + KCN \xrightarrow{K_3Fe(CN)_6} Me—C(CN)(NO_2)—Et \ (\mathbf{21}) \xrightarrow{CH_3(CH_2)_5CH^-NO_2} (Et)(NC)CMe—CH(NO_2)(CH_2)_5CH_3 \quad (77\%) \qquad (47)$$

The Friedel–Crafts acylmethylation of the aromatic ring, using chloroalkyl sulfides (Scheme 40), can also be used to deliver the $^+CH_2CN$ and $^+CH_2COOEt$ synthons. Applicable to a range of substituted aromatics and heteroaromatics, the two-step sequence gives good-to-excellent yields.[1975]

The remaining C–C bond-forming synthon, the cyanide anion, has been used with the α-bromoesters Br—CRR′—COOEt (R,R′ = H or Me) under phase-transfer conditions. The cyano esters were formed in 60–90% yield.[1912]

Nitrogen nucleophiles have been attached to various $^+C—COOR$ systems. Alkali cyanates furnish $^-NH_2$ synthons in reaction with α-bromoesters.[1976] Primary amines combine with glyoxylic acid, resulting in overall mono-carboxymethylation of primary amines.[1977] The nucleophiles associated with Scheme (46) include amides such as potassium phthalimide (yield 68%) or $CH_2F—CONHK$ (yield 62%).[1629] The same route serves also for the preparation of α-alkoxy-substituted *S*-phenyl thioesters (α-MeO, yield 79%; α-*i*PrO, yield 61%).[1629] α-Alkylthio- and α-arylthioesters, lactones, and amides are made advantageously from the corresponding α-halocompounds under phase-transfer conditions.[1907]

5.5. DIFUNCTIONAL SYNTHONS

Just as α-bromocarbonyl compounds function as sources of the $^+$C—CO— synthon, carbonyl compounds having electrophilic α-sites because of other functionalities at that position may be visualized. Thus, for example, glyoxylate esters OHC—COOR could be expected to serve as $^+$CH(OH)—COOR synthon sources. In this section, we will first survey α-hydroxy α-cationic carbonyl synthons and equivalent systems, such as the α-alkoxy, α-alkylthio, and α-amino analogs, and conclude with other difunctional synthons where the "extra" functionality is not directly related to the cationic nature of the α-site.

5.5.1. α-Hydroxy α-Cationic Carbonyl Synthons $^+$C(OH)—CO— and Related Systems

Additions of nucleophiles, such as Grignard or alkyllithium reagents, to various α-dicarbonyl substrates must be run selectively at one carbonyl group only, in order for the overall process to correspond to a $^+$C(OH)—CO— synthon involvement. Alternatively, the carbonyl to be retained can be masked to keep alkylation regioselectivity under control. $^+$CR(OH)—CHO[748,1978] or $^+$CAr(OH)—CHO[825,1978] synthon equivalents have been developed, based on various CHO disguises, for use with Grignards or alkyllithiums. Optically active α-branched α-hydroxyaldehydes are accessible,[748,1978] the key to selectivity residing in the asymmetric nature of the aldehyde mask. A glyoxylate derivative serves as the doubly cationic synthon $^{2+}$C(OH)—CHO, which can be combined with two unlike Grignard alkyl groups, also in a stereoselective manner.[1978]

α-Hydroxyketones, racemic[1979] or nonracemic,[1980] are obtained by similar routes employing organometallics. Stannous triflate promotes the rearrangement of β-ketosulfoxides (Scheme 48)[1981] into

$$\mathrm{MeCOCH_2{-}SO{-}Ar} \xrightarrow[\text{Base}]{\mathrm{Sn(OTf)_2}}$$

$$\underset{\mathbf{22}}{\mathrm{MeCO{-}CH{=}S^+Ar\ ^-OSnOTf}} \xrightarrow{\mathrm{CH_2{=}CPh{-}OSiMe_3}} \tag{48}$$

$$\underset{\mathbf{(91\%)}}{\mathrm{MeCO{-}CH(SAr){-}CH_2COPh}}\ (\longrightarrow \mathrm{MeCOCH_2CH_2COPh})$$

α-thiocarbocations **(22),** which will then accept carbonyl α-anion nucleophiles (the latter being provided by silyl enol ethers). Up to this stage the ketosulfoxides fulfill the role of α-arylthioketone cation synthons, $^+$CH(SAr)—COR, but a final reductive desulfurization step giving 1,4-diketones as end products

would make the overall process correspond to a coupling of $^{-}CH_2COR'$ and $^{+}CH_2COR$.

α-Hydroxycarboxylate synthons $^{+}C(OH)—COOR$ are featured in connection with a number of nucleophiles: Grignards and lithium reagents,[1982–1984] allylic stannanes,[1985] alkenes (ene reaction),[1984] and phenols (Friedel–Crafts)[1986] have been used, the first-mentioned also in enantioselective reactions. Various *O*- and *N*-nucleophiles combine with certain α-aminoester cation[1987,1988] and α-aryloxyester cation[1989] equivalents. Enantioselective utilization of the $^{+}CR(OH)—CONR_2$ synthon in Lewis acid-catalyzed reaction with allylsilane has also been reported.[1990]

5.5.2. Other $^{+}C—CO—$ Synthons Containing Additional Functionality

β-Hydroxyketone[1991] and β-alkoxyketone[1992] synthon equivalents have found rather limited use, the former for C–C bond formation using cuprate nucleophiles. β-(1-Chlorovinyl)-β-propiolactone (**23,** Scheme 49)[1993] is a versatile source for the $^{+}CH_2COCH_2CH_2COOH$ synthon, reacting with cuprates or Cu^{+}-catalyzed Grignards to give 4-chloro-3-alkenoic acids in high yields. The R^{-} carbanion nucleophile may be a primary, secondary, or tertiary alkyl, allyl, vinyl, or aryl group, and may also contain THP and other acetal functionality. The chloroalkenoic acid intermediates may be taken to levulinic acid homologs by $TiCl_4$-mediated hydrolysis. Since the chloroalkenoic acid intermediates may be diverted to 4-oxo-2-alkenoic acids, $^{+}CH_2COCH{=}CHCOOH$ synthon sources are also at hand.

O O Cl $\xrightarrow{R^{-}}$ R Cl COOH $\xrightarrow[H_2O-MeOH]{TiCl_4}$ (49)

(23)

R O COOH

The malonate synthon $^{+}CH(COOEt)_2$ is accessible from the tricarbonyl compound $O{=}C(COOEt)_2$, and provides arylmalonates in a multistep sequence using aryl Grignards or lithiums, or arenes themselves under Friedel–Crafts conditions.[1994] The value of this route is augmented by the ample choice of methods for the $R_2C(COOEt)_2$ to $R_2CH—COOEt$ conversion (for leading references, see Ref. 1995).

5.6. BIDENTATE SYNTHONS

In this section, some of the simpler carbonyl α-cationic synthons, having an additional plus or minus charge elsewhere in the molecule, are discussed. For the $^{+}$C—CO^{-}-type synthons, seen most often in Michael additions of alkyllithiums to ketene dithioacetals, the reader is referred to Section 2.17.5.1. (Chapt. 2), and to a review by Kolb.[491]

Among the doubly cationic synthons, a glyoxylate derivative serves as $^{2+}$C(OH)—CHO, which can be combined with two unlike Grignard alkyl groups, also in a stereoselective manner.[1978] The α,α-dichloroaldehydes Cl_2CR—CHO do not have a broad range of applicability; apparently reactions with thiolate anions only have been reported.[1842] α,α′-Bis(alkylthio)ketones and α,α′-dimercaptoketones are similarly available from the α,α′-dibromoketone.[1996–1998]

The bidentate, doubly cationic synthons are undoubtedly of best value when two unlike nucleophiles can be reacted separately with the synthon source. This has been realized with the $^{+}$CHPh—CO—CH_2^{+} system[573] and $^{+}CH_2$—CO—CH_2^{+} system as shown in Scheme (34).[1932] Similarly, alkyl groups from two different cuprates can be attached to the 2-cyclohexenone 4,6-dication synthon, involving a multistep sequence that starts with 1,3-cyclohexadiene monoepoxide.[1999,2000] Finally, another $^{+}CH_2COCH_2^{+}$ source should be noted. Bridged bicyclic diones can be prepared by making the vinylic sulfoxide (**24,** Scheme 50)[2001] react first with the α-site and then the α′-site of a cyclic ketone, followed by an *in situ* Pummerer reaction and hydrolysis. The overall process is thus a remarkable combination of the dianionic synthon $^{-}$CHR—CO—CHR^{-} with the umpoled analog $^{+}CH_2$—CO—CH_2^{+}.

SOPh
Cl
(24)
+
N
N
COOEt
Et_3N/DMSO
42 %

(50)

SPh
O
N
COOEt

In closing, attention is drawn to a useful approach to synthons of the type $^{+}CHR—CO—CHR^{-}$ and $^{+}CR_2—CO—CR_2^{-}$. Again, α,α′-dibromoketones provide a source, with lithium dialkylcuprate and a reactive alkyl halide employed as the nucleophile and the electrophile, respectively.[2002] Apparently, cyclopropanones are formed first from the dibromoketones, and then undergo ring opening by another equivalent of the cuprate.

6

HOMOENOLATE ANION EQUIVALENTS: $^{-}C—C—CO$

Nick H. Werstiuk

CONTENTS

6.1. INTRODUCTION

Carbonyl compounds form the most important class of compounds used for generating C—C and C=C bonds. This is mainly due to the carbonyl or acyl carbon being electrophilic and its ability to promote facile conversion of substrates into nucleophiles. Furthermore, by umpolung,[6] also called charge affinity inversion,[1640] the normally electrophilic carbonyl or acyl carbons are rendered nucleophilic, thereby further expanding the synthetic utility of car-

bonyl compounds. Homoenolates,[2003,2004] for example those of the β-type, can be considered as charge-affinity-inverted species and are homologous to ambident α-enolates. The noninverted and nonumpoled species ⁺C—C—CO— are encountered in Michael additions and also in the important tandem operations consisting of 1,4-addition of an organocuprate followed by enolate trapping.[2312]

Unlike α-enolization, which is readily carried out under mild conditions to yield high-equilibrium concentrations of α-enolates, vigorous conditions are required to generate even low concentrations of homoenolates.[2003–2006] This problem has been overcome through development of a variety of equivalents formally based on the resonance contributor that bears the charge on carbon. Although a variety of approaches have been used, they are in general of two types: The first includes systems that possess a single nucleophilic site, isolated from a masked carbonyl group but in some cases stabilized by another substituent; in the second approach, attempts have been made to control the ambident nucleophilicity of heteroatom-substituted allylic anions.

This review deals with the preparation and synthetic applications of homoenolate equivalents. It is organized on the basis of types of equivalents, beginning with the propanal β-enolate **(1),** which is the simplest homoenolate. The section concludes with a survey of the few known representatives of γ-enolates.

6.2. ALDEHYDE β-ENOLATES (HOMOENOLATES)

6.2.1. Sources of Propanal β-Enolate

6.2.1.1. Acetal and Ketal Derivatives

The earliest routes to equivalents of the β-enolate **1** involved the preparation of Grignard reagents from β-haloacetals and ketals. Büchi and Wüest[2007] used **2a** in a key step in the synthesis of the aldehyde **3a,** which was readily converted to racemic nuciferal **(3b).**

$^{-}CH_2CH_2\text{-}CHO$

1

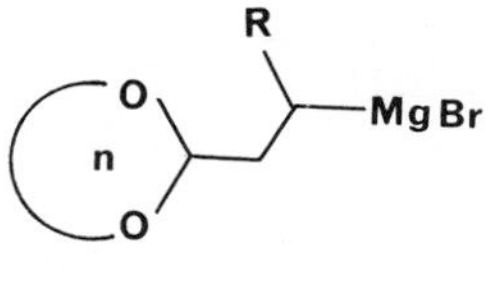

2a R = H, n = 5

2b R = Me, n = 5

13 R = H, n = 6

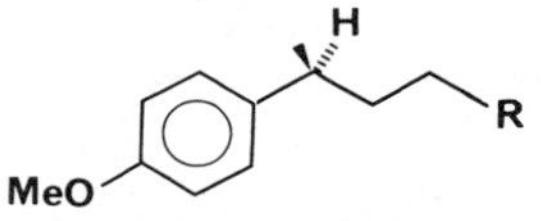

3a R = CHO

3b R = -CMe=CH-CHO

A wide variety of 6,7-substituted benzo[*b*]thiophenes **(5)** were prepared in 60–70% yield by Loozen and Godefroi[2008] by adding **2a** or **2b** to thiophenes

(4) and treating the adducts with 10% refluxing H_2SO_4 to bring about hydrolysis, cyclization, and aromatization. The aldehyde **5a** was obtained by treating **4a** with two equivalents of **2a** and treating the adduct with acid.

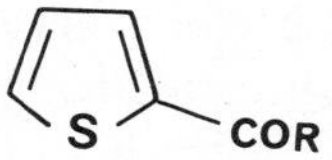

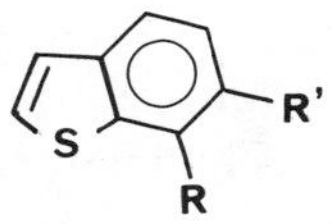

4 R = H or alkyl

4a R = OEt

5 R = H or alkyl, R′ = H or Me

5a R = CH_2CH_2CHO, R′ = H

The 4-substituted benzothiophene **(6)** was obtained after cyclization of the adduct from ethyl thiophene-3-carboxylate and two equivalents of **2a.** This methodology was also used by Loozen and Godefroi[2009] to construct substituted benzimidazoles. Reaction of **2a** or **2b** with the imidazoles **(7)** produced benzimidazoles **(8)** in 26–75% yield after treatment of the adducts with refluxing NaOAc/AcOH.

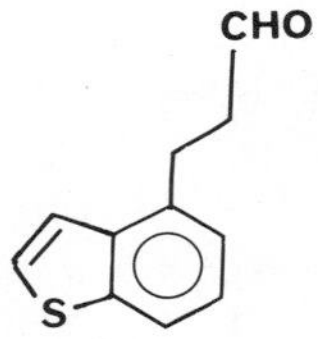

6

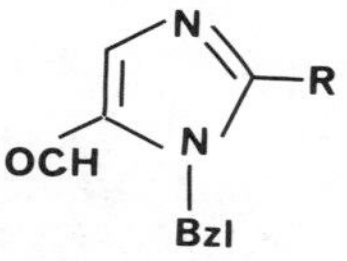

7

8

Similarly, substituted naphthalenes were prepared from **2a** and the corresponding benzaldehydes or acetophenones in 31–95% yield.[2010] Attempts to prepare indoles and benzofurans in the same way failed because of the instability of pyrroles and furans in acidic medium. Loozen and colleagues[2011] used the addition of **2a** to various pyridine and imidazole carbaldehydes to prepare the corresponding lactols **(9)**. The yields in the addition reactions were 70–96% and the lactolization yields in 10% aqueous H_2SO_4 ranged from 74 to 95%.

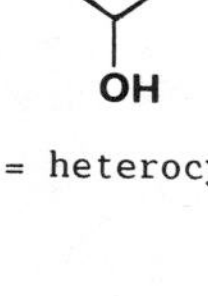

9 R = heterocycle

Eaton and coworkers[2012] used the β-enolate equivalent **2a** as a source of two- and three-carbon units in the synthesis of the peristylane ring system, but because of experimental difficulties the organolithium **10** proved a much better three-carbon source. Helquist and colleagues[302] found that the conjugate addition of **2a** to 2-cyclohexenone, 3-methyl-2-cyclohexenone, 2-cycloheptenone, and 2-cyclopentenone in the presence of catalytic amounts of a cuprous salt followed by acidification yielded the bicyclic enones **11** and the hydroxyketone **12** in 74–87% yield. Finally, alkylations of **2a** with a number of ω-functional primary alkyl bromides, in the presence of LiCl and $CuCl_2$, have also been reported.[2013]

$LiCH_2CH_2CH_2O\text{-}CHMe\text{-}CHOEt$

10

11 R = H or Me

n = 6–7

12

A route to γ-keto aldehydes using the addition of the six-membered-ring acetal **13** to acid chlorides has been documented by Stowell.[2014] 4-Oxodecanal was prepared from heptanoyl chloride in 82% yield. The six-membered-ring acetal was used instead of **2a** because its greater stability allowed for a quick, high-yield preparation of the Grignard reagent in refluxing THF. Fujisawa and co-workers[2015] used **13** as a propanal β-enolate equivalent in the stereocontrolled synthesis of (*R*)(*Z*)- and (*R*)(*E*)-14-methyl-8-hexadecenals, sex pheromones of the *Trogoderma* species. A key intermediate was prepared in 86% yield by alkylation of **13** with 1,4-dibromopropane in the presence of Li_2CuCl_4. Two aldehydes, 7-acetoxyheptanal and 6-methylheptanal (both of value in the synthesis of insect phenomones), have been synthesized by Stowell and King[2016] by coupling **13** with the iodoacetate **14** and the tosylate **15**, respectively, in the presence of CuI. The yields of the aldehydes were 69% from the coupling products after hydrolysis of the acetals.

$^{-}CH_2CH_2\text{-}CHO$

1

2a R = H, n = 5

2b R = Me, n = 5

13 R = H, n = 6

3a R = CHO

3b R = -CMe=CH-CHO

$I(CH_2)_4$-OAc

14

TsO-$(CH_2)_3$-Pr-i

15

PhSO$_2$CH$_2$CH$_2$C(R)(OCH$_2$CH$_2$O)

16a R = H
16b R = Me

Kondo and Tunemoto[2017] utilized the γ-oxosulfones **16a** and **16b** in a convenient route to γ-ketoaldehydes and 1,4-diketones. The dianion of **16a,** prepared by using two equivalents of *n*-butyllithium, was added to a number of aliphatic nonbranched esters, and the resulting ketosulfones were converted with Al—Hg in 10% aqueous *n*-propanol into the corresponding γ-ketoaldehyde ethylene ketals that were easily hydrolyzed to the γ-ketoaldehydes. Ketal **(16b),** when converted to the anion and reacted with methyl *n*-pentanoate, gave the substitution product in 53% yield and the corresponding diketone in 72% yield after reduction and hydrolysis. Epoxides react analogously with **16a**- or **16b**-derived anions and furnish δ-hydroxyaldehydes or -ketones, which normally exist in the cyclic δ-lactol form.[2018]

Grignard reagents react with α,β-unsaturated acetals and ketals at 80–100° and yield mixtures of allylic and vinylic ethers, the latter by way of a net 1,4-addition to the α,β-unsaturated carbonyl compound.[2019] When CuBr is included, addition occurs at lower temperatures (−5 to 30°) to yield predominately vinyl ethers. Acrolein diethylacetal adds various Grignard reagents (RMgX) to give 1-ethoxy-1-alkenes $RCH_2CH{=}CH{-}OEt$ in good yield.[2020] The (*E*)/(*Z*) ratio varies between 1.15 and 3.25, but the stereochemical details are of course lost on hydrolysis to the aldehyde. Similarly, the vinyl ethers **17,** (*Z*)/(*E*) = 2.0, were obtained in 75% yield by adding *n*-BuMgCl to the appropriate acetal. To facilitate the addition of *n*-BuMgCl to the diethyl acetal of (*Z*)-2-butenal, two equivalents of $P(OEt)_3$ were added and a 66% yield of **18,** (*Z*)/(*E*) = 1.2, was obtained after 15 h at 10°.

$BuCH_2$-CMe=CH-OEt

17

Bu-CHMe-CH=CH-OEt

18

6.2.1.2. Heteroatom-Substituted Allylic Anions

It was recognized early that heteroatom-substituted allylic anions could serve as homoenolate anion equivalents, but also that there would be a problem with the regioselectivity of reaction with electrophiles. Consequently a great deal of effort has been expended in attempts to direct the attack to the γ-position and a number of factors that determine selectivity have been recognized.[2021] The nature and size of the groups attached to the heteroatom, the countercation, the solvent and additives, reaction temperature, and reaction time are important. In this connection, Gompper and Wagner suggested

that the concept of allopolarization permits a description of substituent effects in kinetically controlled reactions.[2022] That is, a change in selectivity is related to a change in polarity of ambifunctional anions. Factors affecting the reactivity and regioselectivity of allyl-alkali metal reagents have also been reviewed by Schlosser.[2023] It is clear, however, that heteroatom-substituted allylic anions have received the most attention as equivalents for homoenolates of aldehydes and ketones and provide the widest scope of applications. This section of the review will be structured on the basis of the atomic number of the heteroatom employed in the mask.

6.2.1.2.1. Boron. Although nitrogen-, oxygen-, and sulfur-substituted allylic anions have been used extensively as β-enolate equivalents, a recent report by Pelter *et al.*[2024] indicates that the dimesitylboryl group will be a most useful mask. Readily available allyldimesitylborane **(19)** was converted into the allyl anion by mesityllithium or LDCA and the anion alkylated. Oxidation of the alkenyl boranes with basic hydrogen peroxide gave the corresponding aldehydes in excellent yield (Scheme 1).

$$\underset{\mathbf{19}}{(Mes)_2B—CH_2CH{=}CH_2} \xrightarrow{\text{1) base}\quad \text{2) RI}\quad \text{3) (O)}} RCH_2CH_2CHO \qquad (1)$$

Trimethylsilylation of the anion gave a new synthesis in which the two ends of the three-carbon system **20** are well differentiated. Reaction of the allyl anion of **19** with benzaldehyde followed by oxidation gave the lactol **21** in 70% yield. Hoffmann and Landmann[141] used α-substituted allylboronates **22a–e,** potential β-enolate equivalents, in an attempt to improve the regioselectivity and stereoselectivity of the addition to aldehydes (Scheme 2).

$(Mes)_2B\text{-}CH{=}CH\text{-}CH_2\text{-}SiMe_3$

20 (E)

HO, O, Ph

21

RCHO

$R\text{-}CHOH\text{-}CH_2CH{=}CH\text{-}Y$ (2)

22a Y = Cl

22b Y = Br

22c Y = SEt

22d Y = SBu-t

22e Y = OMe

23 (Z)

24 (E)

The readily available crude undistilled allylboronates were reacted with various aldehydes to yield preferentially the (*Z*)-products of γ-addition **(23)** and minor amounts of the (*E*)-products **(24)**. The sulfur and oxygen analogs **22c, 22d,** and **22e** were prepared by boron-assisted nucleophilic substitution. A rational based on the anomeric effect was presented to account for the (*Z*)-selectivity. Although the addition products undoubtedly can be hydrolyzed to give the corresponding lactols or lactol ethers, the vinyl bromides obtained from hexanal and 2-hexenal, when treated with $Ni(CO)_4/Et_3N$, yielded the corresponding unsaturated δ-lactones, massoia-lactone and tubero-lactone, respectively.

6.2.1.2.2. Nitrogen. Alkylation of 3-metallated enamines obtained by deprotonation of enamines or allylamines has proven to be a useful route to β-enolates. A review by Ahlbrecht[2025] provides an excellent documentation of the preparation of 2- and 3-substituted β-enolates of aldehydes and ketones by the enamine route until 1976. A discussion of the control of regioselectivity, an important factor, was presented. Martin and DuPriest[2026] lithiated allylpyrrolidine, a source of the β-enolate **1,** and studied the reaction of the anion **25** with *n*-butyl bromide, chlorotrimethylsilane and cyclohexylcarboxaldehyde, benzaldehyde, acetone, acetophenone, and cyclohexanone. While the alkylation and silylation proceeded with a high degree of regioselectivity to yield >95% of the γ-adducts **26a** and **26b,** the additions to the aldehydes and ketones yielded approximately equal amounts of α- and γ-products, the latter being cyclized to the aminoacetals **27a.** Treatment of the mixture of products derived from the aldehydes and ketones with ethanolic HCl yielded the corresponding lactol ethers **27b** and the products of α-addition.

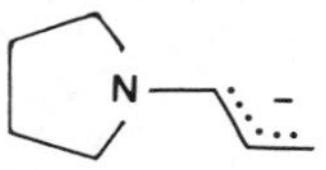

N N R Y O R' R

25 Met = Li^+ or Zn^{2+}

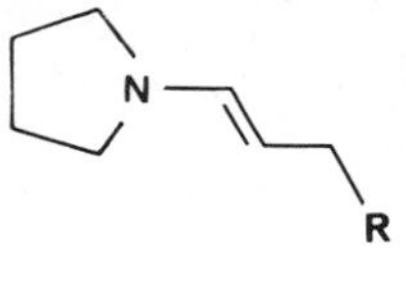

26a R = Bu
26b R = $SiMe_3$

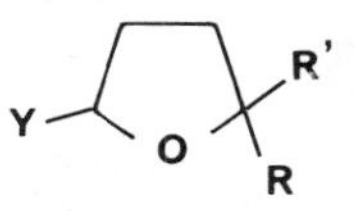

27a Y = $N(CH_2)_4$
27b Y = OEt

In a recent report Ahlbrecht and Sudheendrananth[2027] described the reactions of a variety of 1-aminoallyl anions **28a–d,** sources of the β-enolates **1, 29, 30,** and **31,** with chlorotrimethylsilane.

Ph—NMe—CR^1—CR^2—CHR^3 (⊖)

28

$^-CH_2CH_2CHO$

1

where for **28a,** $R^1 = R^2 = R^3 = H$; **28b,** $R^1 = R^2 = H$, $R^3 = Ph$; **28c,** $R^1 = R^3 = H$, $R^2 = Me$; **28d,** $R^1 = Ph$, $R^2 = R^3 = H$.

$^{-}CH_2$—CHMe—CHO **29** $^{-}$CHPh—CH_2—CHO **30** $^{-}CH_2CH_2$—COPh **31**

The aminoallylsilanes were alkylated further. The reaction showed high γ-regioselectivity, and hydrolysis in dilute aqueous HCl gave the corresponding 3-oxosilanes **(32)** which can be used as protected α,β-unsaturated carbonyl compounds.[2028,2029] Renger and Seebach[1811] prepared anions **33a** and **33b** from allyl *t*-butylnitrosoamine and allyl methylnitrosoamine, respectively, and reacted the anions with alkylating and hydroxyalkylating reagents. While both allylic anions showed kinetically favored α-addition, the addition of the anion **33a** to benzaldehyde and cyclohexanone is reversible and under thermodynamic control the γ-products are favored. The γ-product derived from benzaldehyde, when denitrosated and hydrogenated, gave the aminoalcohol **34.**

R′CO—CH(—)—CHR—$SiMe_3$ **32**

CH_2—CH—CH—NR—NO ($\ominus$ Li^+, delocalized) **33**

*t*BuNH—$(CH_2)_3$—CHOH—Ph **34**

where for **33a,** R = Bu-*t*; **33b,** R = Me.

Compound **35,** the (γ)-adduct of benzophenone, was shown to rearrange to the oxime **36.**

*t*Bu—N(NO)—CH=CHCH$_2$—CPh$_2$—OH

35

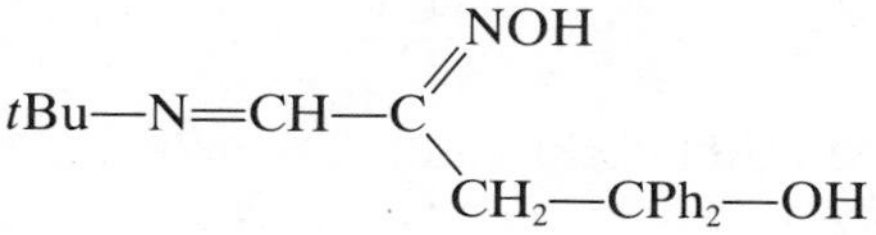

*t*Bu—N=CH—C(=NOH)—CH$_2$—CPh$_2$—OH

36

It appears that *N*-allyl ureas and phosphoramides will be useful sources of **1** as well as substituted analogs. Hassel and Seebach[1388] prepared the anion **37,** accessible from 2,2,6,6-tetramethyl-4-piperidone ethylene ketal, phos-

gene, and allylmethylamine, and reacted the lithium salt or its MgBr analog with electrophiles.

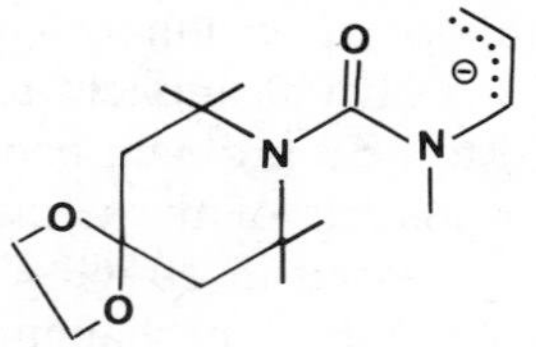

37

The magnesium compound reacts with alkyl iodides, aldehydes and ketones almost exclusively at the γ-position to give (*Z*)-enamides **(38)** that could be converted further to the acetals.

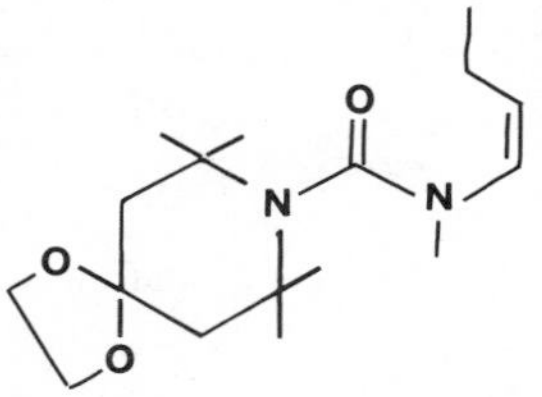

38

Coutrot and Savignac[2030] prepared a number of anions **(39a–e)** that are sources of substituted β-enolates as well as of the β-enolate **1.** Treatment of the starting *N*-methyl-*N*-allylphosphoramides with *n*-butyllithium at −50°, and subsequent reaction of the anions with alkylating agents, resulted in γ-alkylation. Acid-catalyzed hydrolysis yielded the corresponding aldehydes **(40)**.

Me_2N—PO—NMe—CH—CR^1—CR^2R^3 ⊖ Li^+ **39**

OHC—CHR^1—CR^2R^3—El **40**

where for **39a,** $R^1 = R^2 = R^3 = H$; **39b,** $R^1 = Me$, $R^2 = R^3 = H$; **39c,** $R^1 = R^2 = H$, $R^3 = Me$; **39d,** $R^1 = R^2 = H$, $R^3 = Ph$; **39e,** $R^1 = H$, $R^2 = R^3 = Me$.

The $(Me_2N)_2PO$—NMe group is a good mask because it is easy to introduce into allylic structures, it directs attack to the γ-position, and it is easily eliminated. Recently the methodology has been utilized by Coutrot and coworkers[2031] for the preparation of γ-lactols by the addition of the phosphoramido anions

to benzophenone. While the lithium salt gave a mixture of α- and γ-addition, the magnesium compound gave exclusively γ-addition. For example, lactols **41a** and **41b** were prepared in 98 and 72% yields, respectively, by adding the corresponding phosphoramidomagnesium bromides to benzophenone and treating the adducts with 2 *N* HCl at ambient temperature for 15 h. Generalizing the reaction proved to be a problem because of the difficulty in generating and purifying the products of the addition to other carbonyl compounds. Lactols **42** were prepared in 50–60% yield from the appropriate phosphoramides and ketones without exchanging Li^+ for ^+MgBr.

41a R = H

41b R = Me

42a R = H, R′ = R″ = Ph

42b R = Me, R′ = R″ = Ph

42c R = R′ = H, R″ = Pr-i

42d R = Me, R′&R″ = $(CH_2)_5$

6.2.1.2.3. Oxygen. Evans and colleagues[1818] metallated a series of vinyl ethers, sources of the β-enolate **1,** with *s*-butyllithium in THF at −65° and reacted the allylic anions **43** with alkyl halides, 3-methylpropanal, and cyclohexanone. In alkylation reactions with *n*-propyl iodide the following γ:α ratios were observed (column A), and are in sharp contrast with those observed in reactions with cyclohexanone (column B):

R	γ:α (A)	γ:α (B)
THP	54:46	
Ph	63:37	24:76
Et	75:25	30:70
Bu-*t*	89:11	27:73
Me		72:28

The product ratios were cation dependent, with the Zn^{2+} reagent showing a high degree of α-addition. Acid hydrolysis of the γ-adduct of Li-**43** (R = Me) and cyclohexanone gave the lactol **44** in 72% yield. Still and MacDonald[1819] prepared a series of allylic anions **45a–c** and studied the alkylation with methyl, ethyl, *n*-propyl and *n*-hexyl iodide, and *n*-propyl bromide and allyl bromide. The phenoxy derivative **45a** reacted with methyl iodide to yield

71% of γ-product. The anions **45b** and **45c** gave high combined yields (>95%) of α + γ products and showed higher γ-selectivities with primary iodides but lower α-selectivities with cyclohexyl iodide (39% γ) and allyl bromide (78% γ). Apparently only the (*Z*)-enol silyl ethers were formed. In a subsequent publication,[2032] the authors demonstrated that lithium salts **45b** and **45d** in THF/HMPA undergo α-attack with a variety of aldehydes and ketones in a highly regioselective manner (71–99%).

RO-CH-CH-CH $^{-}$ $^{+}$Met

HO O

43

45 (Met = Li)

a R = Ph

b R = $SiEt_3$

c R = $SiMe_2Bu$-t

d R = $SiMe_3$

44

It was concluded that the regiochemistry of allyloxy anions is determined by the nature of the electrophilic species; unsymmetrically substituted allyllithium compounds react preferentially with alkyl halides and protons at the site of highest electron density, and with carbonyl compounds by a rearrangement process that involves the lithium cation. Schlosser and coworkers[2033] prepared **45a** from the (*E*)-vinyl ether and from allyl phenyl ether and alkylated the resulting anions with methyl iodide. Whereas the anion derived from the (*E*)-vinyl ether gave the γ-alkylated product in 84% yield, the anion derived from the allylic ether gave α- (16%) and γ-addition (47%).

As a continuation of their studies on the chemistry of allylsilanes, Sakurai and coworkers[2034] prepared a variety of α-silyloxyallylsilanes **46** and reacted them with a series of saturated and unsaturated acid chlorides RCOCl in CH_2Cl_2 in the presence of $TiCl_4$. The intermediate enol silyl ethers, products of γ-addition, were hydrolyzed to the γ-ketoaldehydes $RCOCH_2CH_2CHO$ in yields ranging between 20 and 80%. In terms of overall yields, silanes **(46)** with $R^1 = R^2 =$ Me and $R^3 =$ Bu-*t* performed best.

Yamamoto and colleagues[2035] prepared **43** (R = Pr-*i*) and studied the effect of additives (Et_3Al and Et_3B) on the regiochemistry of addition to carbonyl compounds. Addition of Et_3Al gives nearly exclusive α-attack. More complex allyloxy masks have been used as well. Mukaiyama and Yamaguchi[2036] reacted 2-allyloxybenzimidazoles **(47)**, yet further sources of the β-enolate **1**, with *n*-butyllithium at −100° in THF and treated the oxyallylic anions with alkylating

agents. The products of α-attack predominated as was also the case in a later study on the effect of additives $ZnBr_2$ and CdI_2 on the regiochemistry of addition.[2037]

47

Recently Hoppe and coworkers[2038] demonstrated that allyl *N,N*-dialkylcarbamates **(48)** provide viable routes to alkyl-substituted β-enolates as well as to the parent enolate **1.** The presence of the *N,N*-dialkylcarbamoyl group increases the kinetic acidity of α-protons by chelation so that α-alkylated derivatives can be deprotonated, which is not always possible with some of the other allyl derivatives. The carbamates **48** can be doubly deprotonated with LDA.

$CH_2{=}CH{-}CH(OSiR^2_2R^3){-}SiR^1_3$ **46**

$R^1CH{=}CR^2{-}CHR^3{-}OCO{-}NR^4_2$ **48**

$MeOCO{-}CR^1(El){-}CR^2{=}CR^3{-}OCONR^4_2$ **49**

$R^1CH{=}CR^2{-}C(COOMe)(El){-}OCONR^4_2$ **50**

When the dianions are reacted with dimethyl carbonate and then a second electrophilic reagent, methyl (*E*)-4-carbamoyloxy-3-butenoates **(49),** together with small amounts of the isomers **50,** are obtained. The method provides a useful and versatile entry to precursors of 4-oxoalkanols and 3-buten-4-olides. The size of the groups at the γ-position of the allyl fragment and on the carbamoyl nitrogen determines the α:γ regioselectivity. In addition to using two fairly bulky substituents on the carbamoyl nitrogen to protect the carbonyl group, Hanko and Hoppe[2039] dideprotonated *N*-alkyl and *N*-phenylcarbamates to yield strongly nucleophilic allyllithium compounds where the carbamoyl carbonyl is protected as the α-enolate. Carbamates **(51)** were dideprotonated with 2.1 equivalents of *n*-butyllithium in THF/TMEDA at −78° to −50°. The dianions react with electrophiles (alkyl iodides, Me_3SiCl, Me_2S_2, carbonates) to give predominantly (*Z*)-configurated γ-adducts. The enol esters are hydrolyzable under acidic conditions to yield the β-substituted carbonyl compounds.

$CH_2{=}CH{-}CHR^1{-}OCONHR^2$ **51**

$R^2R^3C{=}CH{-}CHR^1{-}OCON(Pr\text{-}i)_2$ **52**

where for **52a,** $R^1 = R^2 = R^3 = H$; **52b,** $R^1 = Me$, $R^2 = R^3 = H$; **52c,** $R^1 = R^2 = H$, $R^3 = Me$; **52d,** $R^1 = H$, $R^2 = R^3 = Me$.

In an extension of the work on the deprotonation of allyl-*N,N*-dialkyl carbamates, Hoppe and coworkers[2040] further demonstrated the superiority of the methodology over using 1-oxyallyl anions (R = alkyl, aryl, or trialkylsilyl). High γ-selectivity is observed with aldehydes and ketones as well as alkylating agents. The allyl esters **(52),** prepared from allyl alcohols and *N,N*-dialkylcarbamoyl chlorides, are rapidly lithiated with *n*-BuLi and diethyl ether/TMEDA at −78°. Reaction with a carbonyl compound followed by hydrolysis with $TiCl_4/H_2O$ or CH_3OH afforded the δ-hydroxycarbonyl compounds as lactols or lactol ethers, although no yields were given. Protection of the hydroxy group followed by hydrolysis of the enol esters presents a useful route to γ-acyloxy carbonyl compounds. For example, in this way pivalaldehyde gave the corresponding γ-acetoxyketone in 94% yield (Scheme 3).

$$\textbf{52b} \xrightarrow[\text{3) } Ac_2O \text{ 4) } TiCl_4/H_2O]{\text{1) BuLi 2) } t\text{-BuCHO}} t\text{-Bu—CH(OAc)—}CH_2CH_2COCH_3 \qquad (3)$$

It was postulated that a tight ion pair, in which the lithium cation is held at the α-carbon atom by the complexing carbamoyl oxygen, is the source of the high γ-selectivity. By exchanging the lithium cation for titanium, Hoppe and Brönneke[2041] have shown that *N,N*-diisopropyl 2-alkenyl carbamoyl anions generated from the corresponding esters **52a, 52c, 52d,** and **53** add to aldehydes or acetophenone with high *lk*-diastereoselectivity to form *anti* adducts with greater than 96% ds (Scheme 4).

OCON(*i*-Pr)$_2$

53

52c —(1) BuLi 2) $TiCl_4$; 3) Bu-CHO)→ [HO ... OCON(*i*-Pr)$_2$] (4)

While the enol carbamates could be hydrolyzed to the corresponding lactols or lactol ethers,[2040] thereby providing equivalents of β-enolates **1, 54, 55,** and **56,** treatment with MeOH/MsOH/1% $Hg(OAc)_2$ (or 0.01 equiv. $PdCl_2$), followed by MCPBA/BF_3 etherate in CH_2Cl_2, gave the corresponding γ-lactones in high yields, thus providing the corresponding carboxylic acid β-

enolate equivalents. In this manner the product (Scheme 4) gave the lactone **57** in 90% yield.

$^{-}$CHMe-CH_2-CHO

54

$^{-}$$CMe_2$-$CH_2$-CHO

55

RS-CH-CH-CR'_2 $^{+}$Li

58a R = Ph, R′ = H

58b R = Ph, R′ = Me

58c R = Et, R′ = Me

56

57

6.2.1.2.4. Sulfur. Biellmann and Ducep[1278] prepared lithium thioallyl anions **58a–c** by reacting the corresponding sulfides with *n*-BuLi in THF at −30° in the presence of DABCO. The anions were then alkylated with MeI, CH_2=$CHCH_2Br$, and Me_2C=$CHCH_2Br$. While **58a** yields some γ-addition products (MeI, α:γ = 71:22; CH_2=$CHCH_2Br$, γ:α = 62:29), anions **58b** and **58c,** which are more highly substituted at the γ-position, show a higher α-selectivity (85–100%). In a subsequent publication, Atlani and colleagues[2042] described the preparation and addition of **58b** to acetone and to the above halides in the presence of a variety of complexing agents (DABCO, TMEDA, HMPA, and cryptand [2.2.2]). The results were explained on the basis of the involvement of intimate or solvated ion pairs, solvent-separated ion pairs, and free ions. For example, although in the presence of DABCO **58b** reacts with acetone to yield 100% of the γ-adduct, reactions with alkylating agents yield only small amounts of γ-products (1–12%). At −78° substantial amounts of γ-addition were observed (MeI, [2.2.2], α:γ = 60:40; CH_2=$CHCH_2Br$, (2.2.2), α:γ = 50:50; acetone, no complexing agent, α:γ = 25:75; acetone, TMEDA, α:γ = <1:90; acetone, HMPA, α:γ = 40:60). Hydrolysis of the 1-alkenyl sulfides to the corresponding aldehydes was not carried out in this work, but this can of course be done. The reagents are then, in a limited sense, sources of the β-enolates of propanal **(1)** and butanal **(54).**

Evans and coworkers[1640] studied the effect of chelating groups on the regiochemistry of alkylation of thioallyl anions by preparing a series of heteroatom-substituted anions **(58a, 59a–d, 60b–d** and **61b–c)** and studying their alkylation with alkyl halides at a low temperature (−30° to −65°).

$$RS—CR^1—CR^2—CH_2^{\ominus} \quad Li^+$$

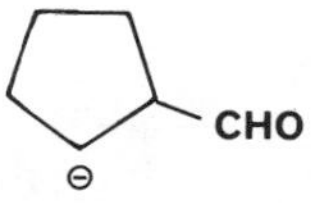

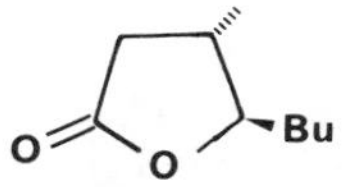

59, 60, 61

where for **59,** $R^1 = R^2 = H$; **60,** $R^1 = H$, $R^2 = Me$; **61,** $R^1 = Me$, $R^2 = H$; and for **a,** R = 4-pyridyl; **b,** R = 2-pyridyl; **c,** R = 2-thiazolyl; **d,** R = 2-imidazolyl.

The yields were generally good to excellent. Highest α:γ ratios were observed for the anions **59b–d** (>99:1); in THF **61b** gave α- and γ-adducts in a ratio of 90:10 and in the case of **63c** the γ:α ratio was 80:20. A dramatic increase in the α:γ ratio was observed when HMPA was added. Ridley and Smal[2043] added the arylthioallyl anions **58a, 62,** and **63,** generated from the corresponding allyl sulfides (*n*-BuLi/THF/ − 70°), to benzaldehyde to study the effect of substituents on the aryl and allyl groups on the α:γ selectivity.

$$\text{ArS—}\overset{\ominus}{\text{CH—CH—CH}_2}\ {}^{+}\text{Li} \qquad \text{PhS—}\overset{\ominus}{\text{CR}''\text{—CR}'\text{—CMe}}\text{—R}\ {}^{+}\text{Li}$$

62 **63**

where for **62a,** Ar = *p*-tolyl; **62b,** Ar = mesityl; **62c,** Ar = *p*-$MeOC_6H_4$; **62d,** Ar = *p*-$O_2NC_6H_4$; **63a,** R = Me, R′ = R″ = H; **63b,** R = R″ = H, R′ = Me; **63c,** R = R′ = H, R″ = Me.

Simple substituents on the aromatic ring (sulfides **62** are sources of **1**) have little effect on the α:γ ratio, and on the ratios of the respective diastereomers. Substitution on the allyl moiety (sulfides **63** are sources of β-enolates **54, 29,** and **64,** respectively) causes large variations.

$$^{-}\text{CHMe—CH}_2\text{CHO} \qquad ^{-}\text{CH}_2\text{—CHMe—CHO} \qquad ^{-}\text{CH}_2\text{CH}_2\text{COMe}$$

54 **29** **64**

Methyl substitution at C-3 yields only products of α-addition. The stereochemistry of the reaction is affected also; there is a change in the ratio of (*E*) and (*Z*) isomers obtained by γ-attack. Binns and coworkers[2044] have studied the effect of HMPA and other complexing agents on the regiochemistry of addition to cyclopentenone of alkyl- and phenylthioallyl, phenylselenylallyl, and phenylthio-stannylallyl anions **58a** and **65.** It is clear that the addition of 1 equiv. of HMPA has an essentially specific α-1,4-directing effect on the reaction of the ambident nucleophile on the ambident electrophile, cyclopentenone. The utility of the reaction as a source of the β-enolate **1** is marginal although reasonable yields of the γ-1,2-addition products are obtained in the absence of HMPA. It is interesting to note that **65e** is a source of the β-enolate **66.**

$$\text{RY—}\overset{\ominus}{\text{CH—CH—CHR}'}\ {}^{+}\text{Li} \qquad ^{-}\text{CHSPh—CH}_2\text{CHO}$$

65 **66**

where for **65a,** R = Ph_3C, R′ = H, Y = S; **65b,** R = Me, R′ = H, Y = S; **65c,** R = Bu-*t*, R′ = H, Y = S; **65d,** R = Ph, R′ = H, Y = Se; **65e,** R = Ph, R′ = $SnBu_3$, Y = S.

Addition of lithium compounds **67a** and **67b,** sources of β-enolates **1** and **68,** respectively, gave the sulfoxides **69a** and **69b.**

$$PhSO\text{—}\overset{\ominus}{CH\text{—}CH\text{—}CHR}\ \ {}^{+}Li \qquad {}^{-}CH(SnBu_3)\text{—}CH_2CHO$$

67 **68**

where for **67a,** R = H; **67b,** R = $SnBu_3$.

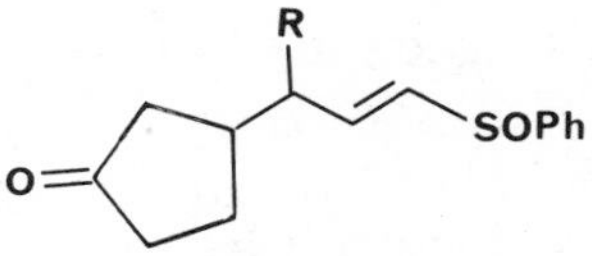

R SOPh O

Cu^+ $\overset{-}{CH_2\text{-}CH\text{-}CH\text{-}SR}$

70

69a R = H
69b R = $SnBu_3$

Yamamoto and coworkers[2045] studied the regiochemistry of the alkylthioallylcopper reagents **(70),** prepared from the corresponding organolithium at −78° by addition of CuI in ether. Treatment of **70** (R = Pr-*i*) with the allylic bromides **71, 72,** and **73** yielded **74** (92%), **75** (88%), and **76** (87%), products of exclusive γ-alkylation by an S_N2' process. In contrast, acetone reacted with a high degree of α-regioselectivity. Although the hydrolysis of the 1-alkenyl sulfides was not reported, the reagents are sources of the β-enolate **1.**

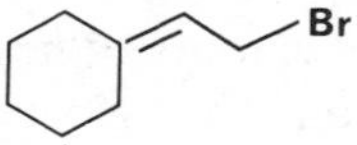

Br $CH_2\text{=}CH\text{-}CH_2Br$ Br

71 72 73

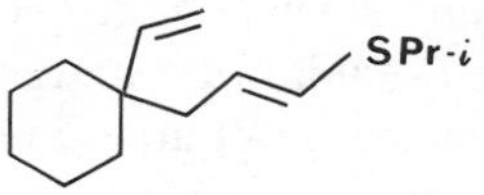

SPr-*i* $CH_2\text{=}CH\text{-}CH_2CH_2\text{-}CH\text{=}CH\text{-}SPr\text{-}i$ SPr-*i*

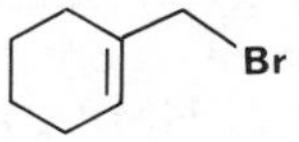

74 75 76

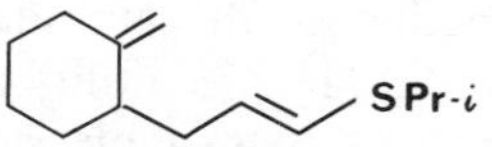

Alkylation of lithium thioalkylborates **(77),** generated by the addition of the appropriate borane to the lithium thioallyl anion **70** (R = Pr-*i*), has been

studied by Yamamoto and colleagues.[2046] Although the borates **77** react with 3,3-dimethylallyl halides **(68a)** regioselectively α,γ′ (92–96%), halides **72** and **78b–c** give significant amounts of γ,γ′ and γ,α′ addition: **77a** + **78b** yield 18% γ,γ′ and 21% γ,α′ products, **77a** + **78c** yield 15% γ,γ′ and 31% γ,α′ products, and **77b** + **72** yield 11% of γ,γ′ + γ,α′ products. Reaction of thioallylanion **70** (R = Pr-*i*) with **78a** yields 55% α,α′ coupling and 45% γ,α′ coupling.

$$\underset{\mathbf{77}}{Li^+\ {}^-YCH_2CH{=}CH{-}SPr\text{-}i} \qquad \underset{\mathbf{78}}{RR'C{=}CH{-}CH_2X}$$

where for **77a,** Y = BBu_3; **77b,** Y = Bu-9-BBN; **78a,** R = R′ = Me, X = Br; **78b,** R = H, R′ = Me, Y = Br; **78c,** R = H, R′ = Me, Y = Cl.

The products of α-γ′ coupling were converted to 1,5-dienes, but the chemistry of the γ-γ′ and γ-α′ addition products that formally are the products of alkylation of **1** and the trialkylboron-substituted β-enolates was not examined.

While thioallylic monoanions clearly react preferentially at the α-position, Seebach and coworkers[2047] established that thioacrolein dianion **(79)** generated from propenethiol and 2.1 equivalents of *n*-BuLi in THF at 0° in the presence of one to two equivalents of TMEDA reacts preferentially at the γ-position (3:1–4:1) with a variety of electrophiles ($^+El^1$ and $^+El^2$) (Scheme 5):

$$\underset{\mathbf{79}}{\overset{2-}{\overline{S{-}CH{-}CH{-}CH_2}}}\ \ 2Li^+ \xrightarrow[2)\ ^+El^2]{1)\ ^+El^1}$$

$$El^1{-}CH_2CH{=}CH{-}S{-}El^2 + CH_2{=}CH{-}CHEl^1{-}S{-}El^2 \qquad (5)$$

Alkyl and silyl halides, epoxides, aldehydes, ketones, and disulfides were used as electrophiles. The yields of addition products ranged from 65 to 95%, and γ:α ratios were close to 3:1. Subsequently the effects of medium on the γ:α ratio were studied and it was established that the addition of one equivalent of *t*-BuOK and 15% HMPA in most cases produced a significant increase in the γ:α ratio.[775]

6.2.1.2.5. Silicon. Although allylsilyl anions were shown to react with electrophiles at the α- and γ-sites[2048–2050] and Chan and Lau[2051] studied the effect of metal halides on the regiochemistry of addition of **80** to carbonyl compounds, it remained for Magnus and colleagues[293] to develop the methodology for converting the γ-adducts **(81),** obtained from addition of **80** to aldehydes and ketones, into lactols and lactones. The anion **80,** prepared from commercially available allyltrimethylsilane and *s*-butyllithium in THF at −76°

containing an equivalent of TMEDA, was reacted with a variety of aldehydes and monocyclic and bicyclic ketones at a range of temperatures (0° to −78°) (Scheme 6).

$$\underset{\textbf{80}}{Me_3Si\text{—}\overset{\ominus}{\overline{CH\text{—}CH\text{—}CH_2}}\ {}^{+}Li} \xrightarrow[65\text{–}97\%]{RR'CO} \underset{\textbf{81}\ (\boldsymbol{E})}{HO\text{—}CRR'\text{—}CH_2CH{=}CH\text{—}SiMe_3} \qquad (6)$$

The vinyl silanes were converted to α,β-epoxysilanes with *m*-chloroperbenzoic acid and further on to the *O*-methyl lactols by dry MeOH/BF_3-etherate. Finally, the γ-lactones **(82)** were obtained from the lactols by Jones' reagent. A wide selection of carbonyl compounds can be reacted: Aliphatic or aromatic aldehydes and all types of ketones respond well, including steroid 17-ones.

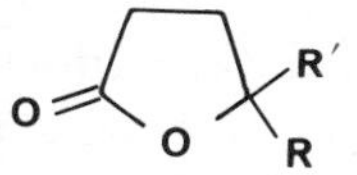

O R' O R

82

Corriu and coworkers[57] prepared the corresponding copper complex of **80** by mixing one equivalent of Cu(I)CN at −78° with the anion generated from allyltrimethylsilane and *n*-BuLi/TMEDA (1.0 equiv.) at 0°. The copper reagent reacts selectively at the γ-site with alkyl and acyl halides; also, a highly selective γ-1,4-addition occurs with α,β-unsaturated ketones and esters. In contrast, α,β-unsaturated aldehydes give γ-1,2-addition. Although the vinyl silanes were not elaborated to lactol ethers, to complete formally the synthon equivalency of $^{-}CH_2CH_2CHO$ **(1),** the vinyl silanes obtained from the reactions with chloromethyl ether, ethyl 2-butenoate, and ethyl cinnamate were converted efficiently to the corresponding α,β-epoxysilanes that are precursors of carbonyl compounds under acidic conditions.[1138]

6.2.1.3. Other Sources of Propanal β-Enolate

Corey and Ulrich[2052] utilized the readily available *cis*- **(83)** and *trans*- **(84)** 2-methoxycyclopropyl lithium, formally equivalents of propanal homoenolate **(1),** in the synthesis of β,γ-unsaturated aldehydes **(85).**

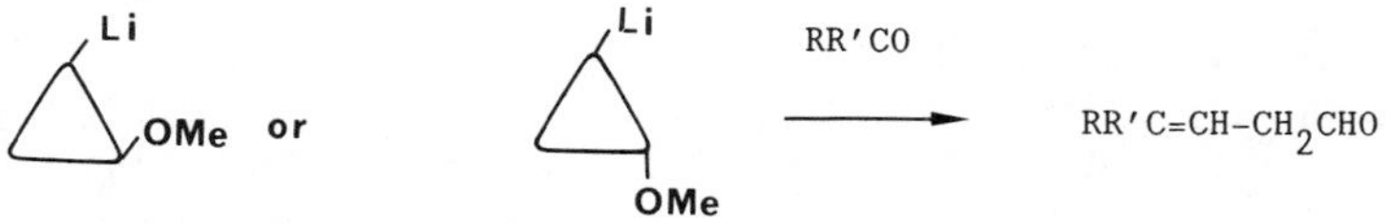

Li OMe or Li OMe RR'CO ⟶ RR'C=CH-CH_2CHO

83 84 85 R and R' = H or alkyl

Martin and Garrison[2053] prepared the phosphonium salt **(86)** by adding triphenylphosphonium bromide to methoxyallene and converted it into the phosphorane **(87)** with *n*-BuLi in THF at −50°. The Wittig reagent was added to a variety of aldehydes and ketones, and the resulting enol ethers were converted to the corresponding α,β-unsaturated aldehydes **(88)** in yields ranging from 40 to 73%. It is interesting to note that Wittig reagents that have protected carbonyl groups can serve as homoenolate anion equivalents in general.

$$\underset{\mathbf{86}}{Ph_3P^+CH_2CH{=}CHOMe\ Br^-} \qquad \underset{\mathbf{87}}{PH_3P{=}CH{-}CH{=}CHOMe}$$

$$\underset{\mathbf{88}}{RR'CH{-}CH{=}CH{-}CHO\ (\text{all } E)}$$

6.2.2. 2-Alkyl-Substituted Propanal β-Enolates

A number of routes to the title synthons have already been documented in Section 6.2.1. Normant *et al.*,[2020] in a study of the addition of Grignard reagents to α,β-unsaturated acetals in the presence of CuBr, added *n*-BuMgCl to **89** and obtained a 75% yield of the vinyl ether **(90)** ($Z/E = 2$).

$$\underset{\mathbf{89}}{CH_2{=}CMe{-}CH(OEt)_2} \qquad \underset{\mathbf{90}}{Me(CH_2)_4{-}CMe{=}CH{-}OEt}$$

$$\underset{\mathbf{91}}{^-CHSiMe_3{-}CHMe{-}CHO}$$

Ahlbrecht and Sudheendranath[2027] generated the 1-aminoallyl anion **(28c)**, β-enolate synthon **(29)**, from the corresponding enamine using *t*-BuOK in *t*-butyl methyl ether followed by *t*-BuLi in pentane. The anion was trimethylsilylated, and could also be alkylated further, thus making it equivalent to the synthon **91.** Coutrot and colleagues[2030,2031] used the lithium derivative of *N*-methyl-*N*-allylphosphoramide **(39b)** as a source of $^-CH_2{-}CHMe{-}CHO$ **(29).** Evans and coworkers[1818] metallated 2-propenyl methyl ether with *s*-BuLi in THF at −65° and studied the reaction of the lithium reagent with electrophiles. The analogous organozinc reagent added to cyclohexanone exclusively α. Hoppe and coworkers[2038] have demonstrated that the allylic *N,N*-dialkylcarbamate **48** ($R^1 = R^3 = H$, $R^2 = Me$, $R^4 = Et$) can serve as the synthon **29.** Ridley and Smal[2043] used the sulfide **63b** as an equivalent of **29.**

A variety of other routes have been described. Tischler and Tischler[2054]

used the dilithiation of *N*-allylcarboxamides **(92)** to prepare the β-enolate synthons **1, 29,** and **93,** as well as the ketone synthon **31.**

$$\text{PhCO—NMe—CHR—CR'{=}CH}_2$$

92

$$\text{PhCONMe—CR{=}CR'—CH}_2\text{R''} \qquad {}^{-}\text{CH}_2\text{—CHPh—CHO}$$

94 **93**

where for **92a,** R = R′ = H; **92b,** R = R′ = Me; **92c,** R = H, R′ = Ph; **92d,** R = Ph, R′ = H. The dilithiated species were generated by the addition of two equivalents of LDA or *n*-BuLi in diglyme at −78°. Subsequent reaction with alkyl halides (R″X) gave predominately the (*Z*)-enamides **(94)** as products of γ-addition in >75% yield. Since the reagents could also be quenched with water, presumably the corresponding β-deuterated carboxamides, and therefore the corresponding β-deuterated aldehydes and ketones could be obtained as well.

Kozikowski and Isobe,[2055] by metalating the dioxane **95** with *s*-BuLi in THF at −78° and reacting the anion with electrophiles, established that **95** is a viable source of the 2-hydroxymethylene analog of **1,** that is, **96.**

95

$$^{-}\text{CH}_2\text{-CH}(\text{CH}_2\text{OH})\text{CHO}$$

96

6.2.3. Butanal β-Enolate

A number of routes to the β-enolate synthon $^{-}$CHMe—CH_2—CHO **(54)** have been described in Section 6.2.1. Loozen and Godefroi[2008] have utilized the Grignard reagent **2b** as an equivalent of **54** to prepare the benzothiophenes **5** (R = Me or Et, R′ = Me) by adding **2b** to the corresponding 2-acylthiophenes **4** (R = Me or Et) and cyclizing the adducts in 10% refluxing H_2SO_4. Similarly, certain benzimidazoles **(8)** were prepared from **2b** and imidazoles **7.** Hoppe and coworkers,[2040] in an extension of their studies on the anions derived from allyl-*N,N*-dialkylcarbamates, prepared the lithium derivative **52c** (an equivalent of **54**) and added it to 2,3-dimethylpropanal to get the corresponding δ-hydroxyenolcarbamate with a *threo*:*erythro* ratio of 85:15.

Additional routes to **54** have been documented. Hayashi and colleagues[2056] reacted the lithium salts of *N,N*-dimethyldithiocarbamates **(97)**—equivalents of β-enolates $^{-}$CHMe—CH_2CHO, $^{-}$CHEt—CH_2CHO, and $^{-}$CH(Pr-*i*)—CH_2CHO—with benzaldehyde to establish the stereoselectivity of the addition. Predominate γ-addition resulted in the formation of *erythro*-(*E*)-**98**

(from *Z*-**97**) and *threo*-(*E*)-**98** (from *E*-**97**) as the major products. Complexation involving a six-membered ring chain conformation was invoked to account for the regioselectivity and diastereomeric induction.

$$\mathrm{RCH{=}CH{-}CH_2S{-}CS{-}NMe_2}$$

97

$$\mathrm{Ph{-}CHOH{-}CHR{-}CH{=}CH{-}S{-}CS{-}NMe_2}$$

98

where for **a,** R = Me; **b,** R = Et; **c,** R = Pr-*i*.

Hoppe and Lichtenberg[2057] have successfully performed the stereocontrolled addition of **54,** $^-CHMeCH_2CHO$, to aldehydes. Transmetallation of the lithium derivative of carbamate **(99a)** with $AlCl(Bu\text{-}i)_2$ gave the (*E*)-(1-carbamoyloxy)crotyl aluminum derivative **99b** that was added to a number of aldehydes. The *threo*-configurated (*E*)-adducts **(100)** predominated. In a subsequent report Hanko and Hoppe[2107] described reactions of the corresponding titanium derivative **99c,** obtained by the transmetallation of the lithium derivative by chlorotris(diethylamino)titanium, to a number of aldehydes.

$$i\text{-}\mathrm{Pr_2N{-}COO{-}\overset{\ominus}{\overline{CH{-}CH{-}CH_2}}\quad Met^+}$$

99

where for **99a,** Met = Li; **99b,** Met = $Al(i\text{-}Bu)_2$; **99c,** Met = $Ti(NEt_2)_3$.

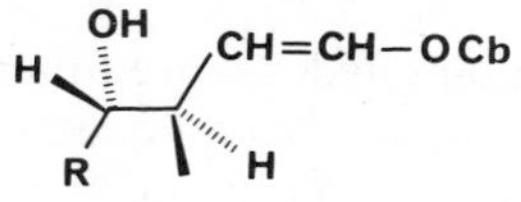

OH, H, R, CH=CH–OCb, H

R = Me, i-Pr, t-Bu, or Ph

100 (E)

101 (Z)

Addition at −78° yields almost exclusively the *threo*-(*Z*)-δ-hydroxyenol carbamates **(101).** Methanolysis gave the corresponding lactol ethers; the dimethyl acetals could be obtained as the acetates.

Sato *et al.*[2058] have shown that the regiochemistry and stereochemistry in the reaction of 1,3-disubstituted allyl anions can be controlled via η^3-allyltitanium compounds. Complexes **102,** prepared at −78° by reaction of the

anions with Et_2TiCl, were added to propanal and 2-methylpropanal to obtain the adducts **103.** In particular, **103b** (R = Et) was formed in 88% yield exclusively in a *threo*- and *cis*-geometry.

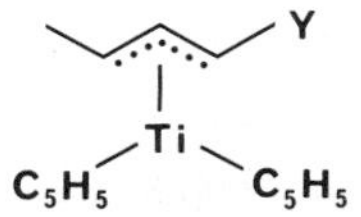

Ti
C_5H_5 C_5H_5 Y

R-CHOH-CHMe-CH=CH-Y

102 a Y = $SiMe_3$ 103

b Y = OPh

c Y = SPh

Pratt and Thomas[2059] used (*E*)-1-methoxymethoxy-but-2-enyl(tri-*n*-butyl)stannane **(104)** as a *threo*-selective equivalent of the β-enolate of butanal **(54).** Heating **104** with aldehydes in toluene under reflux gave the corresponding enol ethers **105** mostly in reasonable yields. Although the ethers were hydrolyzed and oxidized to the *trans*-4,5-disubstituted butyrolactones, the lactols or lactol ethers or other derivatives undoubtedly could be isolated as well.

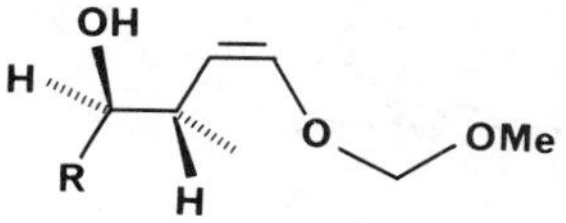

OH H R H O OMe

105

6.2.4. β-Enolates of Other Saturated Aldehydes

A number of routes to β-enolates of other aldehydes—predominately those derived from propanal and butanal—have already been documented in previous sections. Ahlbrecht and Sudheendranath[2027] prepared the 1-aminoallyl anion **28b,** the equivalent of **30,** from the corresponding enamine and allylamine and reacted it with chlorotrimethylsilane. Subsequent alkylation of the silylenamines with EtI gave the corresponding 3-oxosilanes after hydrolysis. Coutrot and Savignac[2030] used the *N*-methyl-*N*-allylphosphoramide **39e** as an equivalent of $^-CMe_2CH_2CHO$ **(55).** Hoppe and coworkers[2038] were able to generate the 3-methoxycarbonyl propanal β-anion equivalents **(106)** by dideprotonating the carbamates **48** and reaction with dimethyl carbonate. Further reaction with an electrophile gives the enol carbamates **49.** An extension of this work[2040,2041] furnished the synthons **55, 56, 106,** and the 3-pinyl-equivalent of propanal β-enolate. Atlani and coworkers[2042] prepared the thioallyl

anion **58b,** the equivalent of **55,** and described its reaction with acetone and a variety of allylating reagents in the presence of complexing agents.

$$MeCH{=}CH{-}CH(SnBu_3){-}OCH_2OMe \qquad {}^{-}CR(COOMe){-}CH_2CHO$$

104 **106**

where for **106a,** R = H; **106b,** R = Me.

Additional routes to β-enolates of other aldehydes have been documented. Ahlbrecht, Enders, and coworkers[2060] have described the preparation of a chiral equivalent of $^{-}CHPh{-}CH_2CHO$ **(30).** Metallated chiral allylamines of the type **107** (Met = Li, K) were prepared and alkylated at the γ-position with alkyl iodides or bromides. Hydrolysis of the resulting enamine yielded the β-substituted aldehydes **(108)** in enantiomeric excess up to 67%. The authors indicated that preliminary studies with metallated phosphoramidates as chiral homoenolate equivalents were promising. Mukaiyama and colleagues[2061] used the anion of the chiral ether **109,** derived from the corresponding chiral alcohol and cinnamyl bromide as a chiral equivalent of **30.** Lithiation of **109** followed by alkylation by a variety of reagents yielded 3-phenylalkanals, in some cases with 75–85% optical purity after hydrolysis of the ethers in aqueous 50% $HClO_4/Et_2O$.

107 108 109

$$MeSCH_2CH{=}CH{-}COOMe \qquad MeSCH{=}CH{-}CRR'{-}COOMe$$

110 (*E*) **111**

A route to the 3-methoxycarbonyl equivalents of propanal and butanal **(106a–b)** has been described by Kende and coworkers.[2062] Treatment of the γ-methylthiobutenoates **110** and **111** (R = H, R′ = Me) with LDA in THF at −78° followed by reaction with the more reactive alkylating agents (MeI, allyl or propargyl halides) in the presence of HMPA gave the corresponding enol thioethers **111** (R = H or Me), generally in good yield. Although direct hydrolysis of the enol thioethers proved difficult, **111** (R = R′ = Me) was saponified [$Ba(OH)_2$, ambient temperature, 24 h] to the free acid (85%) that was hydrolyzed in 1:1 HCl/AcOH (90°, 1 h) to the hemiacylal **112** in 67%

yield. The alkylations led to a number of synthetically useful cyclizations to α,β-unsaturated ketones. The enol thioether **111** (R = H, R′ = HC≡CCH_2) when heated at reflux for 1 h in 1:1 HCl-acetic acid gave the cyclohexenone **113** (93%) that was cyclized quantitatively to the lactone **114** on sublimation at 150°. The ester **111** (R = H, R′ = Me) underwent Michael addition to methyl vinyl ketone (LDA, −78°, THF, 0.2 equiv. HMPA) to produce a 1:1 mixture of the *cis*- and *trans*-enol thioethers. Mild hydrolysis gave the acyl-cyclopentene acid **115** (55%).

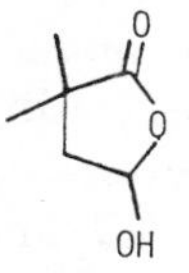

O O OH CH_3 CO_2H O O O CH_3 O CH_3 O CO_2H

112 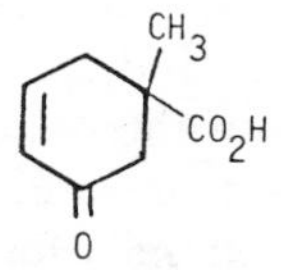 **113** 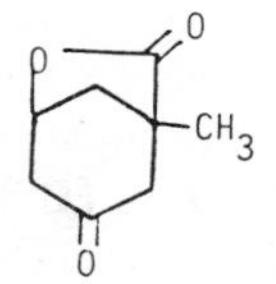 **114** 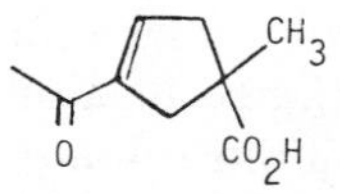 **115**

In a related study Warren and coworkers[2063] prepared a range of γ-phenyl-thiocrotonate esters **(116)** by [1,2] and [1,3]-PhS shifts and reacted the anions, generated using *t*-BuOK/THF at −78°, with alkyl iodides (R″I) to yield the esters **117.** By using two equivalents of base and an excess of alkylating agent, dialkylations could be carried out. Although the enol thioethers were not converted to carbonyl compounds, the authors stated that hydrolysis using documented procedures would yield the 1,4-dicarbonyl compounds. The methodology provides routes to the ketone β-enolates $^{-}CH_2CH_2COMe$ **(64)** and $^{-}CH_2CH_2COEt$ (see Sect. 6.3.1) as well as $^{-}$CHMe—CH_2CHO **(55)** and $^{-}$CH(Bzl)—CH_2CHO, the β-enolate of 4-phenylbutanal.

PhS—CHR—CH=CR′—COOEt PhS—CR=CH—CR′R″—COOEt

116 **117**

PhSCH=CH—CMe_2—CRR′OH

118

where R,R′ = H or alkyl.

Corey and co-workers[2064] used the conjugate addition of the anion of 3-nitropropanal dimethyl acetal to 9-cyano-2-nonenal in the initial step of the synthesis of prostaglandins of the E_1 and F_1 series including 11-epiprostaglandins. Thus, the anion of the acetal is an equivalent of the β-enolate of 3-nitropropanal. Kondo and coworkers[2065] prepared **58b** and reacted this reagent—an equivalent of $^{-}CMe_2$—CH_2CHO **(55)**—with aldehydes and ketones in 70–92% yield. The vinyl thioethers **(118)** were not hydrolyzed, but those from propanal and benzaldehyde were cyclized to cyclopropanes by reaction of the mesylates with MeONa/MeOH and AcOK/*t*-BuOH, respectively. A final hydrolysis to the cyclopropyl aldehydes was also reported.

Cooper and Dolby[2066] used the β-sulfone acetal **119a** to prepare the ketals **119b** and **119c,** the anions of which are equivalents of β-enolates $^{-}CHR—CH_2CHO$, R = Ts, Bu, and $(CH_2)_7CHO$, respectively. Reaction of the two latter anions with acetaldehyde followed by oxidation, sulfinate elimination, and cyclization yielded the cyclopentenones **120.** Reissig and colleagues[2067] developed a flexible route to 4-oxo-carboxylic acid esters by utilizing silyloxycyclopropane carboxylates **(121).**

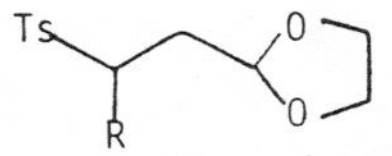

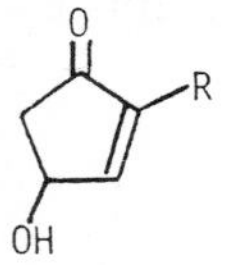

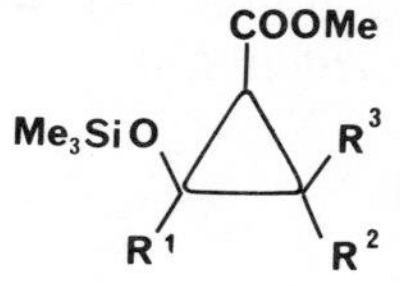

119a R = H

119b R = Bu

119c R = $(CH_2)_6$-$CH(OMe)_2$

120

121

Treatment of the esters with LDA in THF at −78° followed by alkylation (R^4X) and ring-opening by fluoride ion gave the 4-oxo-carboxylic acid esters **(122).** The esters not only provided a route to the β-enolate OHC—C—CH^-COOMe (R^1 = H) but to ketone homoenolate equivalents R^1CO—C—CH^-COOMe as well.

$$R^1CO—CR^2R^3—CHR^4—COOMe$$

122

6.3. β-ENOLATES OF KETONES

A number of routes to the β-enolate equivalent **64** of 2-butanone have been documented in Section 6.2.1. Kondo and Tunemoto[2013] used the anion of γ-oxosulfone **(16b)** as a source of **64** and synthesized undecane-2,5-dione after reaction with methyl pentanoate, reduction, and hydrolysis. Hoppe and coworkers[2038] have demonstrated that the allyl *N,N*-dialkylcarbamate **48** (R^3 = Me) is a good source of the β,β-bisanion **(123).**

$$^{-}CH_2CH_2COMe \qquad ^{2-}CH—CH_2COMe$$

64 **123**

Ridley and Smal[2043] used the allyl anion **63c** to obtain a 38% yield of stereoisomeric vinyl sulfides from addition of the thioallyl anion to benzaldehyde.

Hydrolysis of the vinyl sulfide to the corresponding ketal would be a simple matter. In an interesting approach to the addition of oxygen- and sulfur-substituted allylic anions, Evans and coworkers[2068] converted the lithium reagents into organozinc and organocadmium reagents, which added with α-regioselectivity to acyclic or cyclic α,β-enones. The resulting dienols as their potassium salts (e.g., **124**) were thermally rearranged (oxy-Cope) with varying success (11–93%) to the methyl enol ethers (for example, **125**) of the corresponding 1,6-dicarbonyl compounds. Although not mentioned in the publication, the enol ethers presumably could be converted into 1,6-diketones or ketoaldehydes.

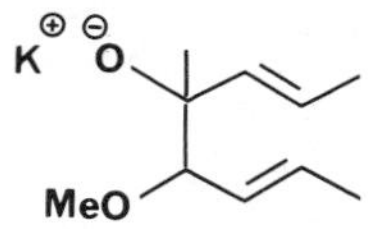

124 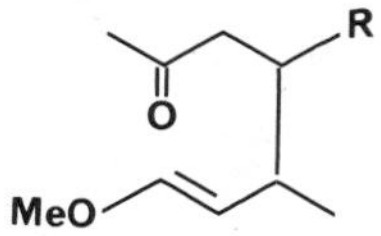 125

The overall reaction corresponds to the addition of homoenolates

$$^{-}CHMe—CH_2COMe \quad \text{and} \quad ^{-}CMe_2—CH_2COMe$$

126 **127**

to an α,β-unsaturated ketone and avoids the persistent problem of the ambident reactivity of enones and allylic anions.

Ponaras[2069] prepared 3,3-ethylenedioxybutylmagnesium bromide **(128)** and established its utility as the C_4 annulation synthon $^{-}CH_2CH_2COMe$ **(64)**, which provides a good route from acyl halides to 1,4-diketones, useful intermediates in the synthesis of cyclopentenones, furans, and other heterocycles. The Grignard reagent **128** is unstable at elevated temperatures and yields the cyclopropyl ether **129.** Fujisawa and coworkers[2070] accomplished the regiospecific ring opening of α-methyl-β-propiolactone with **128** in THF/Me_2S (30:1) in the presence of CuBr (2 mol%). Hydrolysis of the ketal gave 2-methyl-6-oxoheptanoic acid in 60% yield.

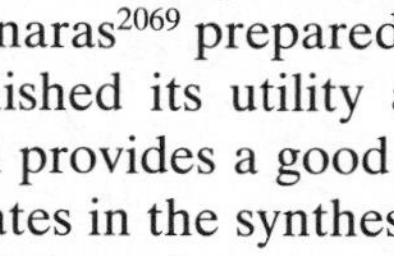

128 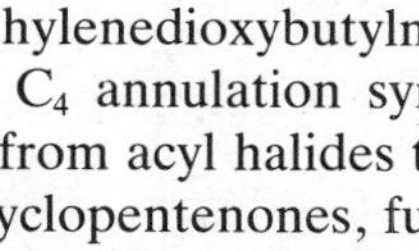 129 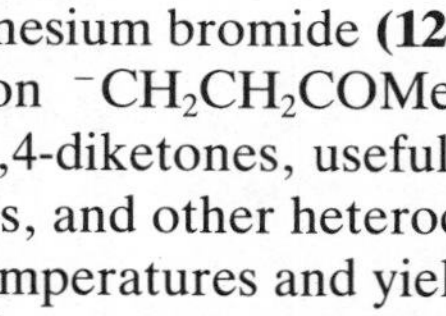 130

A series of β-diphenylphosphinoyl ketals **(130)** that are formally sources of the β-enolates $^{-}CHMe—CH_2COMe$ **(126)**, $^{-}CH_2CH_2COEt$, $^{-}CH_2CH_2COPh$ **(31)**, $^{-}CHMe—CH_2COPh$, $^{-}CH_2—CHMe—COMe$,

$^{-}$CHMe—CHMe—COMe, and the cyclohexanone β-anion **131,** as well as the β-enolate **1,** have been prepared by Warren and coworkers.[2071,2072] The lithium derivatives, generated by treatment with *n*-BuLi in THF at −78°, are stable up to 0° and add to aldehydes and nonaromatic ketones to yield the expected adducts, some of which were reported to eliminate Ph_2PO^- on treatment with NaH in THF to give the Horner–Wittig products **132.** Nakai and coworkers[2073] have reported the oxidation of tertiary 2-alkylcyclopropyl carbinols **133** with pyridinium chlorochromate (5 equiv.) in dichloromethane at ambient temperature, giving the enones **134** in moderate yield. As the carbinols **133** are obtained from the ketones RCOMe and the cyclopropyl organometallic, the latter is here an equivalent of the synthon $^{-}CH_2CH_2COMe$ **(64).**

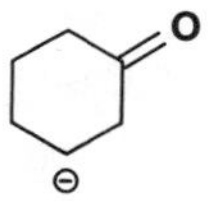

131

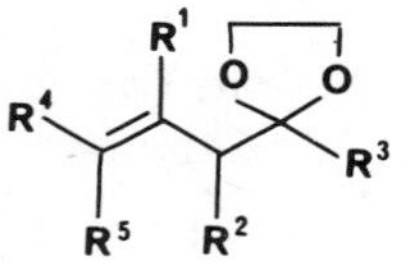

132

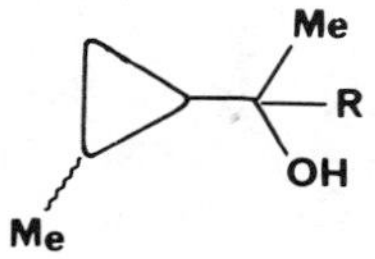

133

MeCO—CH_2CH=CMe—R (MeO)$_2$CMe—CH_2CH_2—SO_2Ph

134 **138**

MeCO—CH_2CH_2—SPh

139

Masaki and coworkers[2074] synthesized (+)-exo-brevicomin **(135),** one of the attractant pheromones of the western pine beetle, in 30% overall yield by transforming **137a** or **137b** to **136,** which was converted to **135.**

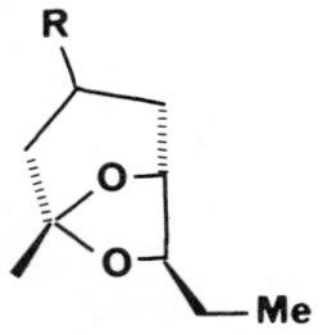

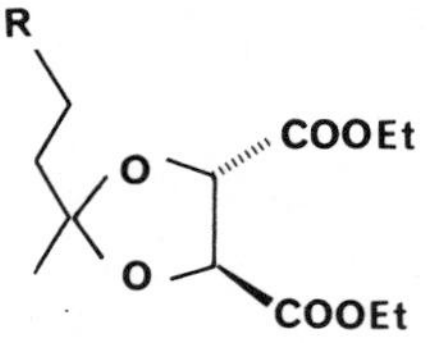

135 R = H

136 R = SO_2Ph

137a R = SO_2Ph

137b R = SPh

The ketals **137** were prepared by treating **138** and **139,** respectively, with diethyl (*S,S*)-(−)-tartrate in benzene at 80–90° in the presence of a catalytic amount of TsOH. In this case the synthon $^{-}CH_2CH_2COMe$ **(64)** is employed

intramolecularly in a substitution reaction. Moreau and Couffignal[2075] have prepared the organolithium reagent **140** from trimethylsilyl 4,4-ethylenedioxypentanoate and LDA in ether at $-60°$.

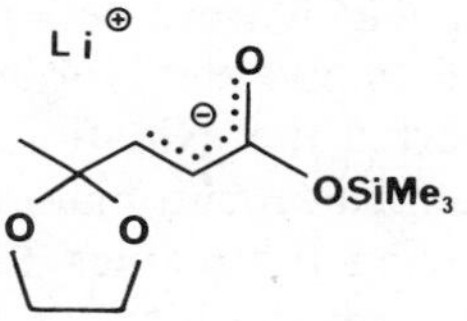

$Li^{\oplus}$ O $\ominus$ $OSiMe_3$ O O

140

Treatment of the anion with diethyl ketone or acetophenone followed by acid hydrolysis and thermolysis at 180° gave the corresponding β,γ-unsaturated ketones **141** in 64 and 69% overall yields, respectively. To prepare the 1,4-diketones **142a** (56%) and **142b** (60%), mixed anhydrides **143** were used in the substitution reactions.

$MeCOCH_2—CH{=}CRR'$ **141**

$MeCOCH_2CH_2COCH_2R$ **142**

$EtOCO—OCOCH_2R$ **143**

where for **141a,** R = R′ = Et; **141b,** R = Me, R′ = Ph; and for **142a,** R = Me; **142b,** R = Pr-*i*.

Rosini and coworkers[2076] have developed 3-ethylenedioxy-1-nitrobutane **144** into the synthon **64.**

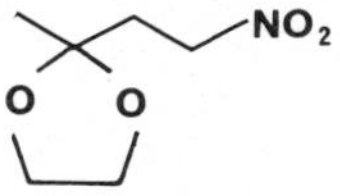

NO_2 O O

144

The ketal, readily prepared from 3-buten-2-one by reaction with $NaNO_2$/AcOH/THF at ambient temperature followed by TsOH-catalyzed ketalization with ethylene glycol in refluxing benzene, was added in a nitro-aldol condensation to a number of aldehydes (RCH_2CHO). Oxidation, denitration by way of *p*-toluenesulfonylhydrazones of the corresponding α-nitroketones and deketalization gave the corresponding 1,4-diketones ($MeCOCH_2$

CH_2COCH_2R, 68–84%), readily cyclized by NaOH/EtOH to 2-substituted 3-methylcyclopentenones. The methodology provides a most convenient route to (*Z*)-jasmone, dihydrojasmone, and other substituted cyclopentenones.

In an interesting approach to ketone β-enolates, Kuwajima and Kato[2077] prepared a series of β-lithiated enol trimethylsilyl ethers **(145)** by rearrangement of lithium salts of the corresponding 1-trimethylsilylallyl alcohols. Alkylation by alkyl iodides R′I gave the corresponding enol ethers **146** (63–74%). High stereoselectivity (*Z*-isomers) and regioselectivity (γ-addition) were observed. It should be a simple matter to convert the silyl anol ethers to the corresponding ketones—the methodology provides a route to synthons $^{-}CH_2CH_2CO$—R, with R = Pr, $PhCH_2CH_2$, and $Me(CH_2)_8$ in this instance.

$Me_3SiO—\overset{\ominus}{CR—CH—CH_2}\ {}^{+}Li$ **145** $Me_3SiO—CR{=}CHCH_2—R'$ **146**

where for **a,** R = Pr; **b,** R = $PhCH_2CH_2$; **c,** R = *n*-C_9H_{19}.

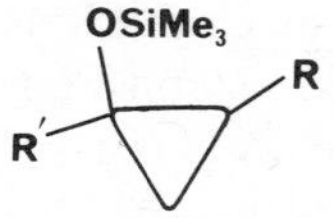

$OSiMe_3$, R, R′

147 R = alkyl or aryl

R′ = H or alkyl

Murai, Seki, and Sonoda[2078] utilized trimethylsilyl cyclopropyl ethers **147** as ketone β-homoenolate equivalents in bromoketone synthesis (*i.e.*, coupling with Br^+). Reaction of the ethers with bromine in CH_2Cl_2 at −70° gave the corresponding bromoketones **148** in high yield after the solvent and Me_3SiBr were removed under vacuum. When two equivalents of bromine were used, the corresponding α,β-dibromoketones were obtained.

RCO—CHR′—CH_2Br **148** $(CH_2{=}CH)_2CY—OSiR_3$ **149, 150**

$CH_2{=}CH—C(OSiR_3){=}CHCH_2—El$ **151**

where for **149,** Y = H; **a,** R = Me; **150,** Y = El; **b,** R = Et.

Oppolzer and coworkers,[2080–2082] in a series of papers, have documented several routes to the unsaturated β-enolate equivalent $^{-}CH_2CH_2CO—CH=CH_2$ using silyloxy-1,4-pentadienes **(149).** Treatment of **149b,** obtained from the corresponding carbinol, with *s*-BuLi in THF at −78°, followed by quenching of the anion with a range of alkyl, epoxy, or carbonyl electrophiles gave the products of α- **(150)** and γ-addition **(151),** the latter being largely in the (3*Z*)-configuration.

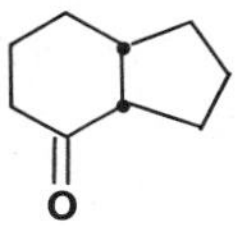

152

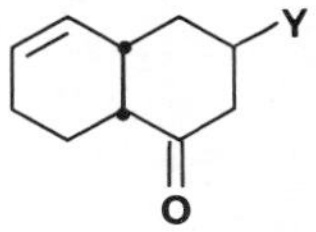

153 Y = H

155 Y = SMe

The γ-products **151** could be directly subjected to intramolecular or intermolecular [4+2]-additions. For example, **151** (El = Me) reacts with *N*-phenylmaleimide to give the [4+2]-adduct. **151** (El = $CH_2=CHCH_2CH_2—$) undergoes an intramolecular cyclization and the ketone **152** (81%) is obtained after cleavage of the enol silyl ether with KF/MeOH; **151** (El = $CH_2=CH—CH=CH—CH_2—$) gives the ketone **153** (78%) in an identical sequence of reactions. It was also established[2082] that the anion of **149b** can be γ-selectively sulfenylated to yield **151** (El = MeS), which on deprotonation (LDA, −78°) and alkylation furnished exclusively the (3*Z*)-γ-adducts **(154)** in high yield.

$$CH_2=CH—C(OSiEt_3)=CH—CHR—SMe$$

154

Again, intramolecular Diels–Alder reactions occur readily if the R group in **154** is suitably olefinic. For example, with R = $CH_2=CH—CH=CH—CH_2$, cyclization and silyl ether cleavage with KF/MeOH at −10 to 0° for 1 h yielded **155** (as an epimeric mixture) in 63% yield. Oxidation of the epimers with $NaIO_4$ in aqueous MeOH at −10°, followed by elimination of sulfonic acid in boiling CCl_4, gave the corresponding α,β-unsaturated ketone that was reduced by $NaBH_4$ in EtOH and then oxidized (1.6 equiv. PCC/NaOAc/CH_2Cl_2) to the ketone **153.** Alternatively, fluoride-promoted desilylethylation provides an efficient route to a variety of α,β-unsaturated ketones.

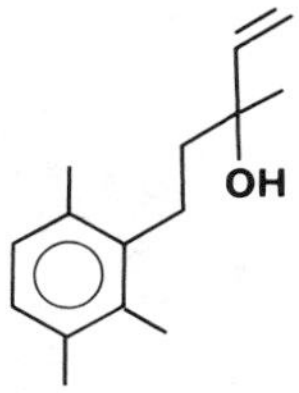

OH

156

In addition, the racemic alcohol **156,** a natural product, was synthesized in four steps (27% overall yield) from **149a** and the dienone **157.** The hydroxy-enone **158,** obtained by the treatment of the addition product with KF/MeOH, was a key intermediate in the synthesis.

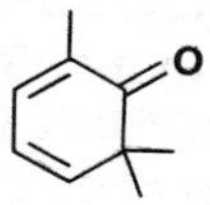

157 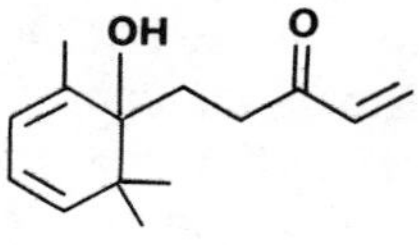 **158**

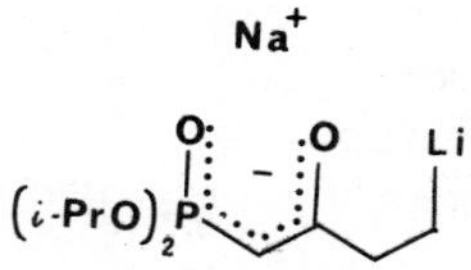

Na^+ Na^+

$(i\text{-PrO})_2P$ O: :O Li $(i\text{-PrO})_2P$ O: :O $SnBu_3$

159 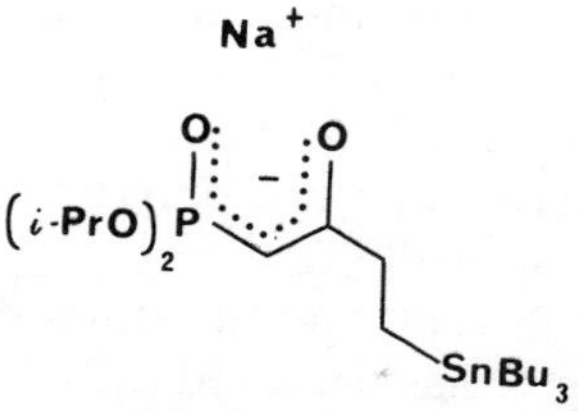 **160**

Goswami[2083] has reported the preparation of a β-ketophosphonate 1,4-dianion **(159)** via tin–lithium exchange of the monoanion **160** derived from diisopropyl 4-tri-*n*-butylstannyl-2-oxobutylphosphonate. Thus, the δ-carbon (β to the carbonyl group) can be functionalized and a series of β-keto-phosphonates **(161)** prepared (R = D, allyl, propyl, trimethylsilyl, acetyl, or HO—CPh_2—). In conjunction with the known reactions of β-ketophos-phonates, **159** serves as a source of vinyl β-enolate equivalents of the type RR′C═CH—$COCH_2CH_2^-$.

$(i\text{-PrO})_2PO—CH_2COCH_2CH_2R$ **161**

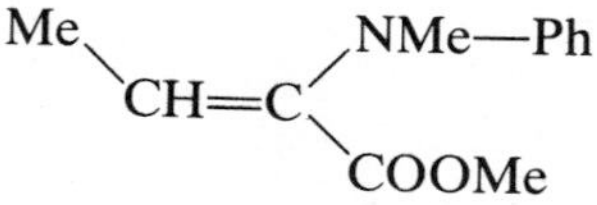

Me—CH═C(NMe—Ph)(COOMe) **162**

Met⁺ ⊖ CH_2—CH—C(NMe—Ph)(COOMe) **163**

RCH_2—CH═C(NMe—Ph)(COOMe) **164**

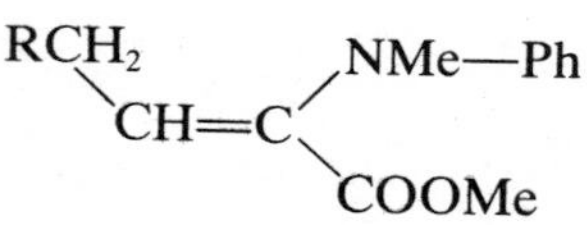

where for **163a,** Met = Li; **163b,** Met = K; **163c,** Met = MgBr.

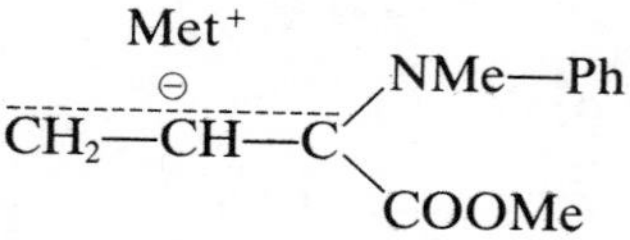

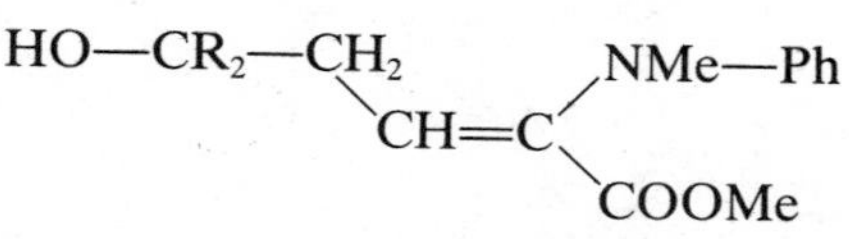

HO—CR_2—CH_2—CH═C(NMe—Ph)(COOMe) **165**

R_2C═CH—CH═C(NMe—Ph)(COOMe) **166**

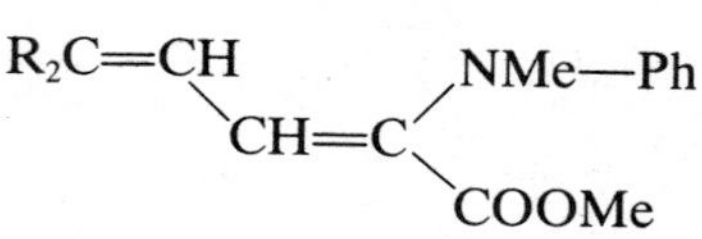

Ahlbrecht and Simon[1153] have shown that the 1-methoxycarbonyl enamine **162** derived from methyl-2-oxobutanoate and *N*-methylaniline quantitatively yields the allyl anions **163a** and **163b** when reacted with LDA/HMPA or KDA/THF at $-78°$. Alkylation (R-I) yields products of α- and γ-addition **(164),** the latter predominating. Acid-catalyzed hydrolysis provides a route to the synthon $^{-}CH_2CH_2COCOOMe$, thus making 2-oxoesters available, such as methyl 2-oxononanoate (from pentyl iodide, 60%). In a following publication,[1154] the authors described the addition of reagents **163a–c** to aldehydes and ketones and obtained the products **165** and **166,** depending on the conditions and the reagent used. Acid-catalyzed hydrolysis of **165** in MeOH yielded the corresponding cyclic ketals.

Murai and coworkers[2084] have studied the coupling of β-transition-metal ketones obtained by electrophilic ring opening of the corresponding silyloxycyclopropanes by $AgBF_4$ or $Cu(BF_4)_2$. The silyloxycyclopropanes formally provide a route not only to $^{-}CH_2CH_2COPh$ **(31)** but to a variety of other species as well.

Other routes to substituted ketone β-enolates have been documented. Reissig[2085] has prepared the trimethylsilyloxy cyclopropyl esters **167** and studied their reaction with carbonyl compounds in CH_2Cl_2 in the presence of equimolar amounts of $TiCl_4$. Reaction of **167a** with benzophenone afforded the unsaturated ketoester **168** (82%) after quenching with MeOH, and the lactol ether **169** (88%) after quenching with water and treatment of the lactol with BF_3-etherate in MeOH.

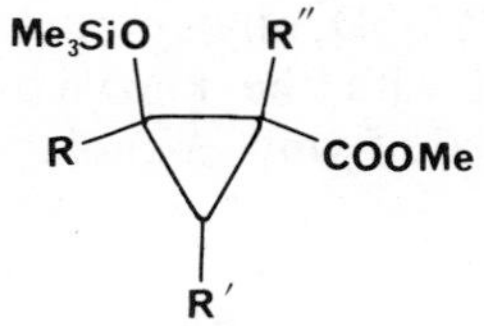

Me₃SiO R″ R COOMe R′

167a R = Bu-t, R′ = R″ = H

167b R = Bu-t, R′ = H, R″ = Me

167c R,R′ = $(CH_2)_4$, R″ = H

$$t\text{Bu—CO—CH}_2\text{—C(COOMe)=CPh}_2$$

168

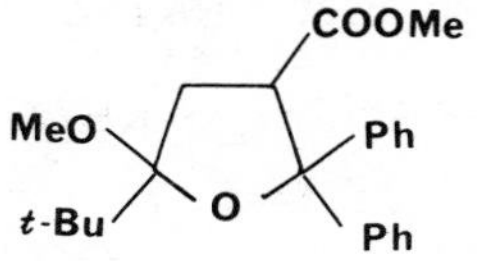

COOMe MeO Ph t-Bu O Ph

169

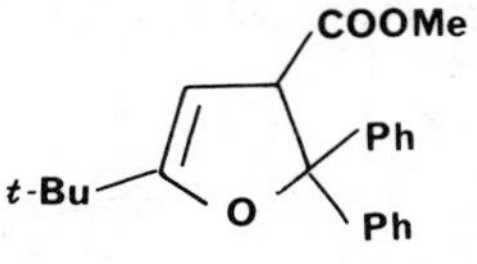

COOMe Ph t-Bu O Ph

170

Thermolysis of the lactol ether **169** at 150° gave the dihydrofuran **170** in 95% yield. Additional examples were documented in the paper. Thus, cyclopropanes **167a–c** provide routes to the ketoester β-enolate equivalents **171–173.**

$$t\text{BuCO—CHR—CH}^-\text{COOMe}$$

171, 172

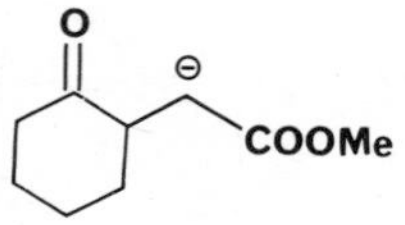

O

COOMe

173

where for **171,** R = H; **172,** R = Me.

Thompson and Huegi[2086] lithiated (*n*-BuLi/THF) indan-1-one pyrrolidine enamine **(174a)** and reacted the anion with MeI to obtain 3-methylindan-1-one (82%) after hydrolysis of the enamine **174b.** Subjected to a second methylation or benzoylation, the enamine **(174b)** gave 3,3-dimethylindan-1-one (95%) and 2-benzoyl-3-methylindan-1-one (19%). The enamine **174a** is thus a source of the homoenolate synthon **175.** In a related study, Trost and Latimer[2079] treated 6-methoxyindan-1-one with two equivalents of LDA in THF at −78° followed by one equivalent of EtI and obtained an 89% yield of 3-ethyl-6-methoxyindan-1-one.

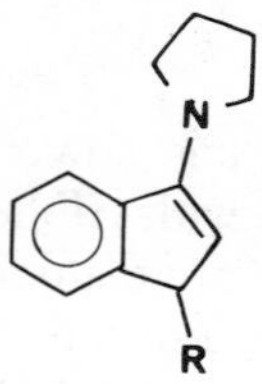

N

R

174 a R = H

174 b R = Me

O

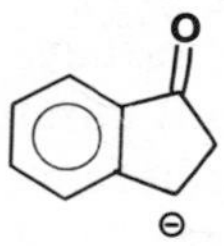

175

6.4. CARBOXYLIC ACID β-ENOLATES

Several routes to equivalents of the homoenolate

$$^-\text{CH}_2\text{CH}_2\text{COOH}$$

176

have already been discussed in Sections 6.2.1.2.3, 6.2.1.2.5, and 6.2.2. Hoppe and Brönneke[2041] have added *N,N*-diisopropyl-2-alkenylcarbamoyl anions to

aldehydes and acetophenone. While the enol carbamates are hydrolyzed to lactols or lactol ethers, reaction with MeOH/MsOH/1% $Hg(OAc)_2$ (or 0.01 equiv. $PdCl_2$), followed by MCPBA/BF_3-etherate in CH_2Cl_2, gave the corresponding γ-lactones in high yields. The anions are thus also equivalents of **176.** Magnus and Ehlinger[293] added the trimethylsilyl allyl anion **80** to a variety of aldehydes and ketones and converted the vinyl silanes **81,** products of γ-addition, into γ-lactones **(82).**

$$Me_3Si—\overset{\ominus}{CH—CH—CH_2}\ {}^{+}Li \qquad MeCH{=}CH—CH(OCH_2OMe)—SnBu_3$$

80 **104 (*E*)**

Oxidation of lactols obtained from carbonyl compounds and ⁻C—C—CHO synthons transforms the overall reaction sequence into one with ⁻C—C—COOH equivalency. For example, Pratt and Thomas[2059] have made *trans*-4,5-disubstituted γ-butyrolactones **(177)** using the *threo*-selective ⁻CHMe—CH_2COOH synthon. In practice, the synthesis involved reaction of the allyl stannane **(104)** with aldehydes RCHO, followed by PCC oxidation of the resulting lactols.

H, Me, R, H, O, O

177

Other routes have been developed as well. Oda and coworkers[2087] established that α,β-unsaturated nitriles and esters react with triphenylphosphine in ethanol or hexane yielding ylides. In the presence of an aldehyde, addition occurs followed by elimination of Ph_3PO to yield β,γ-unsaturated products. Stork and coworkers[2088] used this methodology to prepare 4-(*m*-methoxyphenyl)-3-butenoate, an intermediate used in the synthesis of *d,l*-lycopodine. The method provides a source of homoenolate equivalents of the type ⁻C—C—COOH after the amides and esters are hydrolyzed.

The Wittig reagent **178** has been used[2089] as a source of **176** in the preparation of β,γ-unsaturated acids.

$$Ph_3P{=}CHCH_2COO^-$$

178

Reaction of **178** with NaH (2 equiv.) and *m*-methoxyacetophenone or cyclohexanone in DMSO/THF at 0° afforded the corresponding β,γ-unsaturated acids **179** (*E*/*Z* = 80:20) and **180** in 57–69% and 66% yields, respectively.

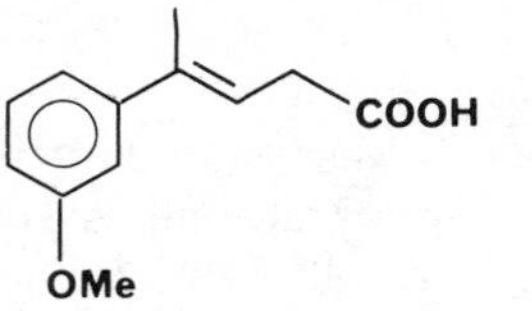

COOH

OMe

179

COOH

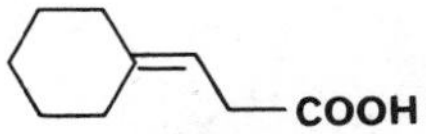

180

As an extension of the secoannulation methodology where the $^{-}CH_2CH_2CO$—Nu unit was introduced into α,β-epoxyketones, Trost and Bogdanowicz[2090] showed that spiroannulation of carbonyl compounds utilizing diphenylsulfonium cyclopropyl anion results in the facile conversion of aldehydes and ketones into γ-butyrolactones **(181)** by way of the spiroepoxides and spirocyclobutanones. Thus, the diphenylsulfonium cyclopropyl anion is an equivalent of **176.** A review by Trost[2091] provides an excellent summary of the methodology.

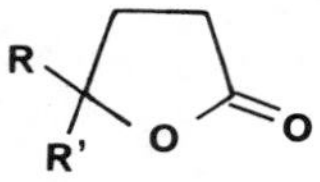

R R' O O

181

In an interesting approach, Sturtz *et al.*[2092] prepared the dianion **182** (2 equiv. *n*-BuLi/THF/ − 50°) and two analogs, one bearing a methyl group at C-2 and another with the methyl at C-3.

$$Me_2N—PO—C(O^-)—CH—CH_2^{-}\ 2Li^+$$

182

Addition of the dianions to a series of aldehydes and ketones led to γ-lactone formation in average yields. Addition to epoxides gave δ-lactones and alkylation with saturated and unsaturated alkyl halides yielded the corresponding carboxylic acids. A key step in the transformations is the hydrolysis of acyl phosphonates or phosphonamides to β- or γ-substituted carboxylic acids. The simple homoenolate dianion **183** has been prepared by Caine and Frobese[2093] from lithium 3-bromopropionate, which was added to lithium naphthalenide in THF at − 70°. The dianion was added to a variety of aldehydes and ketones followed by lactonization of the γ-hydroxyacids to give γ-lactones in reasonable yields.

$LiCH_2CH_2COOLi$ $RCH—CH—\overset{\ominus}{C}Y—CN\ ^{+}Li$

183   **184**

where for **184a,** R = H or Me; Y = NMe_2 or $N(CH_2)_4$; **184b,** R = H, Y = N(Me)Ph.

Continuing their studies on the anions of α,β-unsaturated nitriles as homoenolate equivalents, Johnson and Clader[1302] prepared anions **184a** from the corresponding aminonitriles with LDA in THF at −78°. If the solution is warmed to 0° and a ketone is added, only γ-addition is observed. If the anion is prepared at 0° and anhydrous $ZnCl_2$ is added to the ketone, enolization and hence aldol condensation is reduced. Addition to ketones (mostly cyclic) followed by aqueous work-up and treatment with 0.5 *M* HCl or oxalic acid in 50% dioxane gave 5,5-spirosubstituted γ-lactones. Treatment of the latter with 10% P_2O_5 in MsOH gave cyclopentenones. Costisella and Gross[2094] prepared 1-cyanoenamines that were hydrolyzed to carboxylic acids. Deprotonation of **185a** by LDA in THF at −70° gave the lithium derivative that was alkylated with methyl iodide or reacted with benzaldehyde. In the former case butyric acid was obtained presumably in good yield and in the latter the γ-lactone **186** (R = Ph) was obtained in 46% yield after work-up and distillation.

$$RR'C{=}CR''{-}NMe_2$$

185

where for **185a,** R = Me, R′ = H, R″ = CN; **185b,** R = Me, R′ = H, R″ = $PO(OEt)_2$; **185c,** R = Et, R′ = $PO(OEt)_2$, R″ = CN; **185d,** R = R′ = Me, R″ = $PO(OEt)_2$.

R O O

186

The work on 1-cyanoenamines discussed above was preceded by studies by Ahlbrecht and Vonderheid.[1300] The lithium reagent **184b** was prepared from the corresponding enamine and LDA in THF at −78° and reacted with methyl iodide, benzaldehyde, acetone, or benzophenone. The products of γ-addition **(187)** were obtained in 63–81% yields. Some enamines were hydrolyzed and cyclized to γ-lactones, thus that from benzaldehyde gave **186** (R = Ph) in 67% yield. The size of the alkyl groups on nitrogen affects regiochemistry so that a phenyl or cyclohexyl group in addition to methyl results in exclusive γ-addition when **188** is alkylated by methyl iodide or isopropyl bromide. Two methyls or a piperidine ring give α/γ mixtures.

$$El{-}CH_2CH{=}CPh{-}NMe{-}Ph \qquad Ph{-}\overset{\ominus}{\overline{CH{-}CH{-}C}}(CN){-}NRR'\ {}^{+}Li$$

187 **188**

In a continuation of their study of enamines substituted with anion-stabilizing groups at C-1, Costisella *et al.*[2095] lithiated **185a** and **185b** with LDA in THF at −70° and added the reagents to a number of aldehydes. Hydrolysis of the products of γ-addition gave the corresponding γ-lactones **(186)** (R = alkyl, alkenyl, or aryl) in moderate overall yields. The lithium derivative of enamine **(185c)** when added to benzaldehyde or pentanal gave the corresponding 4,5-disubstituted γ-lactones **189a** and **189b** in 20 and 13% yields, respectively. Ahlbrecht and Farnung[2096] also prepared the lithium derivative of the 1-phosphorylenamine **(185c)** as well as the lithium derivative of **185a** and found that the reagents gave a regio- and stereoselective reaction with alkylating agents (and Me_3SiCl) to yield enamines of the type **190.**

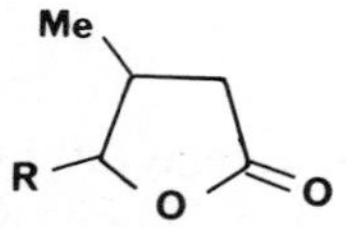

189 a R = Bu

189 b R = Ph

$$RCH_2—CR'{=}C(NMe_2)—PO(OEt)_2$$

190

By hydrolyzing the products of the reaction of *n*-pentyl iodide with **185a** or **185d** to octanoic acid (51%) and 2-methyloctanoic acid (60%), they established that 1-phosphonylenamines can readily be converted to carboxylic acids. Thus the anions are a source of $^-CH_2CH_2COOH$ **(176)** and its 2-methyl analog $^-CH_2—CHMe—COOH$.

Addition of the lithium derivatives of **185a** and **185d** to benzaldehyde and of **185d** to 2,2-dimethyloxirane was also carried out. Multiple alkylations (MeI and Me_3SiCl) of **185d** were performed to yield substituted 1-phosphorylenamines (**191** and **192**). Consequently, this methodology can be used to prepare mono-, di-, tri-, and tetrasubstituted isobutyric acids.

$$(RCH_2)_2C{=}C(NMe_2)—PO(OEt)_2$$

191

$$(RCH_2)(R'R''CH)C{=}C(NMe_2)—PO(OEt)_2$$

192

There has been a considerable number of studies utilizing ketene dithioacetals as β-enolate equivalents of carboxylic acids. Ziegler and Tam[1265] studied the effect of the size of the alkyl groups of the thioacetal and the presence of HMPA on the regiochemistry of alkylation. The authors found that treatment of the lithium derivative of **193a** with CuI/trimethylphosphite yielded

a copper reagent that gave exclusively γ-alkylation. Although the new chain-extended dithioacetals were not hydrolyzed, the authors indicated it would be a simple matter to convert them to esters.

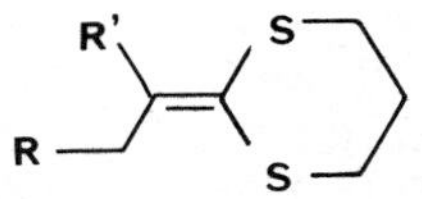

193a R = R′ = H

193b R = H, R′ = Me

193c R = OTHP, R′ = Me

Kozikowski and Chen[502] lithiated ketene dithioacetals **193a–c** with *s*-BuLi or LDA in THF at −78° and reacted the anions with a variety of aldehydes and ketones in fair yields. The γ-addition products were converted in 40–69% yield to γ-lactones in the β-methyl series. Ziegler and coworkers[1273] have reported on the conjugate addition of the lithium or copper derivatives of various dithioacetals **193** to cyclohexenone, cyclopentenone, and 2-methylcyclopentenone. In general, 1,4-addition predominated over 1,2-addition when lithium was the counterion in THF. Use of $CuI/(MeO)_3P$ or HMPA resulted in α-1,4-addition. The dithioacetals were not hydrolyzed, but rather in some cases were ozonized to yield diketoaldehydes that were used in the synthesis of (±)-aromatin and (±)-confertin. The methodology provides a route to the synthons ⁻CHMe—CHMe—COOH and $^-CH_2$—CHMe—COOH, as well as $^-CH_2CH_2COOH$ **(176).** In a following publication,[1272] evidence was provided that the γ-1,4-addition products result from an alkoxy-Cope rearrangement of the first-formed products of α-1,4-addition. Murphy and Wattanasin[1266,1270] lithiated 1,3-dithianes **(194a–b)** and found that the former is alkylated at α only. With **194b,** the γ:α ratio for alkylation correlates with the hardness of both the leaving group and the alkyl group of the alkylating agent, varying from approximately 85:15 in benzylic halides through approximately 30:70 (alkyl halides) to full α-alkylation with dimethyl sulfate and trimethylsilyl chloride. The ketene dithioacetals, products of γ-addition, were not converted to the corresponding acids or esters, but the anion of **194b** nevertheless corresponds to the synthon ⁻CHPh—CH_2COOH.

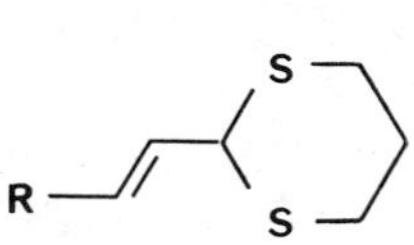

194a R = Me

194b R = Ph

6.5. β-ENOLATES OF ESTERS

As discussed in the introduction, even though the products of reactions of homoenolate equivalents can be formally converted into products of other equivalents—for example, aldehydes to carboxylic acids and carboxylic acids to esters—this review is organized on the basis of equivalent types used in the initial synthon.

Trost's spiroannulation methodology[2091] (see Sect. 6.4.1), which provides a route to γ-lactones by way of spirocyclobutanones, can be diverted to give γ-hydroxyesters by use of alkoxides. Thus, ester β-enolates $^{-}CH_2CH_2COOR$ are formally involved. Nakamura and Kuwajima[2097,2098] prepared 1-alkoxy-1-trimethylsilyloxycyclopropanes **(195)** and established that they add to aliphatic aldehydes in the presence of $TiCl_4$ to yield esters or γ-lactones in good yield in most cases. The authors suggested that the reaction most likely proceeds via the β-titanyl ester.

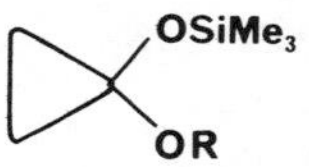

OSiMe₃
OR

195

Chlorination can occur if the intermediates are exposed to the reaction conditions for long periods: for example, benzaldehyde affords the corresponding γ-chloroester **196** in 92% yield. Acetals and ketals can act as electrophiles and give γ-alkoxyesters such as **197.**

$$\mathrm{Ph{-}CHY{-}CH_2CH_2COOEt} \qquad \mathrm{RCH{=}CH{-}CH(OSiMe_3){-}PO(OEt)_2}$$

196, 197 **198 (*E*)**

$$\mathrm{RR'CH{-}CH_2COOR''}$$

199

where for **196,** Y = Cl; **197,** Y = OEt; and for **198,** R = H, alkyl or aryl.

Hata and coworkers[2099] have utilized the products **(198)** of the reaction of α,β-unsaturated aldehydes and diethyl trimethylsilyl phosphite (DTSP) as ester β-enolate synthons. The diethyl 1-trimethylsilyloxyphosphonates **198** were treated with LDA in THF at −78° and reacted with a variety of electrophiles R′X. Treatment of the reaction mixture with TsOH (catalytic amt.) in EtOH or other alcohols (R″OH) gave the esters **199** in good-to-moderate yields.

As an extension of a study of the adducts of organosilanes R_3SiY (Y = CN, PO—R′, SR′) with aldehydes or ketones, Evans and coworkers[2100] prepared and metallated allylic α-silyloxyphosphonamides **(200)** with *n*-BuLi in THF in −65°. The phosphonamides were prepared by the addition of the

phosphorodiamidate $Et_3SiOP(NMe_2)_2$ to the corresponding α,β-unsaturated aldehydes. Addition of alkyl halides (MeI, Bzl—Cl, $C_6H_{13}Br$) or carbonyl compounds (PhCHO, *i*-PrCHO, cyclohexanone) at −78° gave a high γ-regioselectivity (79%) with the bulky alkylating reagents and PhCHO. Hydrolysis of the enol phosphonamides **(201)** with MeONa/MeOH yielded the corresponding esters; the aldehyde adducts gave γ-lactones when treated with tetra-*n*-butylammonium fluoride in THF at 25°. For example, the anion of **200c** yielded **202** (75%) and **203** (71%) when reacted with methyl iodide and benzaldehyde, respectively. Attempts to prepare the binucleophilic synthon $^-CH_2CH^-COR$ by bis-metallating **200a** so that bis-substitution could be accomplished was thwarted by *O*-alkylation. For example, the bis-anion prepared from **200c** with two equivalents of *s*-BuLi in DME at −50°, when reacted with excess MeI followed by methanolysis gave **204** and **205** in a ratio of 17:83. Conversion of the lithium enolate to the tetra-*n*-butylammonium enolate raised the C:O alkylation ratio to 57:43.

RCH═CH—CH($OSiEt_3$)—PO(NMe_2)$_2$

200

RR′CH—CH═C($OSiEt_3$)—PO(NMe_2)$_2$

201 (*E*)

where for **200a,** R = H; **200b,** R = Me (*E*); **200c,** R = Ph (*E*).

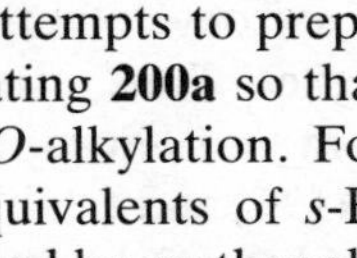

Ph, Ph, O, O

203

Ph—CHMe—CHR—COOMe

202, 204

Ph—CHMe—CH═C(OMe)—PO(NMe_2)$_2$

205 (*E*)

where for **202,** R = H; **204,** R = Me.

Anions derived from α-methoxyallyl phosphine oxides **(206)** (LDA/THF/−70°) have been shown by Maleki and Miller[2101] to react highly regioselectively with electrophiles (MeI, aldehydes) to give adducts of γ-attack **(207)** or α-attack depending on the nature of R and R′. Although the reactions of **207** were not detailed, it should be possible to convert the enol ethers to

the corresponding oxophosphine oxides which, by treatment with alkoxides, should yield the corresponding esters in analogy with phosphonamides.[2100]

$$\text{RCH=CR'—CH(OMe)—POPh}_2 \quad \text{El—CHR—CR'=C(OMe)—POPh}_2$$

206 (*E*) **207**

where R,R′ = H, Me or Ph.

Jacobson and coworkers[2102] have prepared ethoxyethyl-protected cyanohydrins **(208)** and added their lithium derivatives to aldehydes and ketones.

$$\text{RCH=CR'—C(CN)—O—CHMe—OEt}$$

208 (*E*)

where R,R′ = H or Me.

Although addition at −78° is exclusively α, at 0° γ-regioselectivity is observed and the adducts are hydrolyzed routinely in aqueous acid to the γ-lactones. Addition to 2-heptanone or cyclohexanone yielded the γ-lactones **209** (43%) and **210** (60%), respectively. Presumably the corresponding esters can be obtained when alcohols are used as the solvent. The authors pointed out that the methodology is available for converting γ-lactones to cyclopentenones and provided several examples.

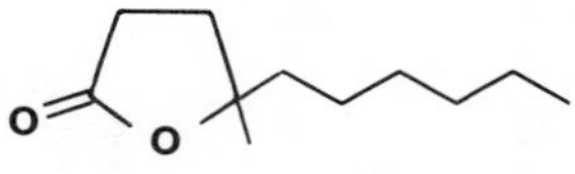

209

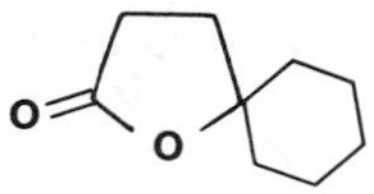

210

Ziegler and Mencel,[2108] continuing a study on the chemistry of anions derived from dithioacetals, lithiated 2-(1-propenyl)-1,3-dithiane (**211,** *E*:*Z* 86:14), reacted the lithium reagent with 2-methyl-2-cyclopentenone and alkylated the resultant enolate with allyl bromide (2.0 equiv./−78° to rt). The ketone **212** was obtained in 46% yield and readily transformed by hydrolysis of the dithioacetal function ($HgCl_2$, 2.4 equiv., 10% aqueous MeOH, reflux) to the ketoester **213a,** further converted into the ketoaldehyde **213b,** a useful

steroid and vitamin D precursor. The anion of dithiane **211** thus corresponds to the synthon $^{-}CHMe—CH_2COOMe$.

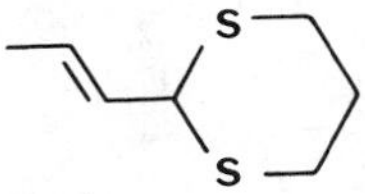

211

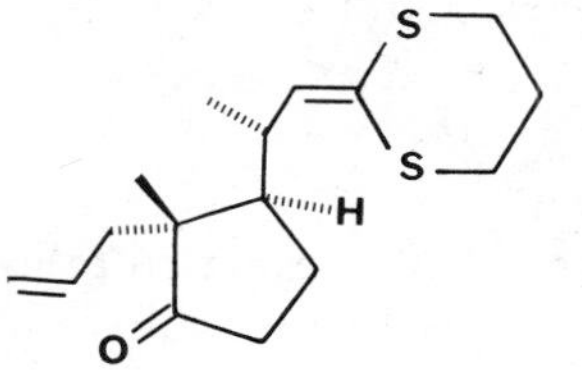

212

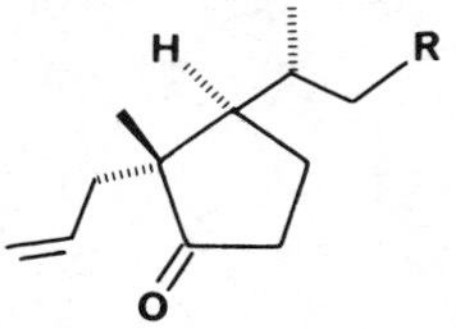

213a R = COOMe

213b R = CHO

The lithiation of acrolein dimethyl and diethyl acetals **214** by *s*-BuLi in THF or THF/Et_2O/pentane at −95° has been carried out. Seyferth and coworkers[2103] reacted **214** with organosilicon or organotin chlorides or allyl bromide and established that γ-addition occurs to give the corresponding ketone acetals. Acid hydrolysis of the adducts affords β-silyl- or β-stannylpropionate esters, or esters of 5-hexenoic acid. The methodology provides a route to the ester homoenolates $^{-}CH_2CH_2COOR$.

$$CH_2{=}CH—CH(OR)_2$$

214

where R = Me or Et.

Stork and coworkers[2088] reacted *m*-methoxybenzaldehyde with ethyl acrylate and triphenylphosphine and obtained ethyl 4-(*m*-methoxyphenyl)-3-butenoate, which was used as the initial intermediate in the synthesis of (±)-lycopodine. Clearly, the phosphonium ylide is an intermediate in the reaction and can be considered as corresponding to the β-enolate $^{-}CH_2CH_2COOEt$.

6.6. β-ENOLATES OF AMIDES

Corey and Cane[2104] prepared and metallated the α,β-ynamine **215** (*t*-BuLi/TMEDA) and reacted the lithium reagent with electrophiles. After isolation of the alkylated (silylated) ynamines, hydrolysis was accomplished by passing the crude mixture over alumina to yield the corresponding amides **(216)**. The ynamine is a source of the amide β-enolate **(217)**.

$$MeC{\equiv}C—NR'_2 \quad RCH_2CH_2CO—NR'_2 \quad {}^{-}CH_2CH_2CO—NR'_2$$

215 **216** **217**

where NR'_2 = 2,2,6,6-tetramethylpiperidinyl.

A report by Goswami and Corcoran[2105] documented the formation of homoenolate dianions of secondary amides by tin–lithium exchange. Treatment of *N*-phenyl- (or *N*-methyl-) 3-(tri-*n*-butylstannyl)propionamide with two equivalents of *n*-BuLi in the presence of DABCO at −78° in THF produced the corresponding dilithio derivate **218,** which reacted with various electrophiles (D_2O, Me_3SiCl, RBr, ketones) to give the appropriate secondary amides **(219).** In a subsequent study,[2106] they reported a novel rearrangement of **218b,** prepared from the stannyl derivative, to **218c.** The latter dilithio derivatives were reacted with same electrophiles to yield amides **(219).** The dilithium compounds are equivalents of amide β-enolates, for example, **218a** corresponds to $^{-}CH_2CH_2COPh$.

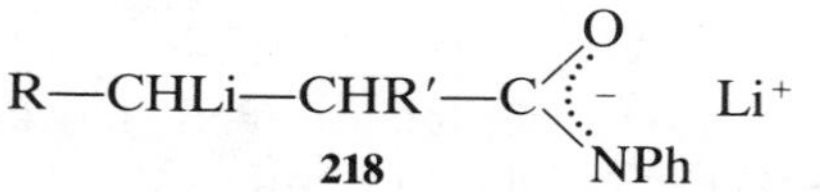

R—CHLi—CHR′—C(O)(NPh)$^{-}$ Li^{+} **218** El—CHR—CHR′—CONHPh **219**

where for **218a,** R = R′ = H; **218b,** R = H, R′ = Ph; **218c,** R = Ph, R′ = H.

6.7. γ- AND FURTHER ENOLATES

Helquist and coworkers[302] prepared the Grignard reagent **220a,** the γ-enolate equivalent $^{-}(CH_2)_3CHO$, and studied its conjugate addition to a number of α,β-unsaturated ketones in the presence of CuBr/Me_2S at −78°. The resultant acetals were hydrolyzed and cyclized in one step to yield the corresponding annulated products **221** and **222.** They also studied the addition of **220a** to hexanoyl chloride and obtained the ketoacetal **223** in 66% yield. Chadha and coworkers[153] used the Grignard reagent **220b,** an equivalent of $^{-}(CH_2)_3COMe$, in the synthesis of (±)-frontalin **(224).** Acetylation by acetic anhydride at −70° gave the ketone **225** (71%), which was converted to frontalin by Wittig methylenation, epoxidation, and acid-catalyzed cyclization.

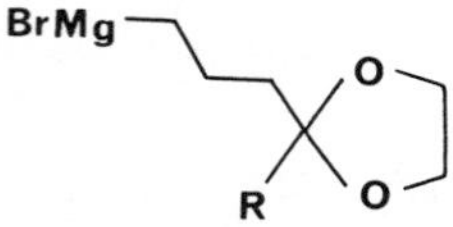

BrMg O O R

220 a R = H

220 b R = Me

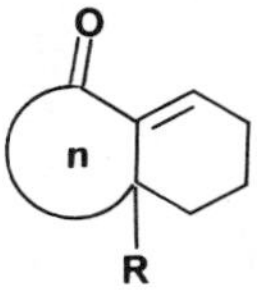

O n R

221

R = H or Me

n = 5 - 7

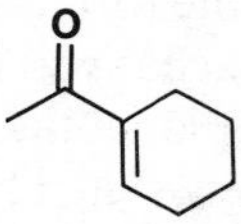

222 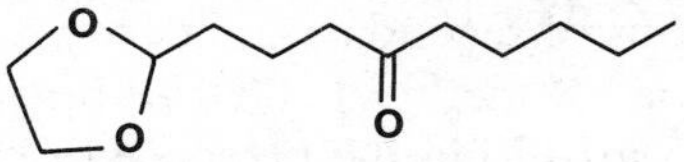 223

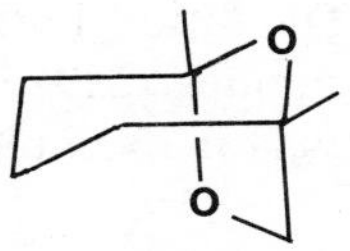

224 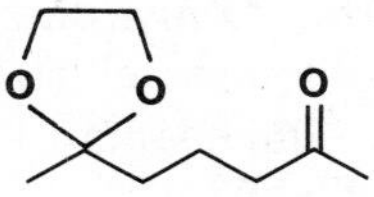 225

Presumably, there is no reason why the basic principle, termed "blocked-carbonyl defensive strategy" by Hoppe,[2310] exploited in developing the β- and γ-enolate anion synthons from the corresponding acetal Grignard reagents (**2a,** and **220a,** respectively) could not be carried further, to furnish synthons with any required distance between the carbonyl and the anionic center. Thus, for example, the synthon $^{-}(CH_2)_8CHO$ is available in the form of the corresponding dimethyl acetal Grignard reagent.[212]

TABULAR SURVEY OF UMPOLED SYNTHONS

INTRODUCTION

The enumeration of the tables follows that of the chapters. Within each group, synthon sources are given first, followed by coded expressions for reaction partners, yields, reaction and unmasking conditions, and references.

Synthon and Synthon Source Formulas. These are listed in order of increasing complexity. Entries having "missing" substituents indicate generality and appear after the fully defined items. Formulas are written in full only when generality is lacking, or was not studied. Thus,

$^{+}$CHMe—CHO means propionaldehyde α-cation.

$^{+}$CHR—CHO means any straight-chain aldehyde α-cation, including propionaldehyde but excluding acetaldehyde.

$^{+}$CH—CHO means any straight-chain or α-branched aldehyde α-cation.

$^{+}$C—CHO means $^{+}CH_2CHO$, $^{+}$CHR—CHO, or $^{+}$CRR′—CHO.

Reaction Partners. A numbered list of electrophiles (for all synthon classes except the carbonyl α-cations) is given below, along with synthon equivalents for each particular electrophilic reagent. A corresponding list of nucleophiles precedes the carbonyl α-cation tables (5.1.–5.6.). A yield indication for the overall reaction, including any demasking steps, is given in parentheses following each electrophile (or nucleophile) symbol:

(A) >80% yield
(B) 60–80% yield
(C) 40–60% yield
(D) 20–40% yield
(E) <20% yield

However, whenever a reaction sequence was not carried to full synthon expression (for example, a dithiane alkylation product was left unhydrolyzed), and the overall yield is therefore unknown, the yield indication carries the

symbol "?". As an example, any particular ^{-}CHO source shown with the entry 4(B) should be understood to furnish C_{n+1} alkanals in 60–80% overall yield from primary C_n alkyl chlorides. On the other hand, 4(B?) means that the alkyl chlorides react in 60–80% yield to give alkylated dithianes or other derivatives as the case may be, but these were not taken to the homologous aldehydes.

Electrophiles that have two potential but unlike sites of reaction are treated as *3*,4(A), for example, indicating that a 1-bromo-ω-chloroalkane will react with the synthon source shown in >80% yield at the —CH_2Br terminus.

Bidentate Synthons. For synthons having both an anionic and cationic center, the nucleophile(s) are given in the "reaction conditions" column, and the electrophiles as usual in the reaction partner column. Synthons having two anionic centers are labeled at these sites and α and β; corresponding characters are shown in the reaction partner column, separated by a semicolon. Yield indications are for the entire reaction sequence.

Reaction Conditions. Any reagents following a semicolon refer to functional group unmasking procedures (for example, ;Hg^{2+}/H_2O for dithiane reaction products).

At the foot of each section, additional references to recent new synthons are noted.

TABLE 1. List of Electrophiles (Y = Leaving Group)

Number	Reagent	Equivalency
1	MeI	Me^+
2	RCH_2I	RCH_2^+
3	RCH_2Br	RCH_2^+
4	RCH_2Cl	RCH_2^+
5	RCH_2OMs	RCH_2^+
6	RCH_2OTs	RCH_2^+
7	Me_2SO_4, R_2SO_4	Me^+, R^+
8	Other RCH_2Y	RCH_2^+
9	R_2CHI	R_2CH^+
10	R_2CHBr	R_2CH^+
11	R_2CHCl	R_2CH^+
12	R_2CHOMs	R_2CH^+
13	R_2CHOTs	R_2CH^+
14	Other R_2CHY	R_2CH^+
15	$R_3C—Y$	R_3C^+
16	$Y(—C—)_nY$	$^+(—C—)_n^+$
17	$CH_2{=}CHCH_2X$	$CH_2{=}CHCH_2^+$
18	C=C—C—X (S_N2)	$C{=}C—C^+$
19	C=C—C—X (S_N2')	$^+C—C{=}C$
20	Other C=C—C—Y (S_N2)	$C{=}C—C^+$
21	Other C=C—C—Y (S_N2')	$^+C—C{=}C$
22	C≡C—C—X	$C{\equiv}C—C^+$
23	$PhCH_2X$	$PhCH_2^+$
24	Ar—C—X	$Ar—C^+$
25	$ROCH_2X$	$RO—CH_2^+$
26	Other RO—C—Y	$RO—C^+$
27	RS—C—Y	$RS—C^+$
28	C=C—X	$C{=}C^+$
29	ArX	Ar^+
30	Epoxide	$HO—C—C^+$
31	Oxetane	$HO—C—C—C^+$
32	RCHO	$HO—CH^+—R$
33	C=C—CHO (1,2-)	$HO—CH^+—C{=}C$
34	C=C—CHO (1,4-)	$^+C—CH—CHO$
35	ArCHO	$HO—CH^+—Ar$
36	Me_2CO	$HO—C^+Me_2$
37	R_2CO	$HO—C^+R_2$
38	C=C—COR (1,2-)	$HO—CR^+—C{=}C$
39	C=C—COR (1,4-)	$^+C—CH—CO—R$
40	C=C=O	$O{=}C^+—CH$
41	ArCOR	$HO—CR^+—Ar$
42	Ar_2CO	$HO—C^+Ar_2$
43	$CH(OR')_2$	$RO—CH^+$
44	$R_2C(OR')_2$	$R'O—C^+R_2$

TABLE 1. (*Continued*)

Number	Reagent	Equivalency
45	$R_2C(SR')_2$	$R'S—C^+R_2$
46	C=NR	$RNH—C^+$
47	$C=N^+R_2$	$R_2N—C^+$
48	$R_2C=N—NHCOOR'$	$^+CR_2—NHNHCOOMe$
49	$R_2C=N—NHTs$	$^+CR_2—NHNHTs$
50	CO_2	^+COOH
51	X—COOR	^+COOR
52	RCOX	$^+CO—R$
53	RCOCN	$^+CO—R$
54	ArCOX	$^+CO—Ar$
55	$(RCO)_2O$	$^+CO—R$
56	$(RO)_2CO$	^+COOR
57	$C(OR)_3$	$^+C(OR)_2$
58	RCOOMe	$^+CO—R$
59	RCOOR′	$^+CO—R$
60	Lactones	$^+CO(—C—)_nOH$
61	C=C—COOR (1,2-)	$^+CO—C=C$
62	C=C—COOR (1,4-)	$^+C—CH—COOR$
63	C≡C—COOR (1,4-)	$^+C=CH—COOR$
64	ArCOOR	$^+CO—Ar$
65	RNH—COOR′	$^+CO—NHR$
66	$Cl—CONR_2$	$^+CONR_2$
67	RN=C=O	$^+CO—NHR$
68	RN=C=S	$^+CS—NHR$
69	DMF	^+CHO
70	$RCO—NR'_2$	$^+CO—R$
71	$ArCONR_2$	$^+CO—Ar$
72	RCN	$^+CO—R$
73	C=C—CN (1,2-)	$^+CO—C=C$
74	C=C—CN (1,4-)	$^+C—CH—CN$
75	ArCN	$^+CO—Ar$
76	ArC≡N → O	$^+CAr=N—OH$
77	H_2O, D_2O	H^+, D^+
78	$B(OR)_3$	$^+B(OR)_2$
79	ArN=NAr	$^+NAr—NH—Ar$
80	$R—NO_2$	$^+NH—R$
81	$Ar—NO_2$	$^+NH—Ar$
82	O_2	$^+O—OH$
83	$R'CO_3R$	$^+O—R$
84	S_8	^+SH
85	RS—X	$^+S—R$
86	R_2S_2	$^+S—R$
87	Ar_2S_2	$^+S—Ar$
88	RSe—X	$^+Se—R$
89	ArSe—X	$^+Se—Ar$
90	Ar_2Se_2	$^+Se—Ar$

TABLE 1. (*Continued*)

Number	Reagent	Equivalency
91	Me_3SiCl	$^+SiMe_3$
92	R_3SiX	$^+SiR_3$
93	R_3SnX	$^+SnR_3$
94	Other metal-X	$^+$Metal
95	Br_2	^+Br
96	$BrCH_2CH_2Br$	^+Br
97	I_2	^+I
98	$CX_3—CX_3$	^+X
99	Other electrophiles	

TABLE 2.1. $^{-}$CHO

Source	Electrophiles (Yields)	Reaction Conditions	Reference
1,3-Dithiane	1(A),2(A),3(A),3,4(A), 4(B),6(C),18(A),29(A), 30(B),33[a],34[a],35(A), 37(B),38[a],39[a],42(A), 46(D),50(D),59(E),65(A), 69(A),70(E),71(B),75(D), 91(B),92(B),94(C)	BuLi/THF; Hg(II)/H_2O	422,424,426, 427,449,452, 453,455–458, 472,475, 476,479,483
1,3,5-Trithiane	1(A),3(A),23(A),42(A), 51(D),70(E),75(A)	BuLi/THF; Hg(II)/H_2O	421,449, 488
See text, Formula **7**	1(B),2(A),3(A),4(C), 5(E),6(E),9(A),10(C), 32(A),35(A),41(A)	BuLi/THF; Hg(II)/H_2O	486,487
See text, Formula **21**	3(B),17(A),18(B), 23(B),30(C)	BuLi/THF;	668,669
EtS—CH_2—SEt	3(B),10(C)	$^{-}NH_2/NH_3$;	632
PhS—CH_2—SPh	1(B),2(A),3(C),30(B), 32(A),35(C),37(B),38[a], 39[a],42(C),77(A),78(B)	$^{-}NH_2/NH_3$ or BuLi/ THF; Hg(II)/H_2O	469,483,632, 648,650–651
MeS—CH_2—S—$CSNMe_2$	2(B),23(B)	BuLi/THF;Hg(II)/MeOH	670
CH_2(S—CS—NMe_2)$_2$	1(A),2(A),19(A)	BuLi/THF;Hg(II)/H_2O	671
MeS—CH_2—SO—Me	1(A),2(B),23(A),24(D), 29(C),35(?),38(C),39(E), 42(B),59(A),64(A),72[b], 75[b]	NaH or BuLi;H_3O^+	703,707–711, 716

EtS—CH_2—SO—Et	2(A),3(A),6(A),10(A),17(A),32(A),37(A),38(A),52(A),54(A),61(A),64(A)	BuLi or LDA/THF;H_3O^+	723–725
1,3-Dithiane-monosulfoxide	1(A),23(D),35(C),37(C),42(A),64(C),77(A)	BuLi or LDA/THF;H_3O^+	726,727
p-TolSCH_2—SO—Tol-*p*	35(A),54(A),59(?)	BuLi/THF;H_3O^+	730,733,734
MeSCH_2—SO_2—Tol-*p*	3(A),23(B)	$^-$OH/PTC;$h\nu$/H_2O	736
1,3-Oxathiane	1(A),2(A),3(D),9(B),15(E),23(D),33(B),38(C),41(C),42(B),72(C)	*s*BuLi/THF; Hg(II)/H_2O	744, 745
See text, Formula **34**	32(A)	BuLi	748
PhS—CH_2OMe	1(A),2(A),3(?),17(B),23(A),30(E),32(B),37(A),54(E),59(C),69(C),70(A),71(A),75(B),91(A)	BuLi/THF; Hg(II)/H_2O	752–754
See text, Formula **39**	1(A),2(C),17(A),23(A),35(B),37[b],41(B),91(A),	BuLi/THF;	784
PhSO_2—CH_2O—CHMe—OEt	2(B),3(C),37(B)	LDA or KDA/THF,HMPA; H_3O^+;HO^-	785
PhS—CH_2SiMe_3	1(A),2(A),3(B),9(C),10(E),23(A),30(B),32[c],35[c],36[c],37[c],41[c],42[c],45[d],46[d],52[b],54[b],55[b],58[b],59[b],64[b],85(B),87(B),91(A)	BuLi/THF; MCPBA/Δ/H_3O^+	765,766, 770
PhSe—CH_2SiMe_3	2(A),3(A),10(D),30(A)	LDA/THF;H_2O_2	839,840
PhTe—CH_2—TePh	2(B)	I_2	1203

TABLE 2.1. (*Continued*)

Source	Electrophiles (Yields)	Reaction Conditions	Reference
Benzothiazole	35(A),39(C),60(B)	BuLi/THF or CsF;H_3O^+	794
Me_2N—CS—S—CH_2CN	1(A),2(A),3(A), 17(A),23(A)	NaOH/Bu_4NI;NBS or NaOH/H_2O	802
TosMIC	1(A),2(A),3(B),9(B),11 (C),17(B),23(A),36(B), 37(B),41(C)	NaOH/Bu_4NOH; H_3O^+	803,811
RS—CH_2—COOH	1(A),2(A),3(A),9(B),10 (A),23(B),32(A),35(A), 36(A),37(A),38(B)	LDA/THF/HMPA; NCS or e^-	815,816
PhCH═N—CH_2COOEt	2(A),3(B),9(A)	LDA/THF/HMPA;$NaIO_4$	853
See text, Formula **59**	1(B),2(B),*3*,4(C), 4(C),23(D)	NaH;$NaBH_4$;H_3O^+	856
Ts—CH_2—py^+	99^e(D)	Et_3N;HBr	855
$(EtO)_2CH$—$SnBu_3$	23(B),35^b,39(C)	BuLi/THF;H_3O^+	1037
CO	18(B),28(A),29(B)	Pd/P cat,Bu_3SnH/THF	1163
$Fe(CO)_5$	2(A),3(B),10(C), 30(C),77(B)	KOH or NaHg/THF/Ph_3P; AcOH	1156,1157, 1161,1162
$CH_2[B(OR)_2]_2$	2(A),*2*,4(A), 3(B),6(B),23(B)	LTMP/THF/TMEDA $NaBO_4/H_2O$	1204
$(EtO)_2PO$—CH_2—$PO(OEt)_2$	32(B),35(B),37(A), 41(B)	NaH;$KMnO_4$; $NaHCO_3/H_2O$/MeOH	1206

For footnotes, see Table 2.11.

TABLE 2.2. $MeCO^-$

Source	Electrophiles (Yields)	Reaction Conditions	Reference
2-Me-1,3-dithiane-monosulfoxide	1(B)	LDA/THF;H_3O^+	727
$CH_2{=}CH{-}OMe$	2(A),18(B),32(B),35(B), 37(A),38(B),59[d],75(B)	*t*BuLi/THF; H_3O^+	1052,1054
$CH_2{=}CH{-}OEt$	28[d],29[d],38(C),39[d]	*t*BuLi/THF;H_3O^+	267, 1056–1058
$CH_2{=}CH{-}SeR'$	1(A),3(A),30(A),32(C), 35(B),36(B),37(B),38(B), 41(B),50(A),51(C),69(B), 77(A),86(B),90(A),91(B)	LDA or KDA/THF; Hg(II)/H_2O	1146,1148 1150

For footnotes, see Table 2.11.

TABLE 2.3. RCO^-: R = Nonfunctional Alkyl Only

Source	Electrophiles (Yields)	Reaction Conditions	Reference
R—CH(SEt)$_2$	3(C)	$^-NH_2/NH_3$;Hg(II)/H_2O	632
R—CS—SEt	1(A),32(C),35(C),36(B), 37(C),50(A),52(B),77(A)	EtMgI;Hg(II)/H_2O	639,673
R—CH(SPh)$_2$	3(B)	$^-NH_2/NH_3$;Hg(II)/H_2O	632,647
See text, 3-R-**(21)**	3(A),17(B),23(C), 30(D)	BuLi/THF or hexane; Hg(II)/H_2O	668,669
See text, Formula **24**	1(A),2(A),30(A),32(A), 35(B),37(B),38(A)	BuLi/THF; Hg(II)/H_2O	667
Me—SO—CH$_2$—SMe	1(A),2(C),3(C),17(A)	NaH or BuLi;H_3O^+	718
Et—SO—CHR—SEt	2(A),3(A),17(A),32(B), 34(A),39(B),52(A), 54(A),62(A)	BuLi or LDA/THF; H_3O^+	723–725
PhSO$_2$—CHR—OCHMe—OEt	2(C),3(C)	LDA/THF/HMPA;H_3O^+;HO^-	785
MeS—CHR—Ts	1(A),3(A),23(A)	NaH/DMF	736
Me$_2$N—CS—S—CHR—CN	3(A),23(A)	HO^-/Bu_4NI;HO^-/H_2O	802
Ts—CHR—NC	2(B),3(B),9(C),17(C), 23(B),86(C),91(C)	NaH/THF or DMSO; H_3O^+	804,807 808
RS′—CHR—COOH	2(B),3(A),60[d]	LDA/THF/HMPA;NCS	816–818
R′Se—CHR—SiMe$_3$	2(A),3(A),32(C), 37(D),91(C)	LDA or BuLi/THF; H_2O_2	842,843

RCH_2—NO_2	35(B),36(C),37(C),39(B),42(C),55(B),58(C),62(B),64(C)	BuLi/THF/HMPA; $TiCl_3$	904–906, 923
PhCH=N—CHR—COOEt	2(A),3(B),39(A),62(A)	LDA/THF/HMPA;LAH/HIO_4	853
*t*BuNH—N=CHR	1(E),2[f],3(B),23(B),32(B),35(C),36(C),37(C)	BuLi/THF;H_3O^+	927
R′—NC	1(A),2(A),9(E),30(A),35(A),50(C),51(B),77(A),91(C)	RLi or RMgX;H_3O^+	930–931
EtO—CHEt—O—CHR—CN	3(A),10(C),17(B),39(C)	LDA/THF/HMPA;H_3O^+	938,964,981
See text, Formula **68**	1(D),2(D),15(D),17(B),2(B),23(A),24(B),62(B),63(B),74(B)	R''_3N;(1)N_2H_4,(2)NO^+, (3)$NaHCO_3$, (4)H_3O^+	1026–1028
MeS—CHR—COOH	2(C),23(B)	LDA/THF/HMPA;NCS	816,1043
Me_3SiO—CR=CR—$OSiMe_3$	2(A),3(B),9(B),23(A)	MeLi/DME;$NaBH_4$/PbTA	1046,1047
RCH=CH—SPh	1(B),3(B),17(A),23(C),27[d],29[d],30(C),32(C),33(C),35(C),37(C),38(C),41(B),50(C),54(C),55(E),69(E),77(B),86(B),87(C),89(A),91(A),93(A)	*s*BuLi or LDA/THF/HMPA; Hg(II)/H_2O	1062,1073, 1075–1081
RCH_2—$C(NR'_2)$=$NNHSO_2Ar$	2(A),16(D)	*t*BuLi/THF;H_2O	1155
CO	32(B),37(B),41(C),59(C),60(?),67(B),68(C),91(D)	RLi/THF/Et_2O/pentane	1170–1173, 1175
$[RFe(CO)_4]^-$	1(B),2(B),*3*,4(A),13(C),23(C),29(D),52(D),54(D)	THF	1178–1180, 1184

TABLE 2.3. (*Continued*)

Source	Electrophiles (Yields)	Reaction Conditions	Reference
$[RCO—Ni(CO)_3]^-$	39(C)	Et_2O	1188
RCOCl	23(C),24(C)	Ni/DME/Δ	1191
RCHO	35(D)	$Pd(PPh_3)_4$/THF	1197
RCOX	37(C),52(C),77(D)	Be	1200
$RCH[B(OR')_2]_2$	1(B),2(B),35(A),58(B)	LTMP/THF/TMEDA	1204
$RCH(OSiMe_3)(PO(OEt)_2)$	1(A),2(B),3(E),9(A),17 (A),18(B),23(A),32(A), 35(A),37(A),41(A),42(A), 86(B),87(A),91(B)	LDA/THF;H_3O^+	1222, 1224–1226
$RCH(OCH_2CH_2SiMe_3)(PO(OEt)_2)$	4(B?),18(C?),51(C?)	*s*BuLi/THF	1227
$R_3Si—C{=}C{<}^g$	1(A),2(A),6(B),9(B), 17(B),23(B),30(C),32(B)	Mg or R'Li/THF; RCO_3H/H_3O^+	1079,1113, 1114,1118

For footnotes, see Table 2.11.

TABLE 2.4. RCO^-: R = Alkyl or Aryl

Source	Electrophiles (Yields)	Reaction Conditions	Reference
2-R-1,3-dithiane	1(A),2(A),2,4(A),3(A), 3,4(B),4(C),6(A),9(C), 18(B),23(A),30(B),31(A), 33[d],34[d],37(B),38[d],39[d], 42(B),46(?),50(B),51(B), 52(C),58(A),59(B),60(C), 61[d],62[d],69(C),75(A),76 (C),77(A),81[d],86(A), 91(B),92(C),94(C)	BuLi/THF; Hg(II)/H_2O	421,422,427, 475,476, 518,521–527, 548,552,559, 565,566, 574–576
R—CH(SPh)$_2$	1(A),2(A),3(B),17(A),23 (C),32(C),36(B),37(B), 41(A),51(D),54(B)	BuLi/THF or TMEDA; Hg(II)/H_2O	646,684
PhS—CR(Li)—SiMe$_3$	1(A),2(B),3(B),10(E)	TMEDA or HMPA	768,782,783
R′Se—CHR—SeR′	3(B),30(A),32(C),35(A), 37(C),91(A)	LDA,LTMP or BuLi/THF; Hg(II)/H_2O	844
R′$_2$N—CHR—CN	2(A),3(A),4(B),6(?), 7(A),17(A),23(B),30(C), 32(C),37(C),39(B)	LDA/HMPA/THF; H_3O^+	867,886–889
RCHO	34(D),35(D),39(D), 62(D),74(D)	Thiazolium salt cat.	1001,1005, 1011
R—CO—CHOH—R	2(C),17(A),18(C), 23(C),24(B)	NaOH/DMSO;H_2O_2/HO^-	1229,1231

For footnotes, see Table 2.11.

TABLE 2.5. $ArCO^-$

Source	Electrophiles (Yields)	Reaction Conditions	Reference
PhS—CHPh—$SiMe_3$	1(A),2(A),3(B),9(D), 10(E),17(B),23(B)	BuLi/TMEDA; MCPBA/Δ/H_3O^+	781
RO—CHAr—CN	3(A),29(D),30(B), 32(C),39(?),76(B)	NaH/THF; H_3O^+	458,940–944, 947
Me_3OSi—CHAr—CN	1(A),2(A),3(B),*3*,4 (A),4(A),6(A),7(B),9(B), 10(C),13(E),15(A),17(A), 23(C),32(A),36(B),37(B), 38^d,39^d,41(A),42(A), 62(C)	LDA/THF; H_3O^+	953–955, 960–962, 965
R_2N—CHAr—CN	1(A),9(A),24(C),29(A), 51(A),62(C),74(B),77(A)	NaH, $^-NH_2$ or LDA/THF; H_3O^+	866,867,874, 877,879–881
ArCH=N—CH_2Ar'	1(A),9(A),10(A), 23(A),51(A),91(A)	BuLi or LDA/THF; H_3O^+	926
Ar—CHO	34(D),39(C),62(C), 74(D)	^-CN/DMSO or DMF	995,996, 998,999
2-Ar-1,3-dioxolane	1(A),2(A),17(A), 32(A),54(B)	BuLi/THF;H_3O^+	1039
Ar—CO—$SiMe_3$	1(C),2(D),17(D),23(A), 35(C),36(C),37(C),77(B)	KF/DMSO or HMPA	1049,1050

For footnotes, see Table 2.11.

TABLE 2.6. $C{=}C{-}CO^-$ **and** $C{\equiv}C{-}CO^-$

Equivalence	Reagent	Electrophiles (Yields)	Reaction Conditions	Reference
$CH_2{=}CH{-}CO^-$	$PhSe{-}CH_2C{\equiv}CH$	1(C),32(B),36(C), 77(B),91(D)	LDA/THF; MCPBA	1236,1237
	$CH_2{=}C{=}CH{-}N(Me){-}PO(OEt)_2$	1(C),23(B)	BuLi/THF; H_3O^+	1262
$RCH{=}CH{-}CO^-$	$RCH{=}C{=}CH{-}OMe$	2(C),3(C),7(A),9(E), 16(C),23(C),28(D),29(C), 32(A),33(E),36(B),37(A), 38(A)	BuLi/THF; H_3O^+	1061, 1240–1243, 1244,1246
	$RCH{=}C{=}CH{-}SR'$	3(B),10(B),36(B),37(B)	$^-NH_2/NH_3$ or BuLi/THF;Hg(II)	1251,1252
	$RCH{=}C(SR')_2$ Dithiane	1(A),2(A),2,4(B),9(A), 11(A),17(A),23(A),39[d], 77(A),86(α:E)	BuLi/THF; Hg(II)/H_2O	499,500, 1264,1266, 1270,1271
	$RCH_2CH{=}C(SPh)(SiMe_3)$	9(C?)	BuLi/THF	1281
	$RCH{=}CH{-}CH(CN)(OSiMe_3)$	1(A),3(B),6(B),9(B),10 (B),17(B),32(C),37(B), 38(C),41(B)	LDA/THF; H_3O^+	965, 1288–1291

TABLE 2.6. (*Continued*)

Source		Electrophiles (Yields)	Reaction Conditions	Reference
$CH_2{=}CR{-}CO^-$	$CH_2{=}CR{-}CH(SPh)(OMe)$	3(B),77(A)	LDA/THF; Hg(II)/H_2O	1282
$HC{\equiv}C{-}CO^-$	$Me_3SiC{\equiv}C{-}CH$	1(B),2(A),3(B),10(D), 15(E),17(B),23(A),30(E), 77(A),91(B),93(A),94(C)	BuLi/THF; H_3O^+/Me_2CO	1234,1235

For footnotes, see Table 2.11.

TABLE 2.7. R-Functional RCO^-: Carbonyl Group on R

Equivalence	Reagent	Electrophiles (Yields)	Reaction Conditions	Reference
$OHC—CO^-$	$(EtO)_2CH—CHO$	39(B),62(D)	Thiazolium salt Et_3N/dioxan	1309,1310
$RCOCO^-$	2-RCO-1,3-Dithiane	1(?),17(C),23(?)	NaH/DMF; Hg(II)/H_2O	448,1308
$OHC—CH_2CO^-$	$MeS—C{\equiv}CCH_2—OMe$	1(B),2(A),3(A),4(C), 36(B),42(B)	LDA/THF; Hg(II)/H_2O	1258
$RCO—CHR'—CO^-$	$RCO—CR'{=}CHSPr$-*i*	62(C)	LTMP/THF;Hg(II)	1313
$OHC—CH_2CH_2CO^-$	Furan	3(B),30(B),32(A),35(A), 37(A),41(A),42(A),50(B), 54(C),59(C),75(A)	BuLi/THF; H_3O^+	1321,1322, 1328

TABLE 2.8. R-Functional RCO^-: COOH or Derivative in R

Equivalence	Reagent	Electrophiles (Yields)	Reaction Conditions	Reference
HOCO—CO^-	$(EtS)_2CH$—COOH	1(A),2(A),3(A),6(A), 9(C),10(C),13(B),17(A), 30(A)	$KN(SiMe_3)_2$;?	1351,1352
	$TsCH_2CF_3$	32(B),35(B),36(B),37(B)	LDA/THF;H_3O^+	1364
EtOCO—CO^-	2-EtOCO-1,3-dithiane	3(A),*3*,4(A),10(B),14 (B),17(B),18(C),23(A), 33^d,34^d,35(A),41(A), 77(B)	NaH/DMF or K_2CO_3/aliquat;	1340, 1342,1344
ROCO—CO^-	$(EtS)_2CH$—COOR	33(E),39(A),62(A), 74(A),77(A)	NaH/DME;NBS	1343,1355, 1356
	$(PhS)_2CH$—COOR	33(E),39(A),62(A),74(B)	NaH/DME;NBS	1357
	2-ROCO-1,3-dithiolane	28(A),39(A),62(A)	NaH/DME;NBS	1353,1359
	$(R'O)_2CH$—COOR	1(C),2(C),3(C),9(D),17 (C),23(B),35(A),37(B), 38(A),41(A),62(A)	LDA/THF;H_3O^+	1354,1360, 1361

	$[(Me_3SiO)(EtO)C{=}C(OEt)(OSiMe_3)]$	15(C),32(A),35(B), 39(B),41(A)	$ZnCl_2/CH_2Cl_2$; H_3O^+	1362
$NC{-}CO^-$	2-cyano-1,3-dithiane	1(A),3(A),9(A), 23(A),30(A)	BuLi/THF;?	1349,1350
$HOCO{-}CH_2{-}CO^-$	$HOCO{-}C{\equiv}CH$	30(D)	LDA/THF; $MeOH/H^+$	1367
$MeOCO{-}CHMe{-}CO^-$	$MeOCO{-}CMe{=}CHSPh$	32(D),33(E),35(D), 52(E),54(D),62(D)	LDA/THF; $Hg(II)/H_2O$	1369
$ROCO{-}CH_2{-}CO^-$	$MeSC{\equiv}CCH(OEt)_2$	1(A),2(A),9(A), 32(B),35(A),36(A)	$LiNEt_2/THF$; $Hg(II)/H_2O$	1259
$HOCO(CH_2)_3CO^-$	1,2-CX-dione	17(C),18(C),22(C)	BuLi/LDA/THF; Al_2O_3;O_2/h	1371

For footnotes, see Table 2.11.

TABLE 2.9. R-Functional RCO^-: Hydroxy or Alkoxy Function in R

Equivalence	Reagent	Electrophiles (Yields)	Reaction Conditions	Reference
$ROCH_2CO^-$	$ROCH_2CHO$	39(C)	Thiazolium salt	1376
	RO—CH=CBr—$SiMe_3$	32(A),36(C),37(C)	*s*BuLi/THF;?	191
HO—C—CH_2CO^-	RO—C—$CH_2CSSR'^h$	19(A),33(A),50(B), 70(B),77(A)	EtMgBr/THF; Cu(II);H_2O	639,640
RO—CH_2CH_2—CO^-	$ROCH_2CH$=C(OCH_2OMe)($SnBu_3$)	32(B)	BuLi/THF; H_3O^+	1373
$HO(CH_2)_3CO^-$	2,3-H_2furan	33(B),36(B)	*t*BuLi/THF;H_3O^+	1066,1067
$HO(CH_2)_4CO^-$	2,3-H_2-4*H*-pyran	2(C),18(C),30(B),32(D),	*t*BuLi/THF;H_3O^+	1066,1067

TABLE 2.10. Bidentate Acyl Anion Synthons: An Additional Cationic Site

Equivalence	Reagent	Electrophiles (Yields)	Reaction Conditions	Reference
$^{+}CH_2CO^{-}$	2-CH_2=dithiane	1(A),2(B),77(A)	RLi;Hg(II)/H_2O	534,535
	CH_2=C(SMe)—SOMe	1(D),77(A)	RLi;H_3O^+	534,716,719
	CH_2=C(SPh)—$SiMe_3$	1(A),2(B),3(B),10(E)	RLi/Et_2O/TMEDA; MCPBA/Δ/H_2O	782
	CH_2=C(CN)—NMePh	1(A),2(B),23(B),77(B)	RLi;H_3O^+	888
$^{+}$CO—CHMe—CO^{-}	MeOCO—CMe=CHSPh	32(D),35(E),52(E), 54(D),62(D)	LDA/THF −78°; Hg(II)/H_2O	1369
$^{+}COCH_2CH_2CO^{-}$	1,2-$(Me_3SiO)_2$-cyclobutene	32(B),37(B)	El^+,Nu^-; KIO_4	1339
$^{+}$CHR—CH=CH—CO^{-}	2-RCH=CH—CH=dithiane	1(B),77(B)	RLi/THF; Hg(II)/H_2O	534

TABLE 2.11. Bidentate Acyl Anion Synthons: An Additional Anionic Site

Equivalence	Reagent	Electrophiles (Yields)	Reaction Conditions	References
$^{-}CH_2^{\alpha}CO^{-\beta}$	Me—CS—SEt	α32;β33,50,77;(B-C)	LDA/THF −78°; $EtOCH{=}CH_2$;EtMgI; Hg(II)/H_2O	639
$^{-}CH^{\beta}{=}CH{-}CO^{-\alpha}$	$CH_2{=}C{=}CH{-}N(Me)P(O)(OEt)_2$	α1;β23;(B)	BuLi;BuLi;H_3O^+	1262
	$CH_2{=}C{=}CH{-}OMe$	α2,91;β2,50,86,91;(C)	BuLi;BuLi;H_3O^+	1248,1379
$^{-}CH^{\beta}{=}CR{-}CO^{-\alpha}$	$CH_2{=}CR{-}CH(SPh)(OMe)$	α3,77;β2-4,77;(C)	LDA/THF;SiO_2; BuLi/TMEDA;$NaIO_4$	1282
$^{-}CR^{\alpha}{=}CH{-}CO^{-\beta}$	$RC{\equiv}C{-}CH_2OMe$	α1,3,7,91;β1,2,3,7,77,91;(A-C)	2BuLi;H_3O^+	1244,1245
$^{-}C_{\alpha}(SiMe_3){-}CH{-}CO^{-}$ β	$Me_3SiC{\equiv}CCH_2OBu\text{-}t$	α1,91;β1,36,91;(C)	BuLi;H_3O^+	1246
	$MeO{-}CH{=}CHCH(SPh)(SiMe_3)$	α2,3,17;β3,4;(C)	BuLi/TMEDA/HMPA/THF; $NaIO_4/H_2O$/dioxan; SiO_2;BuLi/HMPA/THF	1283

$^-\overset{\alpha}{C}(SMe){=}\overset{\beta}{C}H{-}CO^-$	$MeS{-}C{\equiv}C{-}CH_2OMe$	α1-4;β36,42;(C-D)	LDA;$LiNEt_2$	1258
$^-\overset{\alpha}{C}H{=}\overset{\beta}{C}(SePh){-}CO^-$	$PhSeCH_2{-}C{\equiv}CH$	α1,3,9,23,77; β1,32,36,77,91;(B-D)	2LDA;MCPBA	1237
$^-\overset{\alpha}{C}H{=}CH{-}CH{=}\overset{\beta}{C}HCO^-$	$CH_2{=}CHCH_2$ / $RSCS{-}CH_2$ (RSCS bonded to the CH_2 carbon)	α42;β1,2;(A-B)	KH,*s*BuLi/THF; $CaCO_3$/MeI/H_2O	1380

[a]Ratio varies depending on solvent.
[b]Aldehyde derivatives did not result from this condensation.
[c]Gave α-phenylthioketone.
[d]See text.
[e]Maleimides.
[f]Lower alkyl halides gave *N*-alkylation.
[g]Equivalent to >CH—CO^-.
[h]R = 1-ethoxyethyl.

NEW ADDITIONS

(a) ^-CHO, ^-COR, and $^-CO^-$: Ref. 2174 (Li-dithiane with C=C—CN); 2175, 2176 (Li-dithiane with carbohydrate aldehydes); 2177 ($^-CO^-$ for α,ω-dihalides); 2178, 2180, 2186, 2187 (RCO^- and $ArCO^-$); 2179 (RCO^-, stereosel. reaction with aldehydes); 2181, 2182 (Stetter reaction of RCHO with Michael acceptors); 2183 (RCO^- for Michael acceptors); 2184 (RCH_2CO^-); 2185 (RCO^- for R'_2CO, → $RCO{-}CR'_2{-}SR''$).
(b) ^-COR, R = HO-functional: Ref. 2188, 2190 ($^-COCH_2OH$); 2189 ($^-COCH_2CH_2CMe_2OH$).
(c) ^-COR, R = O=C-functional: Ref. 2191 (3-oxocyclohexyl-CO^-); 2192 (^-COCOOH for ArX); 2193 (^-COCOOH); 2194 ($^-COCOOEt$); 2195 ($^-CO{-}CONEt_2$).
(d) ^-COR, other systems: Ref. 2196, 2197 ($^+CH_2CO^-$).

TABLE 3.1. ^{-}CN

Source	Electrophiles (Yields)	Reaction Conditions	Reference
HCN	30(A-B),39(A-B),41(A)	R_3Al	937,1428,
$K_4Ni_2(CN)_4$	28(B)	KCN,ROH	1440
$NaCu(CN)_2$	28(A),29(A)	DMF or HMPA/Δ	1439
R_2AlCN	39(B-D)	HCl	1431
Me_3SiCN	37(A),42(A)		1446
	32(A),33(B),35(B), 36(A),37(A),42(B)	$AlCl_3$	952
	27(A-C),57(A-B)	$SnCl_4$	1447,1448
$(RO)_2PO—CN$	35(B),37(A),38(B), 47(?)	Amine,NH_3 or K_2CO_3	1450–1452
	32(A-C),35(A),37(A-C), 41(A),42(A)	LDA cat., THF	1453
*t*Bu—NC	39(A-B),43(A),44(A), 34(B),60(B)	$TiCl_4$,$EtAlCl_2$ or $AlCl_3$	1451

TABLE 3.2. $^{-}CONH_2$

Source	Electrophiles (Yields)	Reaction Conditions	Reference
Ar—SO—CH_2Li	35(C)	MnO_2;NaOAc;NH_4OH	823
MeSCH_2—SO—Me	72(D)	NaH;Ac_2O;NH_3;(O)	1455

TABLE 3.3. $^{-}$CONHR

Source	Electrophiles (Yields)	Reaction Conditions	References
HCN	39(B)	Et_2AlCl;hydrolysis	1436
KCN	39(B-C)	MeOH;NaOH	1435
MeNC	32(A-B),33(D),35(A-B), 37(A),41(A)	$TiCl_4$,HCl	1475
R—NC	32(A-C),36(A)	Amine or NH_3; Nu	1464
	40(B-C)	RCOOH	1466
	32(A-C),37(A),46(A-B)	Amine,RCOOH	1467
	32(A-D)	Amine,NaOH	1468
	32(A)	Me_2NH,AcOH	1469
	32(A),35(B),36(A-B), 37(A-B)	H_2SO_4	1470
	43(A),44(B)	$TiCl_4$;H_2	1477
	52(A-B),54(B-C)	;H_2O	1465
CO	*28*-64(A),*29*-64(A)	Amine,Pd—P cat.	1478
	81(A-E),*81*-29(D)	$ArNH_2$,Pd—P cat.	1472,1479
	81(A-C)	RNHMgBr,H_2SO_4	1481
	77(C-D)	*t*BuNHLi	1483
	17(D)	$LiCu(NHR)_2$	1484
	1(B),2(D),32(B), 37(B),59(C-D),77(C)	Pr_2NNHPr,BuLi; Li/NH_3	1485
$Fe(CO)_5$	80(B)	RMgX;H_2O	1480
$Na_2Fe(CO)_4$	*3*-37(C),*4*-37(C-D), 23(D-E)	$PhNO_2$	1482
$Co_2(CO)_8$	29(B)	PhH,NaOH,PTC,$h\nu$	1526
Ar—SO—CH_2Li	32(C),35(C)	MnO_2;NaOAc,$PhNH_2$	823

TABLE 3.4. $^{-}CO—NR_2$

Source	Electrophiles (Yields)	Reaction Conditions	References
$HCONMe_2$	32(B),33(C),35(C), 37(B),42(A)	LDA	1505
$HCONPr_2$	23(B),32(B),33(B),35(A), 36(A),37(A),42(A),59(B), 64(B),77(B)	*t*BuLi	1496, 1507
	1(D),2(E),32(C),35(C), 36(D),42(A),64(C),77(B)	LDA	1506
$HCONR_2$	35(B),37(C),38(C),42(A)	LDA	1456
	69(D)	Li, Na or K	1637
$R_2N—COCl$	32(B),37(D)	SmI_2	1509
$Ph—NMe—CH_2CN$	64(C-D)	NaH or KH, $Cu(OAc)_2$	1495
$(EtNCO)_2Hg$	1(D),35(B),42(B),54(B), 58(B),64(D),77(C)	2 BuLi	1504
CO	79(A-C)	$PdCl_2$	1486
	1(D),37(B-C),77(D)	LDA or $LiN(CH_2)_5$	1489
	1(E),17(A-E),29(C),39 (B),51(D),52(B),54(B)	$LiCu(NR_2)_2$	1484
	CO(B)	$LiNR_2$,HMPA	1491
^{11}CO	1(E),2(E),77(E)	$LiN(CH_2)_5$	1490
$Ni(CO)_4$	28(A)	R_2NH	1440
	17(C),18(A),23(B),29(A), 35(B),41(B),42(D), 52(A-B),54(A)	$LiNMe_2$	1493
$Na_2Fe(CO)_4$	3(B)	I_2,R_2NH	1158

TABLE 3.5. $^{-}COOH$

Source	Electrophiles (Yields)	Reaction Conditions	References
CO	28(A),29(B)	$Ni(CO)_4$,Na_2CO_3	1510,1512
	23(A-D)	$Co_2(CO)_8$ or $NaCo(CO)_4$, PTC	1522,1523
	28(A-B),29(A-C)	$Co_2(CO)_8$,PTC,$h\nu$	1525
	2(E),3(E),4(E), 23(D-E),24(B)	$Fe(CO)_5$,PTC	1518,1527
	28(?)	$PdCl_2$	1519
	18,24,28,29	Transition metal(s)	1457
	29(A-D),75(B-E)	NaH,RONa,$Co(OAc)_2$	1521,1524
	23(A)	Pd—P cat., PTC	1520
$Co_2(CO)_8$	28(A-D),29(A-C)	$h\nu$,PTC	1526
$Na_2Fe(CO)_4$	3-4(A-B),4(A),10(C)	O_2 or NaOCl or I_2	1158
MeNC	42(?)	BuLi	1528
$CHCl_3$	35(A-C)	PTC	1553,1554
$CHBr_3$	35(A-E)	KOH	1555
	35(A-B)	KOH,LiCl	1552
$(RS)_3CH$	1(A),9(D),30(D),32(B), 35(B),37(E),42(A), 51(E),77(C),86(A)	BuLi	845
$(MeS)_3CH$	62(A)	BuLi	1536
$(EtS)_3CH$ or $(PhS)_3CH$	3(C-D),10(C-D)	$^{-}NH_2$	632

$(PhS)_3CH$	1(A),4(D),9(D),30(D), 32(A),35(A),51(E), 86(A),91(B)	BuLi	636
$(MeSe)_3CH$	1(A),2(A)	LDA	842
$(EtS)_2CH—CN$	1(B),3(A-B),6(A),8(B), 9(A),17(A),23(A),35(E), 38(A),64(E),*3*-59(A)	KH;NBS,$AgNO_3$	1539
$(PhS)_4C$	1(A),2(B),17(B)	BuLi	635
$MeOCH_2—SPh$	2(?),3(?),4(?),18(?), 32(?),35(?),37(?)	BuLi;Jones	758
R—SO—CH_2Li or Na	35(C),*59*-65(B)	MnO_2;NaOAc;NaOH	823
$(EtO)_2$CH-imidazole	2(A),35(B),37(D),42(B), 50(B),69(A),70(A)	BuLi	1535
Footnote *a*	35(A),39(C),59(B)	CsF;Bu_4NF/SiO_2	794
Alkylthiazoles	30(D),32(A-D),42(D), 50(D),55(A-D)	BuLi or PhLi or RMgBr	1531,1532 1533
2-Br-thiazoles	32(D),35(A-E),41(B-C), 42(A-E),50(B)	BuLi or Mg	1529,1530
Me_2SO	64(C)	*t*BuOK;H^+;KOH or Cu^{2+}	1559
$(RO)_2PO—CN$	32(A-B)	LDA cat.;HCl	1453
o-$(S{=}CS_2)C_6H_4$	1(B?)	MeLi	701

[a] See text, Scheme (49).

TABLE 3.6. $^{-}COOR$

Source	Electrophiles (Yields)	Reaction Conditions	Reference
CO	17(A-E),22(B)	$PdCl_2$,ROH	1561, 1563,1574
	28(?), *35*-59(A-E),*35*-72(B)	$Ni(CO)_4$,ROH	1511,1583
	2(?),30(?)	$KHFe(CO)_4$;I_2,ROH	1594
	2(B),23(B-D)	$Fe(CO)_5$,ROH,K_2CO_3	1518
	17(B)	Pd—P cat., ROH	1575
	3(C),4(C),28(A-E), *29*-64(A),*29*-75(A)	$Pd(OAc)_2$ or Pd—P cat.,Bu_3N	1578
	8(C)	$Hg(OAc)_2$,ROH;$PdCl_2$	1580
$Co_2(CO)_8$	29(A)	hν,PTC	1526
$Fe(CO)_5$	2(A-B),*3*-59(C)	NaOR,THF/NMP	1595
$Na_2Fe(CO)_4$	*3*-59(A),4(A),23(D)	I_2,ROH	1158,1482
$KHFe(CO)_4$	*10*-59(C-D)	I_2,ROH,CO	1160
$Ni(CO)_4$	28(A-B),29(A)	ROH	1440,1582
RNC	34(C),39(B)	Et_2AlCl;K_2CO_3;HCl	1613
$CHBr_3$ or $CHCl_3$	35(C-D)	KOH,MeOH	1596
$CHBr_3$	35(A-E),41(E)	KOH,MeOH	1597
HCOOEt	77(?)	R—CHLi—COOLi	411
$(RS)_3CH$	3(?)	BuLi;$HgCl_2$,HgO,ROH	535
	32(?)	;HgO,HBF_4,ROH	1600

$(MeS)_3CH$	37(E),60(C)	BuLi,Hg^{2+} or MeOH	1601,1602
$(PhS)_3CH$	39(A-B)	BuLi	1598
2-MeS-1,3-dithiane	2(B),23(A),32(B), 37(C),38(D)	BuLi; $HgCl_2$,HgO,ROH	1599
$Me_3SiCH(SR)_2$	35(B),37(A-B),42(B)	BuLi;NCS or NBS,ROH	494
$Cl_2CHCOOR$	35(C-D)	base,$Pb(OAc)_4$	1612
$MeSCH_2$—SOMe	35(B)	PTC;Br_2,ROH	1605
	35(A-C)	PTC	1607
	72(D)	NaH;$CuCl_2$	1608
$MeSCH_2$—Ms	3(B),18(?),19(C), 24(B),35(B-C)	H_2O_2 or SO_2Cl_1, base;ROH	713,1609
$MeSCH_2$—Ts	3(C),17(C),23(C),35 (C),59(C),64(C-D)	H_2O_2 or MCPBA, base;MeOH	1606
$CNCH_2COOR$	32(C),33(C),35(B), 36(C),37(C)	BuLi;HCl	1638
CH_2=CHOMe	39(B-C)	*t*BuLi;CuI_2,Me_2S, H^+, O_3	1054
Alkyloxazolidine	35(C),55(A),77(?)	BuLi	1528
Ph_3P=CHOMe	32(C),37(C),41(D)	$^1O_2$,py	1610
Me_2SO	64(C)	*t*BuOK;NaH,Br_2;ROH	1559
	59(B?)	NaH;I_2,ROH;$SnCl_4$	1604

TABLE 3.7. $^{-}$COSR

Source	Electrophiles (Yields)	Reaction Conditions	Reference
$(RS)_3CH$	32(A?),37(A?)	BuLi;BF_3—Et_2O,HgO	1622
$(PhS)_3CH$	32(C),39(C)	$AgBF_4$,MeCN	1621
$PhCOCH(SR)_2$	77(B-C)	NaH,Br_2 or RSCl;I_2	825
$ClCH_2$—SOPh	1(C),2(C),3(C)	LDA,Me_3SiCl	1625
$PhSCH_2OMe$	32(B),33(A),35(A), 37(A),38(A-C)	*s*BuLi,TMEDA; Me_3SiCl,NaI	756
$MeSCH_2SOMe$	72(D)	NaH;$CuCl_2$ or Ac_2O, H_2O_2 or MCPBA	1455
	72(B)	NaH;$(RCO)_2O$,py	1624
	35(D)	PTC;Br_2,CCl_4	715
$PhSCH_2NO_2$	32(C-D)	KOH;MsCl;O_3	1629
Me_2SO	64(D)	*t*BuOK,NaH,Br_2,H_2SO_4	1559
R—SO—CH_2Li or Na	32(C),35(B),59-65(A-B)	MnO_2;NaOAc	823

TABLE 3.8. $^{-}$CS—N< and $^{-}$CSe—N<

Source	Electrophiles (Yields)	Reaction Conditions	Reference
RNC	32(A)	$R_2NH;Na_2S_2O_3$	1468
HCS—NR_2	1(E),2(C),32(B),35(B), 36(B),37(C),38(C),41 (C),42(B-C),58(B),91(D)	LDA, −110°	2308
CS—NMe_2	1(C),32(A),35(B),37 (A-B),41(B),42(A), 64(A)	LDA	2309
RNC	32(C)	R_2NH,Na_2Se	1468

NEW ADDITIONS

(a) $^{-}$COOH: Refs. 2198–2202.
(b) $^{-}$COOR: Refs. 2203–2210; Ref. 2211 ($^{-}$CO—SPh).
(c) $^{-}$CO—NH_2 etc.: Refs. 2208,2209,2212.
(d) $^{-}$CN: Ref. 2213.

TABLE 4.1. $^-CR—NH_2$: R Saturated and Nonfunctional

Equivalence	Source	Electrophiles (Yields)	Reaction Conditions	Reference
$^-CH_2NH_2$	MeN(NO)—CH(OMe)Me	1(A),2(B-C),23(C), 32(B),36(B)	LDA, −80°; ClCOOMe,H^+	1685
	MeN(NO)CH_2Me	77(B)	^-OD;H^+	1685
	MeNC	3(D?),17(D?),18(B-C?)	BuLi, −70°	1690
		37(C)	BuLi, −60°;H^+	1689
		30(C-D),35(B?),41(D)	BuLi, −60°;H^+	1688
	MeN=CPh_2	37(D),41(D),42(C-E)	LDA, −45°;H^+	1692
		3(A-B?),4(A?),10(A-E?) 23(C),32(C?),37(B-D)	Et_2NLi, −70°;H^+	1693
		3(A-B?),4(B-C?),17 (E?),91(C?),92(C?)	BuLi, −78°; H^+	1693
	$MeNO_2$	37(C?)	Bu_4NF,9Kbar	1697
$^-CHMe—NH_2$	EtN(NO)—CH(OMe)Me	1(C)	LDA, −80°; ClCOOMe,H^+	1685
	$EtNO_2$	32(B?)	Al_2O_3	1696
		37(A-C?)	Bu_4NF,9Kbar	1697
		8(A-D?)	NaH,50–75°	2271

$^{-}CHEt—NH_2$	$PrNO_2$	32(B)	Al_2O_3	1696
		36(A?),37(A-D?)	Bu_4NF,9Kbar	1697
$^{-}CMe_2—NH_2$	*i*$PrNO_2$	32(B?)	Al_2O_3	1696
		36(C?),37(B)	Bu_4NF,9Kbar	1697
		8(A-D?),23(A?)	NaH,rt − 75°	2271
	$Me_2C{=}NOSiMe_2Bu$-*t*	32(D),35(D)	(1)Bu_4NF, − 70° (2)Me_3SiCl;LAH	1698
$^{-}CHPr—NH_2$	Bu—NSO	17(C),18(C)	Ph_3CLi, − 78° or KOBu-*t*,0°;H^+	1700
cycloPr$^{-}NH_2$	cycloPr—NC	3(D?),10(D?), 18(B?),30(B?)	BuLi, − 70°	1668,1690
—″—,2-Ph—	2-PhcycloPrNC	23(B?)	BuLi, − 70°	1690
$^{-}CHPe—NH_2$	Hexyl—NSO	17(C),18(C)	KOBu-*t*,0°;H^+	1700
	$PeCH{=}NOOSiR_3$	32(B-C),35(A-B)	Bu_4NF, − 70°	1698
	Cx—NSO	17(D),18(C)	KOBu-*t*,0°;H^+	1700

TABLE 4.2. $^{-}CR—NH_2$: **R Unsaturated or Aryl**

Equivalence	Source	Electrophiles (Yields)	Reaction Conditions	References
$^{-}CH(NH_2)—CH{=}CH_2$	$CH_2{=}CHCH_2—NC$	23(C),30(D)	BuLi, −60°;H^+	1668,1690
$^{-}CH(NH_2)—C{\equiv}CH$	$Me_3SiC{\equiv}CCH_2N{=}CHPh$	1(C),3(C),14(B)	BuLi, −70°;H^+	1815
$^{-}CH(NH_2)—CH{=}CHMe$	$MeCH{=}CHCH_2NC$	23(B)	BuLi, −70°;H^+	1690
$^{-}CH(NH_2)(CH{=}CHCH{=}CH_2)$	$CH_2{=}CHCH{=}CHCH_2NC$	23(D)	BuLi, −70°;H^+	1690
$^{-}CHPh—NH_2$	$PhCH_2NC$	17(C?),18(A?),23(A?)	BuLi, −70°	1690
		30(B-C),31(B?)	BuLi, −60°;H^+	1668
	$PhCH_2N{=}CHPh$	42(B)	$NaNH_2$, −35°;H^+	2270
	$PhCH_2N{=}C(SMe)_2$	2(D?),23(B?)	KOBu-*t*, −70°	1694
$^{-}CH(NH_2)—CH{=}CHPh$	$PhCH{=}CHCH_2NC$	23(C)	BuLi, −70°;H^+	1690
$^{-}CH(NH_2)$—py-4	4-py—CH_2—NC	3(C),10(C),17(C), 23(C),30(C-D)	BuLi, −70°H^+	1668,1690
$^{-}CH(NH_2)$—py-3	3-py—CH_2—NC	30(D)	BuLi, −60°;H^+	1668

TABLE 4.3. $^{-}CR—NH_2$: Additional Functionality in R

Equivalence	Source	Electrophiles (Yields)	Reaction Conditions	References
$^{-}CH(NH_2)—CH_2OH$	Footnote *ff*	1(D)	LDA, −80°;H^+	1686
$^{-}CH(NH_2)—CHOH—Me$	$HO—CHMeCH_2NO_2$	32(B?)	Al_2O_3	1696
$^{-}CH(NH_2)—CH_2CH_2NMe_2$	$Me_2N(CH_2)_3NC$	30(B?),41(B?)	BuLi, −60°	1668,1688
$^{-}CH(NH_2)—CH_2CH_2OMe$	$MeO(CH_2)_3NC$	30(A?),31(A?)	BuLi, −60°	1668
$^{-}CH(NH_2)—CH_2CH_2COOMe$	$MeOCO(CH_2)_3NO_2$	32(A?)	Al_2O_3	1696
$^{-}CH(NH_2)—CH_2—CMe(OCH_2)_2$	Footnote *v*	32(A?)	Al_2O_3	1696
$^{-}CH(NH_2)—CH_2—OTHF—2$	$THFOCH_2CH_2NO_2$	32(A?)	Al_2O_3	1696
$^{-}CH(NH_2)—CH_2OCHMe—OBu$	$BuO—CHMe—OCH_2—CH_2NO_2$	32(A?)	Al_2O_3	1696

For footnotes, see Table 4.21.

TABLE 4.4. $^{-}CH_2$—NHR

Equivalence	Source	Electrophiles (Yields)	Reaction Conditions	References
$^{-}CH_2NHMe$	Me_2NNO	23(B?)	$NaNSi_2Me_6$, −80°	2272
		41(B?),42(B-C?)	KOBu-*t*, −20°-rt	1667
		1(B?),2(B?),16(C?), 17(D),32(B?),35(A?), 36(B?),37(B-C?),42 (B?),52(C?),53(C-D?), 54(D?),57(B),86(A?), 89(A?),90(B?),93(B?)	LDA, −78° to −20°; HCl	1667,1705, 1706, 2273–2275
	Bu_3SnCH_2NNOMe	32(C-D?)	2 *N* HCl	1667
	Me_2NCONR_2, R = *i*Pr or Cx	42(C-D?)	*s*-BuLi,TMEDA −78°	1728
	Me_2NCOBu-*t*	42(A?),77(A?)	*s*BuLi,TMEDA,0°	1828
	Me_2NCONR_2[b]	1(B?),2(B?),32(B?), 35(A?),36(A?),42(A?)	*s*BuLi,TMEDA 0°	1728
	Me_2NCONR_2[c]	2(B?),35(B?),42(B?)	*s*BuLi,TMEDA −80°	1728
	Me_2NCOAr[a]	1(A?),2(A?),17(C?), 35(B?),36(B?),37(C?), 77(A?),91(C?)	*s*BuLi,TMEDA, −78°	1724,1725
	Me_2NCO—CPh_3	2(A?),35(B?),37(B?)	*s*BuLi,TMEDA −78°	1724
	Me_2N—CS—Bu-*t*	1(B?),2(B?),3(B?),23 (C?),32(D?),35(B?), 37(E?),42(D?)	*s*BuLi, −78°	1731

	Me$_2$NCH=N—R R = *t*Bu or Cx	1(A?),2(A?),9(?),32 (C),35(B),37(C),41(B)	*t*BuLi, −78°; KOH—MeOH	1716
	Me$_2$NCH=N—Bu	23(C),35(B)	*t*BuLi, −78°;KOH—MeOH or HCl/MeOH	1716
	Me$_2$N—PO(NMe$_2$)$_2$	32(A?),35(C?),37(B?)	*s*BuLi, −78°	1733
$^-$CH$_2$NH—Et	Et—NMe—NO	32(B?),51(A?)	LDA, −80°	2274,2275
	Me$_2$NCH=N—Cx	1(B?)	*t*BuLi, −78°	1716
$^-$CH$_2$NH—Pr-*i*	*i*Pr—NMe—NO	23(A),32(B?),42(B)	LDA, −80°;HBr	2274
$^-$CH$_2$NH—Bu-*t*	MeNH—Bu-*t*	16(B?)	LDA, −78°	1706,2273
	*t*Bu—NMe—NO	4(B?),23(A?),32(A?), 35(B),36(B),51(A), 53(B-C?),54(C?),86 (A?),91(B?),93(A?)	LDA, −80°; HCl	1705,2274
		37(A?),42(O)	KOBu-*t*, −20°-rt	1667
	Me$_3$SnCH$_2$N(NO)Bu-*t*	32(B-C?)	2*N* HCl	1667
$^-$CH$_2$NH—Cx	Cx—NMe—NO	42(A?)	LDA, −80°	2274
$^-$CH$_2$NHPh	PhNMe—CH=NBu-*t*	1(A),2(B)	*t*BuLi, −78 to −30°; KOH—MeOH or N$_2$H$_2$ or LAH	1720
$^-$CH$_2$NH—Bzl	Bzl—NMe—NO	32(C?)	LDA, −80°	2274
$^-$CH$_2$NH(CH$_2$)$_2$Ar[d]	Ar(CH$_2$)$_2$NMe—NO	23(B)	NaNSi$_2$Me$_6$, −80°; HCl, urea	2272
$^-$CH$_2$NH—C$_{11}$H$_{23}$	ArCO—NMe—C$_{11}$H$_{23}$[a]	35(A?)	*s*BuLi,TMEDA, −78°	1724

For footnotes, see Table 4.21.

TABLE 4.5. $^-$CR—NHR′: R Saturated and Nonfunctional

Equivalence	Source	Electrophiles (Yields)	Reaction Conditions	Reference
$^-$CHMe—NHMe	Et—NMe—COAr[a]	35(C)	*s*BuLi,TMEDA, −78°	1724
	Ph_2PO—CHMe—NMe(—COOEt) (EtOCO on N)	35(C)	LDA, −78°;(1) Δ, (2) H_2/PtO_2,(3) HO^-	1814
	PhCO—NMe—CHMe—CN	35(B)	LDA, −78°;$NaBH_4$	862
$^-$CHMe—NHEt	Et_2N—NO	35(B)	LDA, −78°;HCl	1706,2273
	Et_2N—COBu-*t*	77(A?)	*s*BuLi,TMEDA,0°	1828
	Et_2N—COCEt_3	2(B?),35(C),77(A?)	*s*BuLi,TMEDA, −78°; (1) HCl(2) KOBu-*t*	1669
$^-$CMe_2—NHPr-*i*	*i*Pr_2N—NO	32(D?),51(C?)	LDA, −78°	1667,2275
		1(C?),32(C?)	KDA, −78°	1667
$^-$CHPr—NHBu-*i*	*i*Bu_2N—NO	51(B?)	LDA, −78°	2275
$^-$CHPe—NH—Hex	Hex_2N—NO	23(A),35(B)	LDA, −78°;HCl	1706,2273

For footnotes, see Table 4.21.

TABLE 4.6. $^{-}$CR—NHR′: R Unsaturated or Aryl

Equivalence	Source	Electrophiles (Yields)	Reaction Conditions	Reference
$^{-}$CH(NHMe)—CH=CH_2	CH_2=$CHCH_2$—NMe—NO	1(A?),35(A-C?), 37(A?),42(B?)	LDA, −78°	1811
		35(A?),42(A?)	KOBu-*t*, −20°	1811
$^{-}$CH(NHBu-*t*)CH=CH_2	CH_2=$CHCH_2$—NBu—NO	1(B?),2(B?),9(D?), 17(A?),35(C?), 36(C?),37(B?)	LDA, −78° to −30°	1811
$^{-}$CHPh—NHMe	Bzl—NMe—NO	51(B?)	LDA, −78°	2275
	Bzl—NMe—PO$(NMe_2)_2$	41(C?)	BuLi, −78°	1732
$^{-}$CHPh—NHBu-*t*	BzlN(NO)Bu-*t*	23(B?)	LDA, −78°	1706, 2273
$^{-}$CHPh—NH—Bzl	Bzl_2N—NO	1(A?),2(A?),3(D?), 35(B),50(A?)	LDA, −78°	1722
		51(A?)	LDA, −78°	2275
	Bzl_2N—COPh	1(A?),36(A?), 50(A?),55(A?)	LDA, −78°	1722

TABLE 4.7. $^-$CR—NHR′: Cyclic Amines

Equivalence	Source	Electrophiles (Yields)	Reaction Conditions	Reference
$^-$CHNH(CH$_2$)$_2$	CH$_2$NY—(CH$_2$)$_2$[e]	35(B?)	tBuLi, −40°	1726
	(CH$_2$)$_2$N(NO)CH$_2$	30(B)	LDA, −78 to −30°; HCl	1706,2273
$^-$CHNH(CH$_2$)$_3$	CH$_2$NY—(CH$_2$)$_3$[e]	42(C?)	sBuLi,0°	1726
		32(D?)	tBuLi, −40°	1726
	CH$_2$NY—(CH$_2$)$_3$[f]	32(A),35(A),36(D), 42(D),77(A?)	sBuLi,TMEDA,0°; (1) H$^+$(2) KOBu-t	1828
	CH$_2$N(COAr)(CH$_2$)$_3$[a]	35(D?),42(D?) 77(A?)	sBuLi,TMEDA, −78°	1669
	(CH$_2$)$_3$N(NO)CH$_2$	2(C?),23(B?),42(B), 50(D?),51(B?), 53(B?),86(B?)	LDA, −78°; Raney-Ni,H$_2$	1667,1705, 1706,1710, 2273,2275
		41(B)	LDA, −78°;R—Ni	2277
	(CH$_2$)$_4$N—CH=N—Cx	35(C)	tBuLi, −78°; KOH—MeOH	1716
$^-$CHCH=CHCHR—NH	R = H or Bu	2(B-D?),3(B?)	LDA, −78°	1813,2306
$^-$CHNHCHEt(CH$_2$)$_2$	(CH$_2$)$_3$CHEt—NNO	2(C)	LDA, −78°;HCl	2276
$^-$CHNHCHBzl(CH$_2$)$_2$	(CH$_2$)$_3$CHBzl—NNO	23(A)	LDA, −78°;HCl	2276
$^-$CHNH(CH$_2$)$_4$ (*1*)	CH$_2$NH—(CH$_2$)$_4$ (*2*)			
	(2) Y = H	2(D-E?)	KDA, −78°	1667
	(2) Y = COCPh$_3$	35(E?),42(D?)	tBuLi, −40°	1726

	(*2*) Y = COBu-*t*	32(B),35(B),36(B), 77(A?)	*s*BuLi,TMEDA,0°; (1) H^+ (2) KOBu-*t*	1828
	(*2*) Y = $COCEt_3$	32(D),77(A?)	*s*BuLi,TMEDA, −78°; (1) HCl (2) KOBu-*t*	1669
	(*2*) Y = NO	2(C),17(C?),23(C?), 32(A-B?),35(C), 37(C),42(C)	LDA, −78°; Raney-Ni,H_2	1706, 1710, 2273
	$(CH_2)_4N—CHPOPh_2$ (ring), COOEt	32(C)	LDA, −78°;(1) Δ, (2) H_2/PtO_2 (3) HO^-	1814
	$(CH_2)_4N—CH—CN$ (ring), COPh	32(B),35(B)	LDA, −78°;$NaBH_4$	862
(*1*) 3-Me	Y = NO	2(C?)	KDA, −78°	1667
(*1*) 3-,4- or 6-Me	Y = NO	42(D)	LDA, −78°;HCl	2278
(*1*) 4-Bu-*t*	Y = COAr	35(C?),42(D?), 77(A?),92(D?)	*s*BuLi,TMEDA, −78°	1669
(*1*) 6-Pe	Y = NO	2(C)	LDA, −78°;R—Ni/H_2	2278
(*1*) 4-Ph	Y = $COCEt_3$	2(B?),35(C)	*s*BuLi,TMEDA, −78°; (1) HCl (2) KOBu-*t*	1669
(*1*) 4-OH	Y = NO	2(C),42(C?)	LDA,78°;HCl	2278
		1(B)	KDA, −78°;R—Ni	1667
$^-CHNH(CH_2)_5$ (ring)	$(CH_2)_5N(NO)CH_2$ (ring)	42(A)	LDA, −78°;HCl	1706,2273
$^-CHNHCH_2NRCH_2CH_2$ (ring)	*N*-Nitroso	R = Me,32(C?)	LDA, −96 to −60°	1706,2273
		R = Ph,50(C?)	MeLi, −78°	1722

TABLE 4.7. (*Continued*)

Source		Electrophiles (Yields)	Reaction Conditions	Reference
Footnote *x*		1(A),2(B),35(B)	*t*BuLi, −78 to −30°;KOH—MeOH or NH_2NH_2 or LAH	1720
Footnote *y*		1(B),2(A-B)	*t*BuLi, −78 to −30°;KOH—MeOH or NH_2NH_2 or LAH	1720
Footnote *aa*	Y^{bb} = COPh	23(B?),35(B?)	*t*BuLi, −40°	1726
	Y = COBu-*t*	1(B),2(A?),3(A?), 4(A?),9(A?),24(B?), 35(A?),37(B),41(A?), 91(A?),93(A?)	*t*BuLi,TMEDA, −40° $NaAlH_2(OCH_2CH_2OMe)_2$	2279
		32(A?)[g],35(A?)[g]; $MgBr_2$, −78° 30(B),35(C)	MeOH or NH_2NH_2—AcOH	1730
	Y = CH=NR[h]	1(A)[h],3(A)[h],23(A)[h]	LDA, −78°; NH_2NH_2—AcOH	2280
	Y = CH=NR[i]	1(B-C)[i],2(A-C)[i], 3(B)[i],17(B)[i]	LDA, −100°; NH_2NH_2—AcOH	1718,2280

Footnote *aa*, 1-Me	Y^{bb} = COBu-*t*	1(C?)	*t*BuLi,TMEDA, – 40°	2279
	Y = $PO(NMe_2)_2$	1(E),23(D)	BuLi, – 78°; HCl—MeOH	1729
Footnote *aa*, 1-Bzl	Y^{bb} = CH=NBu-*t*	1(C)	*s*BuLi, – 78 to – 30°; LAH	1717
—″—, 6,7-Me_2	Y^{bb} = COPh	23(B?)	*s*BuLi, – 40°	1726
—″—, 6,7-$(MeO)_2$	Y^{bb} = CH=NBu-*t*	1(C)	KDA, – 78°; NH_2NH_2—AcOH,53°	1717
	Y = NO	23(C),35(C?)	LDA, – 78°;LAH,R—Ni	1710,2278
	Y = CH=NR^i	$1(C)^i$,$2(B)^i$,$24(B)^i$	LDA, – 78°; NH_2NH_2—AcOH	1718
Footnote *aa*, 4-OH-6,7-$(MeO)_2$	Y = NO	24(B)	LDA, – 78°;R—Ni	2277
Footnote *cc*		1(B),3(C),9(B), 32(A?)	BuK, – 78°; H^+,60°	1719,1721
Footnote *dd*		1(C?)	KDA, – 78°	1667

For footnotes, see Table 4.21.

TABLE 4.8. $^{-}C—NR_2$

Equivalence	Source	Electrophiles (Yields)	Reaction Conditions	Reference
$^{-}CH_2NMe_2$	$Me_4N^+Cl^-$	42(C)	PhLi, −70°;Δ	1738a
	$Bu_3SnCH_2NMe_2$	35(B)	BuLi,0°	1737
		35(A)	BuLi,TMEDA,rt	1681
$^{-}CH_2NEt_2$	$Et_2N—CH_2CN$	2(A?),30(C?)	LDA, −78°;$NaBH_4$	886
$^{-}CH_2N(CH_2)_4$	$Me_2N^+(CH_2)_4Cl^-$	42(C)	BuLi, −70°;H^+	2281
$^{-}CH_2N(CH_2)_5$	$Bu_3SnCH_2N(CH_2)_5$	35(B)	BuLi,0°	1737
$^{-}CH_2N(Pr\text{-}i)_2$	$Pr_2N—CH_2CN$	32(C)	L(K)DA, −78°;Δ	887
$^{-}CH_2NR_2$[gg]	Bu_3SnCH_2N	35(D)	BuLi,0°	1737
$^{-}CHR—NR'R''$[hh]		35(C-D),77(B?)	*t*BuLi, −70–0°; Si_2Cl_6	2282

$^{-}CH_2NMe—Ph$	$Bu_3SnCH_2NMe—Ph$	35(A)	BuLi,0°	1737
$^{-}CH_2NPh_2$	$Bu_3SnCH_2NPh_2$	35(C)	BuLi,0°	1737
$^{-}CHMe—NMe_2$	$Me_2N—CHMe—CN$	32(B?),35(B)	LDA, −78°;$NaBH_4$	862,887
$^{-}CHMe—NEt_2$	$Et_2N—CHMe—CN$	32(B?),33(B?)	L(K)DA, −78°	887
$^{-}CHMe—\overline{N—CH}(CH_2)_3$ with CH_2OMe on CH	$MeCH(NC)—\overline{N—CH}(CH_2)_3$ with CH_2OMe on CH	35(A?)	L(K)DA, −78°	887
$^{-}CHEt—NMe_2$	$Me_2N—CHEt—CN$	3(B?)	KNH_2,NH_3;$NaBH_4$	2284
		32(B-C?)	L(K)DA, −78°	887
$^{-}CHEt—NEt_2$	$Et_2N—CHEt—CN$	32(C?)	L(K)DA, −78°	887
$^{-}CHPr—NMe_2$	$Me_2N—CHPr—CN$	32(C?)	L(K)DA, −78°	887

For footnotes, see Table 4.21.

TABLE 4.9. $^{-}C—OH$

Equivalence	Source	Electrophiles (Yields)	Reaction Conditions	Reference
$^{-}CH_2OH$	$ArCOOMe^{j}$	1(A?),2(B-D),41(B?), 42(D),77(A-B?),93(B?)	sBuLi,TMEDA, −75°; LAH	1642,1724, 2283
	B-Me-9-BBN	77(B?)	LTMP, −78°	1744
	Mes_2BMe	2(B),23(B)	MesLi,0°;H_2O_2,HO^-	1643
	Bu_3SnCH_2OH	3(C),4(C),23(C),29(C), 32(C),35(C),37(C)	BuLi,0°	179
	$(iPrO)_2SiMe$—CH_2MgCl	3(A),18(B), 28(A-B),29(B)	CuI or Ni/P cat; H_2O_2	1742
	$(iPrO)_2SiMe$—CH_2ZnCl	29(A-B)	Pd/P cat;H_2O_2	1742
$^{-}CHMe—OH$	$ArCOOEt^{j}$	2(C?),14(C?),17(B?), 32(B?),36(A?),37(D?), 51(B?),77(A?),92(A-B?), 93(C?)	sBuLi,TMEDA, −78°	1642
	$Mes_2B—Et$	2(B)	MesLi,0°;H_2O_2,HO^-	1643
$^{-}CHEt—OH$	$ArCOOPr^{j}$	2(B-D),16(B?),27(C?), 52(B?),60(B?),79(C?)	sBuLi,TMEDA, −78°; H^+	1642,2285

$^{-}$CHBu—OH	ArCOOPe[j]	77(B?)	sBuLi,TMEDA, −78°	1642
$^{-}$CHHex—OH	$ArCOOC_7H_{15}$[j]	17(B?),36(C?), 77(B?),92(B?)	sBuLi,TMEDA, −78°	1642
$^{-}$CHPh—OH	ArCOOBzl[j]	1(A?)	sBuLi,TMEDA, −78°	1642
	$Me_2NCOOBzl$	55(A?)	LDA, −78°	1743
	$Et_2NCOOBzl$	1(A?),91(C?)	LDA, −78°	1743
$^{-}$CHOH—C_6H_4OMe-*p*	*i*$Pr_2NCOOCH_2Ar$	2(A?),32(A?), 33(A?),36(A?)	LDA, −78°	1743
Ph^{-}CMe—OH	$Et_2NCOOCHMe$—Ph	1(B?)	LDA, −78°	1743
$^{-}$CHOH—CHMe—Ph	Ph—CHMe—CH_2(—OOCAr[j])	1(B?)	sBuLi,TMEDA, −78°	1642
H_4-1-naphthol-1-anion	*N,N*-Et_2-carbamate	91(A?)	LDA, −78°	1743
$^{-}$CHOH$(CH_2)_{14}$Me	$Me(CH_2)_{15}O$(—COAr[j])	36(E?),77(E?)	sBuLi,TMEDA −78°	1642
$^{-}$CHOH—CH_2NMe_2	$Me_2NCH_2CH_2$(—OOCAr[j])	2(A?),77(B?)	sBuLi,TMEDA −78°	1642

For footnotes, see Table 4.21.

TABLE 4.10. $^{-}C{-}OR$

Equivalence	Source	Electrophiles (Yields)	Reaction Conditions	Reference
$^{-}CH_2OR$ (OAr)	X_3SnCH_2OR (OAr)	35(A),	BuLi, −78°	1650
$^{-}CH_2OMe$	Me_2O	35(C),91(D)	BuK, −78°	1646
	Me_3SiCH_2OMe	32(?),37(?)	*s*BuLi, −78°	1220
$^{-}CH_2OBu$-*t*	MeOBu-*t*	32(A),35(A),37(A), 39(A),41(B)	*s*BuLi,KOBu-*t*, −78°	1645
		35(B),91(B)	BuK, −78°	1646
$^{-}CH_2{-}Y^m$	Bu_3SnCH_2Y	37(A)	BuLi, −78°	1649
$^{-}CH(OMe){-}Pr$	$PrCH(OMe){-}SnBu_3$	35(A),91(B)	LN or LDMAN, −78°	1741
$^{-}CMe_2{-}OMe$	$PhS{-}CMe_2{-}OMe$	35(B)	LDMAN, −63 to −78°	1741
$^{-}CH(OMe)CH{=}CH_2$	$MeOCH_2CH{=}CH_2$	37(A)	(1) *s*BuLi, −65° (2) $ZnCl_2$	1818

1-MeOcycloPr$^-$	1-MeO-1-PhS-*cyclo*Pr	32(A)	LDMAN, −63 to −78°	1741
$^-$C(Me)(Et)(OCH$_2$OY)[k]	Bu$_3$SnC(OCH$_2$OY)Me—Et[k]	7(?)	BuLi, −78°	1746
$^-$CH(OEt)CH=CHMe	EtOCH$_2$CH=CHMe	37(A)	(1) *s*BuLi, −65° (2) ZnCl$_2$	1818
$^-$CHY—C$_6$H$_{13}$-*n*[m]	Bu$_3$SnCHY—C$_6$H$_{13}$	37(A)	BuLi, −78°	1649
$^-$CHPh—OMe	PhCH$_2$OMe	3(A),42(B-C),59(D)	BuLi,TMEDA, −10°	1747
$^-$CHY—(3-furyl)[m]	Bu$_3$SnCHY(furyl)	1(B)	BuLi, −78°	1649
$^-$CHY—Ph[m]	Bu$_3$Sn—CHY—Ph	1(A),37(B)	BuLi, −78°	1649
2-THF$^-$	THF	91(B)	BuK, −78°	1646
2-Me-2-THF$^-$	2-Me—THF	91(D)	BuK, −78°	1646
2-THP$^-$	THP	91(B)	BuK, −78°	1646
	2-PhS—THP	35(B),91(B)	LN or LDMAN, −78°	1741
2-Et-2-THP$^-$	2-Et—THP	91(C)	BuK, −78°	1646

For footnotes, see Table 4.21.

TABLE 4.11. $^{-}$C—SH: Saturated Systems

Equivalence	Source	Electrophiles (Yields)	Reaction Conditions	Reference
$^{-}CH_2SH$	MeSCOAr[n]	1(A?),17(A?),23(C), 35(B?),41(B?),50(C?), 77(A?),92(C?)	BuLi, −78°;LAH	1725, 1751
$^{-}$CHMe—SH	EtS—COAr[n]	23(B),35(A?)	*s*BuLi,TMEDA, −98°	1751
	EtS—COAr[p]	42(A?)	*s*BuLi,TMEDA, −95°	1725
	EtS—$CONMe_2$	2(C),23(C),32(B?), 35(C),77(A?)	*s*BuLi,TMEDA, −98°; HO^-	1750
$^{-}CMe_2$—SH	*i*PrS—COAr[p]	1(D?),2(B?),3(C?), 18(C?),32(B?),35(C?), 50(D?),51(B?),55(E?), 64(C?),77(B?),92(D?)	*s*BuLi,TMEDA, −98°	1750
$^{-}$CH(SH)—C_6H_{13}-*n*	C_7H_{15}SCOAr[p]	35(B?),36(D?)	*s*BuLi,TMEDA, −98°	1750
4-*t*Bu—Cx-thiol-1-anion	R_2CH—SCOAr[p]	1(B?)	*s*BuLi,TMEDA, −98°	1750
$^{-}$CH(SH)—CH_2NMe_2	$Me_2NCH_2CH_2$SCOAr[p]	1(B?)	*s*BuLi,TMEDA, −98°	1750

For footnotes, see Table 4.21.

TABLE 4.12. $^{-}$C—SH: Unsaturated Systems

Equivalence	Source	Electrophiles (Yields)	Reaction Conditions	Reference
$^{-}$CH(SH)—CH=CH$_2$	CH$_2$=CHCH$_2$S-het[q]	1(B?),3(C?), 17(C?),23(B?)	BuLi, −60°	1759
	CH$_2$=CHCH$_2$SCOAr[a]	1(C?),2(C?)	BuLi or LDA, −98°	1750
	CH$_2$=CHCH$_2$SCSNMe$_2$	1(A?),2(A?), 9(A?),23(A?)	LDA, −60°	1758
$^{-}$CH(SH)—C≡CH	HC≡CCH$_2$S-het[q]	1(C?),22(C?), 23(B),87(C?)	BuLi, −60°	1759
$^{-}$CH(SH)—CH=CHR	RCH=CHCH$_2$SCONMe$_2$	1(C?),2(?), 86(?)	LDA, −78°	1263, 2286
$^{-}$CH(SH)—CMe=CH$_2$	CH$_2$=CMeCH$_2$SCSNMe$_2$	1(A?),2(A?) 4(A?),9(A?),	LDA, −60°	1758
	CH$_2$=CMeCH$_2$S-het[q]	2(B?),3(A-B?), 23(B?)	BuLi, −78°	1640, 1759
$^{-}$CMe(SH)—CH=CH$_2$	CH$_2$=CHCHMe—S-het[q]	2(C?),3(C?),23(D?)	BuLi,HMPA, −60°	1759
$^{-}$CH(SH)—C≡CMe	MeC≡CCH$_2$S-het[q]	22(C?),23(C?)	BuLi, −60°	1759
$^{-}$CH(SH)—CMe=CHR	RCH=CMeCH$_2$SCONMe$_2$	86(?)	LDA, −78°	1263
$^{-}$CH(SH)—CH=CMe—R	RCMe=CHCH$_2$SCONMe$_2$	86(?)	LDA, −78°	1263
$^{-}$CH(SH)—CH=CHPh	PhCH=CHCH$_2$S-het[q]	23(B?)	BuLi, −60°	1759
$^{-}$CH(SH)—C≡CPh	PhC≡CCH$_2$S-het[q]	18(C?),23(C?)	BuLi, −60°	1759

For footonte, see Table 4.21.

TABLE 4.13. $^{-}$CR—SR′: R Saturated

Equivalence	Source	Electrophiles (Yields)	Reaction Conditions	Reference
$^{-}CH_2SMe$	Me_2S	3(D),32(C),35(A), 92(C)	BuLi,TMEDA, −20°	1755
	$ClCH_2SMe$	2(B),14(B)	Mg,0°	1752
		35(B)	Mg, −78°	1752
$^{-}CH_2SPh$	MeSPh	77(A)	*t*BuLi,HMPA, −78°	1644
		9(C),36(B), 42(A),77(A)	BuLi,DABCO,0°	580
$^{-}CH_2S$-het[q]	MeS-het	1(C),3(C-D),10(D), 17(D),23(B),24(B), 32(B),35(B)	LDA or BuLi −20° or −78°	241,1754, 1756, 1786
$^{-}$CH(*i*Pr)—SPh	*i*BuSPh	2(B),3(B),17(A),35 (A),64(C),75(B), 77(A),92(A),	*t*BuLi,HMPA, −78°	1644
	*i*PrCH(SPh)—SePh	38(C),39(D)	BuLi,HMPA, −78°	1752
*i*Pr^{-}CMe—SPh	*i*PrC(Me)(SPh)—SePh	38(A)	BuLi, −78°	1752
$^{-}$CHPe—SPh	CH_2═CH—SPh	50(C)	BuLi, −20°	1071
$^{-}$CH—SPh	CH_2SPh			
$CHMe(CH_2)_3OMe$	$CHMe(CH_2)_3OMe$	77(A),92(A)	*t*BuLi,HMPA, −78°	1644

For footnotes, see Table 4.21.

TABLE 4.14. $^{-}$CR—SR′: R Unsaturated or Aryl

Equivalence	Source	Electrophiles (Yields)	Reaction Conditions	Reference
$^{-}CH(SPr\text{-}i)CH{=}CH_2$	$CH_2{=}CHCH_2SPr\text{-}i$	17(C),19(B),32(B), 35(A-B),37(B)	(1) *s*BuLi, −78°, (2) R_3B, −78°	1823,2295
$^{-}CH(SR)CH{=}CH_2$	$CH_2{=}CHCH_2SR$	38(A?)	LDA,HMPA, −50°	1820
$^{-}CH(S\text{-}2\text{-}py)CH{=}CH_2$	$CH_2{=}CHCH_2S\text{-}2\text{-}py$	3(A?),23(B)	PhLi, −25°	2294
$^{-}CH(SCH_2COOR)(CH{=}CH_2)$	$CH_2(SCH_2COOR)(CH{=}CH_2)$	2(C),3(B-C), 4(C),10(C)	*s*BuLi,HMPA, −78°	1821
$^{-}CH{—}SCH_2CH_2CH{=}CH$ (ring)	$CH_2SCH_2CH_2CH{=}CH$ (ring)	2(B),3(B),4(B),18 (A),23(B),30(A-D)	*s*BuLi, −78° or BuLi,TMEDA, −49°	1822,2292
$^{-}CHPh{—}S{—}CS{—}NEt_2$	$PhCH_2SCSNEt_2$	32(A)	LDA, −78°	1757
$^{-}$CHPh—S-het[q]	$PhCH_2S$-het	35(B),36(A)	LDA, −78°	1757
$^{-}$CHPh—S-2-pyrrole	$PhCH_2S$-het	35(B)	BuLi, −78°	1757
$^{-}CH(SPh)CH{=}CHPe$	$PeCH{=}CHCH_2SPh$	39(A?)	BuLi,HMPA, −78°	2293

For footnotes, see Table 4.21.

TABLE 4.15. $^{-}$C-Halogen

Equivalence	Source	Electrophiles (Yield)	Reaction Conditions	Reference
$^{-}CH_2Cl$	$BrCH_2Cl$	50(E)	BuLi, −110°	1763
		26(C),37(B),41(C) 59(C),64(C)	*s*BuLi/LiBr, −110°	1764
	PhSO—CH_2Cl	2(A-B?)[r],32(B-C) 29(A-B?)[r],37(A?)[s]	BuLi, −78°;	1767–1769
$^{-}CH_2Br$	$(Ph_2As)_2CH_2$	2(B),3(B-C?),32 (B?),35(B?),37(B?)	BuLi, −40° to rt; Br_2	1765, 1766
$^{-}$CHBu—Cl	Br—CHBu—Cl	58(B-C),69(D)	BuLi, −155°	1648,1762
$^{-}$CHBu—Br	Bu—$CHBr_2$	58(B-D),69(D)	BuLi, −115°	1648,1762

For footnotes, see Table 4.21.

TABLE 4.16. $^{-}C—SeR$ **or** $^{-}C—SeAr$

Equivalence	Source	Electrophiles (Yield)	Reaction Conditions	Reference
$^{-}CH_2SeMe$	$(MeSe)_2CH_2$	2(C),3(C),32(B) 35(A),38(A)	BuLi, −78°	1783,2287, 2290,2291
$^{-}CH_2SePh$	$(PhSe)_2CH_2$	32(B),35(B),37(B) 41(A-B),42(A),77(A)	BuLi, −78°	1653,1782, 2285,2288
	$PhSeCH_2Br$	3(A),32(B),36(B)	BuLi, −78°	1790
$^{-}CHMe—SeMe$	$(MeSe)_2CHMe$	33(A),35(A),37(A), 38(B),41(A),92(B)	BuLi, −78°	1783,2289, 2291
$^{-}CHMe—SePh$	$(PhSe)_2CHMe$	32(B-C),37(B), 41(B-C),42(C)	BuLi, −78°	1782,1783, 2287,2288
	PhSe—CHBr—Me	32(B),36(B)	BuLi, −78°	1790
$^{-}CMe_2—SeMe$	$(MeSe)_2CMe_2$	2(A),3(A),35(A), 37(C),38(B-C), 41(B),92(A)	BuLi, −78°	1783,2289, 2291
$^{-}CMe_2—SePh$	$PhSe—CMe_2—Br$	32(B)	BuLi, −78°	1790
	$(PhSe)_2CMe_2$	32(B),33(B),37(B), 38(B),41(B),51(B), 52(C),91(B)	BuLi, −78°	1782, 1788–1790, 2291,2296
$^{-}CHEt—SePh$	$(PhSe)_2CHEt$	32(C),35(A)	BuLi, −78°	1782,2288
$^{-}CH(CH{=}CH_2)—SePh$	$CH_2{=}CHCH_2SePh$	3(B)	LDA, −78°	1780
1-MeSe-cycloPr^{-}	1,1-$(MeSe)_2$-cyclopropane	2(B),3(B),32(B-C), 37(B-C),41(B)	BuLi, −78°	1784,1786, 2297

TABLE 4.16. (*Continued*)

Source	Electrophiles (Yields)		Reaction Conditions	Reference
1-PhSe-cycloPr$^-$	1,1-$(PhSe)_2$-cyclopropane	2(B),3(C-D), 32(B)	BuLi, −78°	1786, 2297
1-MeSe-cycloPe$^-$	1,1-$(MeSe)_2$-cyclobutane	32(A),37(B)	BuLi, −78°	1785
—″—, 2-pentyl		32(B)	BuLi, −78°	1785
$^-$CH(SePh)—CH=CHMe	MeCH=$CHCH_2SePh$	3(B)	LDA, −78°	1780
$^-$CH(SePh)—CH=CMe_2	Me_2C=$CHCH_2SePh$	92(B)	LDA,0°	1780
$^-$CH(*s*Bu)—SeMe	$(MeSe)_2$CHBu-*s*	32(A),35(A),37(C)	BuLi, −78°	1783
$^-$CH(*i*Bu)—SePh	CH_2=CH—SePh	36(B),77(A),91(A)	*i*PrLi,0°	1656,1781
$^-$CH(*s*Bu)—SePh	$(PhSe)_2$CHBu-*s*	87(B)	BuLi, −78°	2296
$^-$CHPe—SePh	CH_2=CH—SePh	1(A),3(B),35(B), 36(C),41(C),75(C), 89(A),91(A)	BuLi,0°	1656, 1781
$^-$CH(CH_2Bu-*t*)SePh	CH_2=CH—SePh	77(A)	*t*BuLi,0°	1656,1781

MeSe-cyclohexane-1-anion	1,1-$(MeSe)_2Cx$	2(A),3(A),32(B),35 (A),37(C),50(B), 51(C),54(B),55(D)	BuLi, −78°	1783,1789, 2290, 2298
4-*t*Bu-1-PhSe—Cx^-	4-*t*Bu—$Cx(SePh)_2$	51(D),69(C)	BuLi, −78°	1789
$^-$CHHex—SeMe	$(MeSe)_2CH$—Hex	2(A-B),3(A-B),17(C), 32(A),37(C),38(B),51 (D),54(B),91(A),92(A)	BuLi, −78°	1783,1789, 2287, 2290–2291
$^-$CHHex—SePh	$(PhSe)_2CH$—Hex	2(D),3(D),32(A-B)	BuLi, −78°	2287,2290, 2299
	PhSe—CHBr—Hex	32(B)	BuLi, −78°	1790
$^-$CHPh—SePh	Bzl—SePh	3(A),10(B), 23(B),30(B)	LDA,rt	1778, 1779
$^-CH(SeMe)C_8H_{17}$-*n*	$(MeSe)_2CH$-octyl	32(B)(CH_2O)	BuLi, −78°	2299
$^-$CH(SePh)—CH=CHPh	PhCH=$CHCH_2SePh$	36(C)	LDA, −78°	1780
$^-$CH(SePh)(CH=CCl—Me)	Me—CCl=$CHCH_2SePh$	3(A),23(B) 30(B),92(C)	LDA, −78°	1780

TABLE 4.17. $^-C—SiR_3$ **or** $^-C—SiAr_3$

Equivalence	Source	Electrophiles (Yield)	Reaction Conditions	Reference
$^-CH_2SiMe_3$	$ClCH_2SiMe_3$	30(B),32(?),35(?), 36(?),42(?),50(B), 52(?),91(B)	Mg	773, 2301,2302
	$BrCH_2SiMe_3$	50(C)	BuLi, −23°	2300
		17(B),35(E), 36(C),55(C)	Mg	1795
$^-CH_2SiPh_3$	$BrCH_2SiPh_3$	35(A),42(A),50(B) 60(A-B)	BuLi, −78°	2300
$^-CH_2SiMePhNp$-1	$BrCH_2SiMePhNp$	50(B)	BuLi, −78°	2300
$^-CHMe—SiPh_3$	$Br—CHMe—SiPh_3$	50(D)	BuLi, −78°	2300
$Ph_3SiC^-{=}CH_2$	$Ph_3Si—CBr{=}CH_2$	50(B)	BuLi, −23°	2300
$^-CHEt—SiMe_3$	$PhS—CHEt—SiMe_3$	32(B),37(D)	LDMAN, −78°	257

$^{-}CMe_2$—$SiMe_3$	PhS—CMe_2—$SiMe_3$	32(B)	LDMAN, −78°	257
$Me_3SicycloPr^{-}$ and subst'd.	1-PhS-1-Me_3Si-cyclopropanes	32(A),33(A),35(A), 37(A)	LDMAN, −78°	257
$^{-}CHPe$—$SiPh_3$	CH_2=$CHSiPh_3$	50(E),77(B)	BuLi,rt	1797
$^{-}CH(SiMe_3)$—CH_2Bu-*t*	CH_2=$CHSiMe_3$	77(D)	*t*BuLi, −40°	2303
Me_3Si—Cx^{-}	1-Me_3Si-1-PhS-cyclohexane	32(C)	LDMAN, −78°	257
$^{-}CHPh$—$SiMe_3$	Br—CHPh—$SiMe_3$	92(D)	Mg	1795
$^{-}CHPh$—$SiPh_3$	$PhCH_2SiPh_3$	1(B),50(C),92(B)	BuLi,rt	2300
$^{-}CHBzl$—$SiPh_3$	CH_2=$CHSiPh_3$	50(B-E),77(A-B) 92(D)	PhLi,rt	1654,1797 2300
$^{-}CH(SiPh_3)CH_2Ar^{t}$	CH_2=$CHSiPh_3$	77(D-E)	$ArLi^{t}$,rt	1797
Ph_3SiC^{-}=CHPh	Ph_3Si—CBr=CHPh	50(B)	BuLi,rt	2300
$^{-}CPh_2$—$SiPh_3$	Ph_2CH—$SiPh_3$	1(A),77(A)	BuLi, −78°	2300

For footnotes, see Table 4.21.

TABLE 4.18. $^{-}C—GeAr_3$, $^{-}C—SnAr_3$, and $^{-}C—PbAr_3$

Equivalence	Source	Electrophiles (Yield)	Reaction Conditions	Reference
$^{-}CH_2GeMePhNp$-1	$BrCH_2GeMePhNp$	50(B)	BuLi,rt	2300
$^{-}CH_2GePh_3$	$BrCH_2GePh_3$	50(B)	BuLi,rt	2300
$^{-}CH_2SnPh_3$	ICH_2SnPh_3	32(D),35(B), 37(D),41(C)	BuLi, −70°	1799
$^{-}CH_2PbPh_3$	$CH_2(PbPh_3)_2$	35(D),37(C),42(C), 92(B),Ph_3GeBr(B)	BuLi, −100°	1799, 1652

TABLE 4.19. $^-C—PR_2$, $^-C—PAr_2$, $^-C—SbAr_2$, and $^-C—TeAr$

Equivalence	Source	Electrophiles (Yield)	Reaction Conditions	Reference
$^-CH_2PMe—Ph$	$Me_2P—Ph$	50(D)	*t*BuLi,rt	1800
$^-CH_2PHex_2$	$Hex_2P—Me$	50(D)	*t*BuLi,rt	1800
$^-CH_2PPh_2$	$MePPh_2$	42(B),50(C), 77(B),94(C)	*t*BuLi,rt	1802
$^-CHPe—PPh_2$	$CH_2=CHPPh_2$	50(C),60(C)	BuLi,rt	1655
$^-CH(PPh_2)—CH_2Bu$-*t*	$CH_2=CHPPh_2$	50(A)	*t*BuLi,rt	1655
$^-CHPh—PPh_2$	Bzl—PPh_2	23(?)	PhLi,Δ	1801
$^-CH_2SbPh_2$	$(Ph_2Sb)_2CH_2$	2(E),3(E),32(B), 35(D),37(C)	PhLi, −70°	1766
		2(B-C)	(1) PhLi, −70°, (2) CuCl, −50° to rt	1766
$^-CH_2TePh$	$(PhTe)_2CH_2$	35(C),42(C),77(?)	MeLi or BuLi, −78°	1791

TABLE 4.20. $^-CY—N<$; Y = Activating Group[a] (See also Tables 2.1.–2.5.)

Equivalence	Source	Electrophiles (Yield)	Reaction Conditions	Reference
$^-CH(COOH)NH_2$	$Ph_2C{=}NCH_2COOEt$	1(A),2(A-B),3(A), 9(B-C),23(B)	LDA, −78° or $Bu_4N^+HSO_4^-$,HO^-; H^+,Δ	1805
$^-CH(COOEt)NH_2$	PhCONH—CH_2COOEt	1(C?),9(D?), 23(B?),55(A?)	LDA,TMEDA	1806, 2304
	$EtOCOCH_2$—NC	28(C?)	NaOEt,rt;H^+/Δ	1807
	$EtOCOCH_2$—N=$C(SR)_2$	1(B?),2(A?),9(C?), 23(A?)	KOBu-*t*, −70°; H_2O_2/HCOOH,0°	1694
$^-CMe(COOEt)NH_2$	EtOCO—CHMe—NC	28(C?)	NaOEt,rt;H^+/Δ	1807
$^-C(COOEt)(Bu\text{-}i)(NH_2)$	*i*BuCH(COOEt)—N=$C(SR)_2$	1(B?)	KOBu-*t*, −70°; H_2O_2/HCOOH,0°	1694
$^-C(COOEt)(CH_2C_6H_4Br\text{-}p)(NH_2)$	EtOCOCHArN=$C(SMe)_2$	2(A)	KOBu-*t*, −70°; H_2O_2/HCOOH,0°	1694
$^-C(COOH)(CH_2OH)(NH_2)$	Footnote *ii*	1(B),2(B?),17(C?), 23(C?),77(A?)	LDA, −78°; H^+/Δ	1808
$^-CH(SMe)$—NHBu-*t*	$MeSCH_2N(NO)$—Bu-*t*	23(A?),35(A?), 38(B?),86(A?)	LDA, −78°	1705
^-CH—$S(CH_2)_nNH$ (ring) n = 5–6	*t*BuN=CH—N<	3(C?),42(C?)	*t*BuLi, −78° to −25°	2305

For footnotes, see Table 4.21.

TABLE 4.21. α-Heterosubstituted Carbanions, Other Than Nitrogen, with an Additional Activating Group (See also Tables 2.1.–2.5.)

Equivalence	Source	Electrophiles (Yield)	Reaction Conditions	Reference
$^-$CHOH—COOH	$ROCH_2$-het[ee]	2(B-C),9(D), 23(B)	BuLi, −78°; H^+/Δ	1817
$^-$CH(SH)—OR	$ArCOSCH_2OMe$[a]	1(B?),77(A?)	*s*BuLi,TMEDA, −98°	1750
$^-$CPh(SEt)—CN	EtS—CHPh—CN	17(B)	LDA	1447
$^-$CMe(SeMe)—COOMe	MeSe—CHMe—COOMe	39(B)	KDA, −100°	1824
$^-$CH(SPh)—$SnPh_3$	Ph_3SnCH_2SPh	35(A)	LDA, −50°	1827
$^-$CH(SPh)—$PbPh_3$	Ph_3PbCH_2SPh	35(C)	LDA, −50°	1827
$^-$CHI—$MetPh_3$	Ph_3Met—CHI_2 Met = Si,Ge, or Sn	35(B-C)	PhLi, −90°	1825

[a]Ar = 2,4,6-tri-isopropylphenyl
[b]HNR_2 = 2,2,6,6-tetramethylpiperidine.
[c]HNR_2 = 4,4-ethylenedioxy-2,2,6,6-tetramethylpiperidine.
[d]Ar = 3-benzyloxy-4-methoxyphenyl
[e]Y = CO—CPh_3.
[f]Y = CO—Bu-*t*.
[g]Diastereoselective.
[h]Chiral R; products are in (*R*) configuration at C-1.
[i]Chiral R; products are in (*S*) configuration at C-1.
[j]Ar = 2,4,6-tri-R-phenyl (R = *t*-Bu or *i*-Bu) or Ar = 2,6-bis(Me_2N)-3,5-($iPr)_2$phenyl.
[k]Y = (−)-α-MeO-α-trifluoromethylphenylacetyl; the reagent is in (*R*) configuration at C(Sn); the synthon is in (*S*) configuration.
[m]Y = O—CHMe—OEt.
[n]Ar = 2,4,6-triethylphenyl.
[p]Ar = 2,4,6-triisobutylphenyl.
[q]het = a thiazolidine or oxazolidine.

TABLE 4.21. (*Continued*)

[r]Pyrolysis gives vinyl chlorides.[1768]

[s]Treatment of the adduct in basic medium gives the α-epoxysulfoxide.[1769]

[t]Ar = *o*- or *p*-tolyl, *p*-Me_2Nphenyl or 1-naphthyl.

[u]Various $^-$CHR(CN)—NR′$_2$ and $^-$CHR(CN—NHR′ synthons are shown in Tables 4.5. and 4.8.

[v]2-Methyl-2-(2-nitroethyl)-1,3-dioxolane.

[x]Dihydroindole 2-anion; source is *N*-(*t*BuN=CH—)—H_2-indole.

[y]Tetrahydroquinoline 2-anion; source is *N*-(*t*BuN=CH—)—H_2-quinoline.

[aa]Tetrahydroisoquinoline 1-anion.

[bb]*N*-*Y*-tetrahydroisoquinoline.

[cc]See **1** below; source is *N*-(*t*BuN=CH—)-derivative.

[dd]See **2** below; source is *N*-nitroso derivative.

[ee]Source is a camphor-based oxazoline (see text, Formulas **80–82**); enantioselective alkylation.

[ff]See text, Scheme (23), Formula **5**.

[gg]NR_2 = morpholino.

[hh]$^-$CHR—NR′R″ = quinuclidine 2-anion; source is quinuclidine *N*-oxide.

[ii]See **3** below.

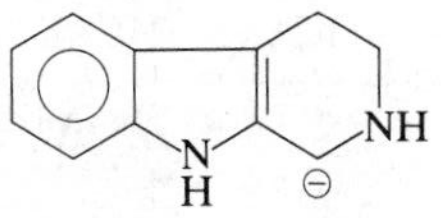

1 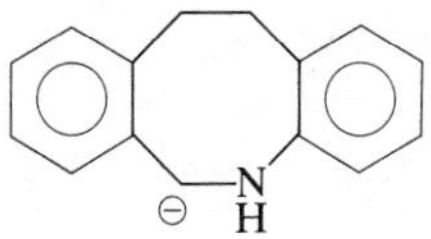 **2** 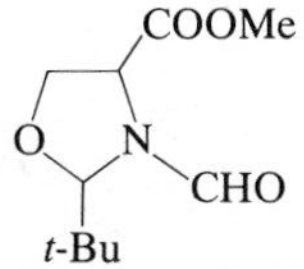 **3**

NEW ADDITIONS

(a) $^{-}$C—N<: Refs. 2180, 2214–2220; Ref. 2221 (assymmetric alkylation).
(b) $^{-}$CW—N<, W = additional activating group: Refs. 2222–2226; Ref. 2227 ($^{-}$CHOH—COMe for aldehydes).
(c) $^{-}$C—OH, $^{-}$C—OR: Refs. 2229–2231; Ref. 2228 (1-alkoxycyclopropyl anion); 2232 ($^{-}CH_2OMe$ for ArBr); 2233 ($^{-}CH_2OMe$ & $^{-}CH_2OEt$); 2234 ($^{-}$CHOEt—CH_2C═C); 2235, 2236 ($^{-}CH_2OH$); 2237 ($^{-}CH_2OSiR_3$); 2238 ($^{-}CR_2$—OR′).
(d) $^{-}$CW—OH or $^{-}$CW—OR, W = additional activating group: Refs. 2239–2241; Ref. 2242 ($^{-}$CHOR—CH═CH_2, diastereoselective addition to aldehydes); 2243 ($^{-}$CHOMe—COOR).
(e) $^{-}$C—SH, $^{-}$C—SR: Refs. 2244, 2245; Ref. 2246 ($^{-}$CHSEt—CH═CH_2, diastereoselective addition to aldehydes).
(f) $^{-}$C—X: Ref. 2247; Ref. 2248 ($^{-}$CHCl—CH═CH_2 for epoxides); 2249, 2250 ($^{-}$CxMe—R, X = Cl or Br, R = H or alkyl).
(g) $^{-}$C-metal: Refs. 2251, 2252 ($^{-}$C—SnR_3); 2253 ($^{-}$C—AsR_2).

List of Nucleophiles

Reagent	Equivalency
1. RMgX	R^-
2. Allylic or benzylic MgX	$C{=}C{-}C^-$, $Ar{-}C^-$
3. ArMgX	Ar^-
4. RLi	R^-
5. ArLi	Ar^-
6. R_2CuLi, etc.	R^-
7. Ar_2CuLi, etc.	Ar^-
8. Allylsilane or stannane	$C{=}C{-}C^-$
9. C=CLi or MgX	$C{=}C^-$
10. Acetylide	$C{\equiv}C^-$
11. Enol silyl ether	$RCO{-}C^-$, etc.
12. Enolate anion	$RCO{-}C^-$, etc.
13. Malonate, diketone, or ketoester anion	$^-C(COOR)_2$, $^-C(COR){-}COOR'$
14. Nitronate anion	$^-C{-}NO_2$
15. Sn enolate	$RCO{-}C^-$, etc.
16. Zn enolate (Reformatsky)	$RCO{-}C^-$, etc.
17. Enamine	$RCO{-}C^-$, etc.
19. Lithio imine, imidate	$RCO{-}C^-$, etc.
20. $Ph_2PO{-}C^-$	
21. $(RO)_2PO{-}C^-$	
22. $RN(NO){-}CH_2^-$	$^-CH_2{-}NHR$
23. Dithioacetal anion	^-CHO, RCO^-
24. Alkene (ene reaction)	$^-C{-}C{=}C$
25. ArH (Friedel–Crafts)	Ar^-
30. Amine	RNH^-, etc.
31. Isocyanate anion	$^-NH_2$
32. Isothiocyanate anion	
33. Azide anion	$^-NH_2$
34. Amide or $RCO{-}NR^-$	RNH^-
35. ROH, ArOH	RO^-, ArO^-
36. Alcoholate, phenolate	RO^-, ArO^-
37. Carboxylate	$RCOO^-$, HO^-
38. RSH, ArSH	RS^-, ArS^-
39. Thiolate (incl. $RCO{-}S^-$)	RS^-, ArS^-
40. Thiocyanate anion	
41. Phosphine or phosphite	
42. Cyanide anion	^-CN, ^-COOH
43. Other	

YIELDS

A >80%
B 60–80%
C 40–60%
D 20–40%
E <20%

TABLE 5.1. $^{+}$C—CHO

Equivalence	Source	Nucleophiles (Yields)	Reaction Conditions	Reference
$^{+}CH_2$—CHO	CH_2=CH—SAr=NTs	1(B-C?),3(B?) 13,30,36,39(A-C)		1636,2109
	CH_2=C(SMe)—SO—Me	12(A)[i]	;HBF_4	1845
		12(A-B),13(A)	;$HClO_4$	718
	CH_2=CH—SOMe	12(A?)		1850
	CH_2=CH—SOAr	13(A-C?),14(A?), 30(A?),36(A?), 38(A?)		1851,1852
		6(B-D?)		1631
	CH_2=C(CN)—NMe—Ph	12(A-B),13(A) 19(A-C)	;H_3O	1630
	$BrCH_2CH(OEt)_2$	23(?),42(A?)		428
$^{+}$CHMe—CHO	MeCH=$CHNO_2$ (*E*) or (*Z*)	12(A-D?) diastereoselective addition		1861
	MeCH=C(SMe)—SOMe	12(A)	;$HClO_4$	720
$^{+}$CHBzl—CHO	PhCH=$CHCH(OEt)_2$	4(A-B?),5(C?)		1843
$^{+}$CHR—CHO[c]	RCH=CH—NO_2	12(A-C?),22(B)		474
		23(A-D)		474
		12(A-D?)[b]		1860
	RCH=C(SMe)—SOMe	12(A),13(A-B) 17(A)	;$HClO_4$	717
$^{+}$CHR—CHO	R—CBr=$CHOSiMe_3$	2(C),3(B)	Ni/P cat;H_3O^+	1844
	RCH=C(SMe)—SOMe	12(D),19(C)	HBF_4	719
$^{+}$CHPh—CHO[k]	ArSO—CH=CHPh	13(B?)		1856,2110
$^{+}$C—CHO	Cl—C—CHO	39(A-C)		1842

For footnotes, see Table 5.6.
Note that $^{+}$C—CHO synthons result when $^{+}$C—CO^- synthons (Sect. 2.17.5.1. in Chapt. 2; Table 2.10.) are quenched with a proton source.

TABLE 5.2. $^{+}C—COR$ **and** $^{+}C—COAr$

Equivalence	Source	Nucleophiles (Yields)	Reaction Conditions	Reference
$^{+}CH_2—COMe$	$ClCH_2COMe$	34(B)		2112
		41(B)		2113
	$BrCH_2COMe$	13(A-B)		1919,1921
		37(C)		1878
		41(C)		2115
		$R_3BC{\equiv}CR'$(B)		2114
	$BrCH_2—CMe{=}N—OMe$	25(A-B)	$AgBF_4$	1962
	$BrCH{=}CMe—OSiMe_3$	1(B-C),3(B-D)	Ni/P cat;H_3O^+	1844
	$CH_2{=}CMe—NO_2$	11(A-D)	$TiCl_4$	1944,2111
		12(D?)		474
		12(D?)[b]		1860
	$CH_2{=}C(NO_2)CH_2OCOBu\text{-}t$	1(C?),4(B-D?), 5(A-B?),9(B-C?), 12(A-C?),23(A?), 30(A?),39(B?)		1932
	$HC{\equiv}C—CH_2OH$	30[g](C-D)	$Zn(OAc)_2—Cd(OAc)_2$	1964
	$MeS—CHCl—Me$	25(A)	$SnCl_4$;Zn—AcOH	1961
	$XCH_2—CX{=}CH_2$	20(B),21(B-C)	;H_3O^+ or Hg^{2+}	1959,1960
	$ClCH_2—CCl{=}CH_2$	11(B-C)	Pd/P cat;$Hg(OAc)_2$	1946
	$AcOCH_2—CCl{=}CH_2$	13(C-D)	Pd/P cat;H^+/Hg^{2+}	1951
	$BrCH_2C{\equiv}CSiMe_3$	13(B)		1924
	$ClCH_2—C(OSiMe_3){=}CH_2$	13(A-B), 19(A-B)	$(COOH)_2$	1950
$^{+}CH_2—COEt$	$ClCH_2COEt$	37(E)		1876
$^{+}CH_2COCH_2R$	$CH_2{=}C(OEt)—CHR—OAc$	13(B-C)	DBU, Pd/P cat	1952
$^{+}CH_2CO—CMe{=}CH_2$	$ClCH_2CH{=}CMe—CH_2SePh$	34(C),36(C-D)	NaH,HMPA;MCPBA	95

$^+CH_2$—COBu-*t*	$ClCH_2COBu$-*t*	42(A)		1913
$^+CH_2$—COR	$ClCH_2COR$	33(A)		1894
		34(B)		1897
		39(A-B)		1905
	$BrCH_2COR$	30(A-D)		1890
		37(A-B)		1975
		38(A-C)	Et_3N	1898,1899
		$^-$SeCOR (A-B)		1908
		42(B)		1910
		R_4B^-	KOBu-*t*	2116
	CH_2=CR—SO—Ph	12(C),13(B)	F_3—Ac_2O,F_3AcOH	1855
	X—CH_2COR	30(A-C),34(B)		1889
		39(A-C)		1896
$^+CH_2$—COPh	MeS—CHCl—Ph	25(B?)	$SnCl_4$	1961
	$BrCH_2COPh$	30,32,34,41,43		1892
		41(A-B)		2117
		42		1912
$^+CH_2$—COAr	$ClCH_2COAr$	H_2O(C-E)		2119
		30(C)		1889
	$BrCH_2COAr$	30(A-D),37(D)		1873
		37(B-E)		1877
		39(A-B)		1873,1904
		42(C)		1915
		$^-$SeCOR(A-B)		1908
	BrCH=CPh—$OSiMe_3$	1(A-B),3(A)	Ni/P cat;H_3O^+	1844
	CH_2=CAr—$OSiMe_3$	11(B-C)[a],35(B)	PhIO—BF_3	1943
$^+$CHMe—COMe	Cl—CHMe—COMe	42		1914
		HO^-		2118

TABLE 5.2. (*Continued*)

Equivalence	Source	Nucleophiles (Yields)	Reaction Conditions	Reference
	Br—CHMe—COMe	36(A)		1887
	MeO—CHMe—COMe	20(C-D)	TFA;H_3O^+	1959
	MeCH=CMe—NO_2	12(D)	;$NaNO_2$—PrONO	1947
$^+$CHMe—COR	Cl—CHMe—COR	34(A)		1895
$^+$CHMe—COPh	Br—CHMe—COPh	37(B)		1874
$(CH_2)_3CH^+C{=}O$	$(CH_2)_3CBr{=}C$—$OSiMe_3$	3(D)	Ni/P cat;H_3O^+	1844
$(CH_2)_4CH^+C{=}O$	Footnote *f*	4(A?),5(A?)		573
		23(B?),30(B?)		573
	α-Cl—CX-one	13(C)		1920
	Footnote *w*	36(A),38(A)		1965
$^+$CHBzl—COMe	PhCH=CHCMe$(OEt)_2$	4(A?)		1843
$^+$CHR—$COCHR_2$	RCH_2CO—C(Y)R_2	7(B-C)[e]	LDA;H^+	1942
$^+$CHR—COR′	Cl—CHR—COR′	HO^-(B)		1864
		5	−50°	1926
		13(B-D)		1917,1918
		36(A-D)		1882–1885
		39(A)	PTC	1902
	Br—CHR—COR′	HO^-(B-D)		1863, 1869–1871
		17(A-E)		1890,1916
		30(A)		1891
		36(A),39(A-E)		1871
		39(A-B)		1906

		R_4B^-(B-E)		2120
	X—CHR—COR′	39(A-C)		1903
	RCH—CR′—OAc (epoxide, O bridging)	6(C-D)		1941
	PhN=N—CR′=CHR	38(A-B?)		1966
	Ts—NMe—N=CR′—CH_2R^m	6(C)	H_3O^+	1940
	RCH=CR′—NO_2	13(A-E)	KF	1949
$^+$CHR—COCH=CR'^d	RCH_2COCH—CR′ (epoxide, O bridging)	6(AB),7(A-C)	LDA;H^+	1942
		9(A)		
$^+$CHR—COAr	Cl—CHR—COAr	3(A-D)		1925
	Br—CHR—COAr	34(C-D)	Amberlite	1896
		36(?)		1886
$^+$CHPh—COMe	PhCH=C(NO_2)(CH_2OCOBu-*t*)	4(A?)		573
	Br—CHPh—COMe	25(C)	$AlCl_3$	1931
$^+$CHAr—COMe	ArCH=CMe—NO_2	12(B?),23(B-C?)		474
	Cl—CHAr—COMe	HO^-		1872
$^+$CHPh—COR^n	BrCHPh—CR=NNH—Y^o	13(B-D?)	PTC	1957
$^+$CHAr—COR	Cl—CHAr—COR	35(A)	Lutidine	1882
$^+$CHPh—COPh	Cl—CHPh—COPh	36(B)		1881
	Br—CHPh—COPh	25(A),35(A-B)	$AgSbF_6$	1888
$^+$CHAr—COAr	Cl—CHAr—COAr	HO^-(A)		1866
$^+$CH—COR	CH=CR—NO_2	11(A-C)	Lewis acid	2122
		12(A-E)	HMPA;H_3O^+	1948
	CHBr—CR=N—NHTs	6(B-C?),7(D?)		1936
		7(A-B)	BF_3—Et_2O	1939

TABLE 5.2. (*Continued*)

Equivalence	Source	Nucleophiles (Yields)	Reaction Conditions	Reference
$^{+}$CH—CO—CH	Br—CH—C(SePh)=C	25(B-C?)	$AgClO_4/MeNO_2$	1963
$^{+}$CH—COPh	Br—CH—COPh	15(C-D)	Ru cat	1922
$^{+}$CMe$_2$—COPh	Br—CMe$_2$—COPh	HO$^-$(B-C)	liq. NH_3 or amine	1868
$^{+}$CR$_2$—COR′	Cl—CR$_2$—COR′	4(B-E)	*i*PrMgBr	1927
	Br—CR$_2$—COR′	4(A-E)	CuI	1928
		6(A-E)		1929
		33,40		1893
		42	PTC,DMSO	1912
	X—CR$_2$—COR′	39		1900,1901
$^{+}$CR$_2$—COPh	Cl—CR$_2$—COPh	HO$^-$(A)		1865
$^{+}$CPh$_2$—COMe	Br—CPh$_2$—COMe	HO$^-$(A)		1867
$^{+}$CPh$_2$—COPh	Br—CPh$_2$—COPh	H_2O(A),35(B)		2121
$^{+}$C—COR	Br—C—COR	15(A-E)	Pd cat.	1922
		6(A)		1930
	X—C—COR	38(A)	PTC	1907
	C=C—NO_2	11(A-E)	$TiCl_4$;H_3O^+	1944

For footnotes, see Table 5.6.

TABLE 5.3. $^{+}$**C—COOH and Derivatives**

Equivalence	Source	Nucleophiles (Yields)	Reaction Conditions	Reference
$^{+}CH_2COOH$	OHC—COOH	30(A-C)		1977
$^{+}CH_2COOMe$	PhS—CHCl—COOMe	11(B-C)	$ZnBr_2$;RaneyNi	1971
$^{+}CH_2COOEt$	MeS—CHCl—COOEt[h]	25(A-B)	$SnCl_4$;RaneyNi	1975
$^{+}CH_2COOR$	$BrCH_2COOR$	16(A-D)		1972
$^{+}$CH—COOH	X—CH—COOH	31(C)	H_3O^+	1976
$^{+}$CH—COOMe	PhS—C(Cl)—COOMe	11(B-C)	$TiCl_4$ or $ZnBr_2$; Raney-Ni	1970
$^{+}$CH—COOR	X—CH—COOR	38(A-B)	PTC	1907
$^{+}$C—COOEt	Br—C—COOEt	42(A-B)	PTC	1912
$^{+}$CHR—CO—SPh	RCH=C(NO_2)—SPh	13(B),34(B-C), 36(B),$ArSO_2^-$(C)	O_3, −78°	1629
$^{+}CH_2$—CN	Br—CH_2CN	2(B-D)		1968
$^{+}CR_2$—CN	O_2N—CR_2—CN	14(A-C)	KH,DMSO	1974

For footnotes, see Table 5.6.

TABLE 5.4. Vinylogous Systems

Equivalence	Source	Nucleophiles (Yields)	Reaction Conditions	Reference
$^+CH_2CH{=}CHCHO$	$(EtO)_2CHCH_2CH{-}CH_2$ (epoxide, O bridging)	1(B),3(B-C)	;HCl	1862
$^+CH_2CR{=}CH{-}COMe$	Footnote *j*	13(B)	DBU, Pd/P cat	1952
	$Me{-}CR{=}CH{-}COMe$	13(A-B)	$PdCl_2$	1953
2-Cyclohexenone-4-cation	MeO-diene $Fe(CO)_3$-complex (review)	13		1954
$^+CH_2CH{=}CHCOOMe$	$BrCH_2CH{=}CHCOOMe$	12(B),13(D)	HMPA	1973
$^+CH_2CH{=}CHCOOR$	$BrCH_2CH{=}CHCOOR$	16(C)		1972
	$MeCH{=}CHCOOR$	13(A-B)	$PdCl_2$	1953
$^+CH{-}CH{=}CH{-}COO$ (ring-closed)	Footnote *aa*	1(?)	Retro-D-A	1969

For footnotes, see Table 5.6.

TABLE 5.5. Difunctional Synthons

Equivalence	Source	Nucleophiles (Yields)	Reaction Conditions	Reference
$^{+}CR(OH){-}CHO$	$RCO{-}CH(SR')(OR'')$[q]	1(C-E)[k]	$NCS, AgNO_3$	748
	RCOCHO	1(A-B)[k],12(A-C)[k]		1978
$^{+}CPh(OH){-}CHO$	$PhCOCH(SR)_2$	1(A-C),3(B)	;I_2	825
	PhCOCHO	1(A-B)[k],12(A-C)[k]		1978
$^{2+}C(OH){-}CHO$	MeO—CHOH—COOMe	1&1(A-C)[k]		1978
$^{+}C(OH){-}CO(CH_2)_n$ (ring)	Footnote *r*	1,3,4 or 5(A-C)[k]		1980
$^{+}CR(OH){-}COR'$	R′COCOR[v]	2(A-E)		1979
$^{+}CH(SAr){-}CO{-}R$[u]	$RCOCH_2{-}SO{-}Ar$	11(A-D)		1981
$^{+}CH(CH_2OH){-}CHO$	Footnote *z*	36(C)	;H^+;IO_4^-	2312
$^{+}CH(CH_2OMe)(COR)$	$RCO{-}CHBr{-}CH_2OMe$	Iminoester		1992
$RCO{-}C^{+}{-}C{-}OH$	Footnote *y*	6(A-B)	;$TiCl_3$	1991
$^{+}CH_2COCH_2CH_2COOH$	Footnote *p*	1,2,3,6 or 7 (B-E)	;$TiCl_4$/MeOH	1993
$^{+}CH_2COCH_2Z$[l]	$BrCH_2C(OEt){=}CHZ$	12(A),13(B)	HMPA;HCl	2124
$^{+}CH_2COCH{=}PPh_3$	$BrCH_2COCH{=}PPh_3$	12(D-E),13(D)		2123

TABLE 5.5. (*Continued*)

Equivalence	Source	Nucleophiles (Yields)	Reaction Conditions	Reference
$^{+}CH(NH_2)$—COOEt	Footnote *s*	30(B),35(A-B), 37(B),38(A)		1987
	Ph_2C=N—CH(OAc)COOEt	6 or 7(B-C)		1988
$^{+}$CH(OAr)—COOMe	BrCH(OAr)COOMe	33(A),36(B-C) 40(A-B)		1989
$^{+}$C(OH)—COOR	α-Ketoester	25(A-C)	Lewis acid	1986
	OHC—COOR	1,4 or 24(A-B)[k]		1982–1984
$^{+}$CMe(OH)—COOMe	MeCO—COOMe	8(A)		1985
$^{+}CH(COOEt)_2$	O=$C(COOEt)_2$	3 or 5(A-E),25(A-C)		1994

For footnotes, see Table 5.6.

TABLE 5.6. Bidentate Synthons (For $^{+}C—CO^{-}$-type synthons, see Table 2.10.)

Equivalence	Source	Nucleophiles (Yields)	Reaction Conditions	Reference
$^{2+}CR—CHO$	$Cl_2CR—CHO$	39(A-D)		1842
$^{\pm}CR—COR'$	$R'CO—CHX—R$	$CH_2{=}SMe_2$(C-E)		2126
$^{+}CH_2COCH_2^{+}$	$CH_2{=}C(NO_2)CH_2OAc$	14(C-E?)		2125
	$ClCH_2COCH_2Cl$	HS^{-}(A),39(B),38(B)		1998
	$CH_2{=}C(SO_n—Ph)(CH_2Cl)$ n = 1 or 2	[f](C-E?)		2001
	$CH_2{=}C(NO_2)(CH_2OCOBu\text{-}t)$	α4,12,23,25 β12		1932
$^{+}CHPh—COCH_2^{+}$ (α) (β)	$PhCH{=}C(NO_2)CHOCOBu\text{-}t$	α4;β11		573
		6(C),7(D)		1999,2000
$^{+}CHR—CO—CHR^{+}$	$Br—CHR—CO—CHBr—R$	39(A)		1996
	$TsNH—N{=}C(CHR—Br)_2$	6 or 7(A-C?)		1936
$^{+}CR_2—COCH^{+}R$	$R_2CBr—COCHBr—R$	^{-}SH(B-E)		1997
$^{+}CHR—CO—CHR^{-}$	$Br—CHR—CO—CHBr—R$	6(A-C)[x]		2002

[a]Products are symmetrical 1,4-diarylbutane-1,4-diones.
[b]Nucleophiles are β-hydroxyester enolate enantiomers.
[c]R = H, alkyl, or aryl.
[d]In cyclopentenone or -hexenone.
[e]Y = Cl or OCOPh.
[f]2-Nitro-3-pivaloyloxy-cyclohexene.
[g]Secondary only.
[h]MeS—CHCl—CN reacts similarly.

TABLE 5.6. (*Continued*)

[i]Nucleophiles are enolates of MeCO—CHR—SMe.
[j]CH_2=C(OEt)—CH(OAc)—CR=CH_2.
[k]Enantioselective reaction.
[l]Z = $PO(OMe)_2$.
[m]R′ is vinylic, *i.e.*, parent system is steroid 4-en-3-one.
[n]R = Me, Bzl, or Ph.
[o]Y = Ts or DNPH.
[p]β-(1-chlorovinyl)-β-propiolactone; products can also be diverted to RCH_2COCH=CHCOOH (*E*).
[q]The SR′ + OR″ system is a chiral 1,3-oxathiane.
[r]O=C—C(NR_2)=CH—$(CH_2)_{n-1}$, n = 3–6.
[s]Ph_2C=N—CH_2COOEt or Ph_2C=N—CH(OAc)—COOEt.
[t]Cyclic ketone α,α′-dianion.
[u]R = alkyl, phenyl, or methoxy.
[v]Either R = Me or Ph.
[w]1-Chlorocyclohexene epoxide.
[x]MeI and $PhCH_2Br$ were reported as electrophiles.
[y]α,β-Epoxycyclohexanone oxime.
[z]See Formula **A.**
[aa]See text, Scheme (43), Formula **19.**

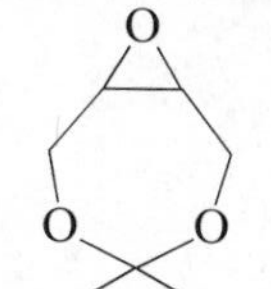

NEW ADDITIONS

Ref. 2255 ($MeCOCH_2^+$ and $^+CH_2COCH_2^+$); 2256 ($^+$C—C=C—CHO); 2257 ($^+CHNH_2$—COOMe, ass. reaction with malonates).

TABLE 6.1. $^{-}CH_2CH_2CHO$

Source	Electrophiles (Yields)	Reaction Conditions	Reference
$BrCH_2CH_2$-diox[e]	3(A-C),18 + 19(C), 24(?),33(B),35(A-C), 39(A-B),41(A-D),64(?), Ar^+(A),tropone(C), RCOSAr(?)	Mg	302, 2007–2011, 2013, 2130–2138
	20(C),39(B-C)	Mg, CuBr	2139–2141
$BrCH_2CH_2CH(OR)_2$	2(B),3(?),6(B),52(A) 67(C)	Mg	2014–2016 2142–2145
	39(B-C)	Mg, CuBr	2146,2147
$ClCH_2CH_2CH(OR)_2$	34(D)	Mg, CuI	2148
$ArSO_2CH_2CH_2$-diox[e]	2(?),3(B),10(C) 23(A),30(?),59(?)	BuLi −78°	2017,2066,2127 2018,2149
$ArSO_2CH_2CH_2CH(OEt)_2$	2(C),9(C),10(C), 23(C)	BuLi −78°	2149
$PhSO_2CH_2CH_2CH(OR)_2$	6(?)	BuLi, −20°	2074
$Mes_2BCH_2CH{=}CH_2$	2(A),35(B)	MesLi or LDCA	2024
$CH_2{=}CHCHX{-}B(OR)_2$	32(?), 35(?)[a]		141
diox-$CH_2CH_2NO_2$[e]	32(C-D)	Amberlyst	2129
	24(?),70(?)	NaH or LiOEt	2150
$(MeO)_2CHCH_2CH_2NO_2$	34(?)	Base	2064,2160

TABLE 6.1. (*Continued*)

Source	Electrophiles (Yields)	Reaction Conditions	Reference
$CH_2{=}CHCH_2NR_2$	3(?),32(B),35(C), 36(C),37(C),41(B), 91(?)	*s*-BuLi, −10°	2026
$CH_2{=}CHCH_2NRR'$	32(?),37(?) diastereoselective	BuLi, −78° $(Et_2N)_3TiCl$	2128
$PhCONMeCH_2CH{=}CH_2$	2(?), 77(?)	LDA or BuLi −78°	2054
$PhNMeCH_2CH{=}CH_2$	1(?)[b],3(?)[b],30(?) 36(?),77(?),91(?)	BuLi BuLi/*t*-BuOK	2027 2151
$Me_2NPO\text{-}NMe\text{-}CH_2CH{=}CH_2$	3(B),18(B),23(B), 32(C),42(B)	BuLi, −50°	2030,2031
t-$BuN(NO)CH_2CH{=}CH_2$	35(?),37(?)	BuLi	1811
$CH_2{=}CHCH_2{-}NMe{-}CONR_2$	1(?),2(B),32(B), 35(C),37(B),42(?)	BuLi	1388
$CH_2{=}CHCH_2{-}NAr_2$	4(C),11(?),20(?) 23(?),36(?),42(A)	BuLi, −15°	2152
$ROCH_2CH{=}CH_2$ (R = Me or Bu-*t*)	2(?),3(?),9(?),86(?) 32(?),37(?)	*s*-BuLi, −65°	1818
$PhOCH_2CH{=}CH_2$	39(?) 1(?)[b]	*s*-BuLi, −65° BuLi or BuLi + *t*-BuOK, −30°	2068 1819,2033

PhOCH=CHMe	1(?)	BuLi/*t*-BuOK −30°	2033
$R_3SiOCH_2CH{=}CH_2$[c]	1(?),2(?),17(?)	*s*-BuLi	1819
$R_2NCOOCH_2CH{=}CH_2$	1(?),2(?),9(?),35(?), 36(?),41(?)[b],56(?), 86(?),91(?)[b]	BuLi, −78°	2039–2041, 2153
$CH_2{=}CHCH_2OH$	29(A)	$Pd(OAc)_2$	2156
$PhSCH_2CH{=}CH_2$	1(?),2(?), 18(?),35(?)[b] 39(?)	BuLi/*t*-BuOK BuLi, −30 to −70° *s*-BuLi, −65° Cope	1278,1640, 2033,2043 2068
i$PrSCH_2CH{=}CH_2$	19(?)	BuLi, −78° *s*-BuLi	1279,2045
$CH_2{=}CHCH_2SH$	3(?),4(?),11(?), 30(?),32(?),35(?), 37(?),41(?),42(?), 77(?),86(?),91(?)[b]	BuLi 0° *t*-BuOK/BuLi BuLi/$ClTi(OR)_3$	775,2047 2154
$PhS(O)CH_2CH{=}CH_2$	39(?)	LDA, −78°	2044,2155
$CH_2{=}CHCH(SiR'_2OR'')$—SiR_3	54(B-D)	$TiCl_4$	2034
$R_3SiCH_2CH{=}CH_2$	37(A-D), 38(?)	*s*-BuLi −78°	239,293, 2051
	25(?),32(?),33(?), 39(?),41(?),42(?), 52(?),62(?),93(?)	BuLi,0°	57
	1(?),30(?)[b],50(?)[b], 77(?)[b],91(?)	*t*-BuLi, −78° BuLi	260,2051 1136, 2048,2049

TABLE 6.1. (*Continued*)

Source	Electrophiles (Yields)	Reaction Conditions	Reference
$R_3SiCH{=}CHCH_2Br$	91(?)	Mg	2049
$Me_3SiCHXCH{=}CH_2$ X = Cl, Br	32(?),39(?),43(?), 52(?)	BF_3, $TiCl_4$ etc.	240
$CHCl{=}CHCH_2X$	47(?),91(?),93(?), 94(?)	Li—X-exchange	237

For footnotes, see Table 6.9.

TABLE 6.2. $^{-}C—C—CHO$ with Alkyl or Aryl Substituents

Equivalence	Source	Electrophiles (Yields)	Reaction Conditions	Reference
$^{-}CH_2CH(Me)CHO$	$PhNMeCH_2CMe{=}CH_2$	91(?)	BuLi	2027
	$CH_2CMe{=}CH_2$	2(?),3(B),18(B),	BuLi, −50°	2030,2031
	$NMe—PO—NMe_2$	37(C),42(B),77(?)	LDA, −78°	2054
		4(?),11(C),23(?)	BuLi, −15°	2152
		42(?)	−78°	
	$EtOCH_2CMe{=}CH_2$	2(?),18(?),37(?)[b]	s-BuLi, −65°	1818
	$ArSCH_2CMe{=}CH_2$	2(?)[b],35(?)[b]	BuLi, −30°	1640,2043
	CH_2CHMe-diox[e] | $SO_2—Ar$	23(C),35(C)	BuLi	2157
	$Me_3SiCHXCMe{=}CH_2$ X = Cl,Br	32(?),43(?)	$TiCl_4$, −78°	240
$^{-}CHMeCH_2CHO$	diox-$CH_2CHBrMe$[e]	41(B-D)	Mg	2008,2009
	$CH_2CH{=}CHMe$ | $NMe—PO—NMe_2$	3(B),23(B),10(B)	BuLi, −50°	2030
	$CH_2CH{=}CHMe$ | $OCON(Pr\text{-}i)_2$	32(?)[b]	BuLi, −78°	2040,2041,2153
		33(?),35(?)	BuLi + iBu$_3$Al	2057,2107
			BuLi + "Ti"[d]	
	$MeCH{=}CHCH_2—NAr_2$	4(?),11(?)	BuLi, −15°	2152
	$PhOCH_2CH{=}CHMe$	32(?)	t-BuLi, −78° + Et_3TiCl or Cp_3TiCl	2058
	$CH_2{=}CMeCH_2SH$	35(?),42(?)	BuLi/t-BuOK	775
	$PhSCH_2CH{=}CHMe$	32(?)	t-BuLi, −78° R_3TiCl	2058

TABLE 6.2. (*Continued*)

Equivalence	Source	Electrophiles (Yields)	Reaction Conditions	Reference
	$CHMeCH_2diox^e$ (with SO_2—Ph on CH)	3(?),10(?),18(?)	BuLi, −78°	2149
	$MeCH{=}CHCH_2SCONMe_2$	32(?)[b],35(?)[b]	LDA, −70°	2153
	$Me_3SiCH_2CH{=}CHMe$	32(?)	*t*-BuLi, −78° R_3TiCl	2058
	$MeCH{=}CHCH(SnBu_3)(OCH_2OMe)$	32(?),33(?),35(?)		2059
$^-CHRCH_2CHO$	$RCH{=}CHCH_2SSCNMe_2$ R = Me,Et,*i*Pr	35(?)	LDA	2056
	$ArSO_2$—$CHRCH_2diox^e$ R = Bu or $(CH_2)_6CHO$	32(?)	EtMgBr	2066
$^-CMe_2CH_2CHO$	$CH_2CH{=}CMe_2$ (with NMe—PO—NMe_2 on CH_2)	23(C)	BuLi	2030
	$Me_2C{=}CHCH_2$—NAr_2	4(?),11(?),20(?)	BuLi, −15°	2152
	$CH_2CH{=}CMe_2$ (with O—CON(Pr-*i*)$_2$ on CH_2)	32(?),35(A)	BuLi, −78°	2040,2041, 2153
	$PhSCH_2CH{=}CMe_2$	1(?)[b],18(?)[b],32(?), 35(?),36(?),37(?)	BuLi, −30°	2042,2065

Footnote *g*	ROCON-*i*Pr_2	35(?),37(?)	BuLi	2040,2041
$^-CH_2CH(Ph)CHO$	$PhCONMeCH_2CPh{=}CH_2$	2(?),77(?)	LDA or BuLi −78°	2054
$^-CHPhCH_2CHO$	$PhNMeCH_2CH{=}CHPh$	1(?),30(?),35(?), 91(?)	BuLi BuLi/*t*-BuOK	2027 2158
	$CH_2CH{=}CHPh$ | NMe—PO—NMe_2	4(C),7(B)	BuLi	2030
	$PhCH{=}CHCH_2NRR'$	1(D),2(E),9(B), 17(D) (enantiosel.)	*t*-BuLi 0° to −100°	2060
	$PhCH{=}CHCH_2OR$	1(C),2(D),6(?),7(?) 8(?),9(D),17(E)	*s*-BuLi, −100° LDA −78°	2061
$^-CHRCH_2CHO$	$RCH{=}CHCHO$ R = Ar, Pr	29(C-D)	$Pd(OAc)_2$	2159

For footnotes, see Table 6.9.

TABLE 6.3. $^{-}$CR—C—CHO: R = Functional

Equivalence	Source	Electrophiles (Yields)	Reaction Conditions	Reference
$^{-}CH(CH_2CHO)(COOMe)$	$R_2NCOOCH_2CH{=}CH_2$	1(?),23(?),56(?)	LDA	2038
$^{-}CMe(CH_2CHO)(COOMe)$	$R_2NCOOCH_2CH{=}CHMe$	23(?)[b],56(?)	LDA	2038
$^{-}CH(CMe_2CHO)(COOMe)$	Footnote *h*	1(B),17(B)	LDA, −78°	2067
$^{-}CR(CH_2CHO)(COOMe)$	$PhSCH_2CH{=}CRCOOR'$ R = H, Bzl, Et	1(?),2(?)	*t*-BuOK, −78°	2063
	R = H or Me	1(?),18(?),20(?)	LDA, −78°	2062
$^{-}CH(CH_2CHO)(SiMe_3)$	$CH(SiPh_3){=}CHCH_2SiMe_3$	91(?)	BuLi	2048

$^{-}C(SiMe_3)$—C—CHO	$CHR(SiMe_3)$—CR′=CH—NMePh, R = H or Ph, R′ = H or Me	2(B),3(A-B),77(B)	BuLi	2027
$^{-}CH(SnBu_3)CH_2CHO$	$CH(SnBu_3)$=$CHCH_2SOPh$	39(?)	—	2044
$^{-}CH_2CH(CH_2OH)CHO$	Footnote *i*	1(?),30(?)[b],35(?), 91(?)	*s*-BuLi, −78°	2055
$^{-}CH(NO_2)CH_2CHO$	$O_2NCH_2CH_2CH(OR)_2$	32(?),34(?)	Base	2129,2150, 2160,2064
$^{-}CH(SO_2—Ph)CH_2CHO$	$CH_2(SO_2—Ph)CH_2CH(OR)_2$	2(B),9(C),10(C) 23(A),30(C)	BuLi	2018,2066, 2149
$^{-}C(SO_2—Ar)$—C—CHO	$CHR(SO_2—Ar)$—CHR-diox[e]	3(B),10(D),18(B), 23(?),35(?)	BuLi, −78°	2149,2157
$^{-}CH(SMe)CH_2CHO$	$MeSCH=CHCH_2SMe$	19(?)	*s*-BuLi, −78°	1279

For footnotes, see Table 6.9.

TABLE 6.4. $^{-}CH_2CH_2COR$: **R = Alkyl**

Equivalence	Source	Electrophiles (Yields)	Reaction Conditions	References
$^{-}CH_2CH_2COMe$	$ArSO_2(CH_2)_2diox^f$	9(?),15(?),23(C), 30(?),35(C),58(?), 59(?)	BuLi, −78°	2013,2017, 2018,2070, 2157,2161
	$R_2NCOOCHMeCH{=}CH_2$	1(?),32(B)[b],33(?)[b], 36(?)[b],56(?)	LDA, BuLi, −78°	2038–2040, 2153
	$ArSCHMeCH{=}CH_2$	2(?),3(?)[b],35(?)[b]	BuLi −30° to −78°	1640,2043
	$BrCH_2CH_2diox^f$	18(?),52(A),60(?)	Mg	2065,2070, 2134,2144
	$O_2NCH_2CH_2diox^f$	32(?)	Alumina	2076
	$MeCCl{=}CH{-}CH_2X$	2(?),18(?),47(?), 91(?),93(?),94(?)	Li—X exchange	237
	$CH_2{=}CH{-}CHMe{-}NAr_2$	4(B),11(?),23(?), 42(?)	BuLi, −15°	2152
	$Me_3SiCH(Cl)CH{=}CHMe$	43(?)	$TiCl_4$	240

$^{-}CH_2CH_2COEt$	$CH_2{=}CHCHEt{-}OH$	29(?)	$Pd(OAc)_2$	2156
$^{-}CH_2CH_2COBu$-*t*	Footnote *j*	95(?)	−70°	2078
$^{-}CH_2CH_2COR$	$MeCH{=}CR{-}OSiR'_3$	1(B),2(B)	*t*-BuLi, −10°	2162
	$RCOSiMe_3$	34(?),39(?),62(?)	$CH_2{=}CHMgBr$ and BuLi	2163
		1(?),2(?)		2077
	$Ph_2PO(CH_2)_2$-diox[f]	32(?),35(?),37(?)	BuLi	2071,2072
$^{-}CH_2CH_2COCH{=}CH_2$	$(CH_2{=}CH)_2CHOSiMe_3$	1(?),2(?),3(?), 5(?),6(?),9(?), 17(?),18(?),20(?), 23(?),32(?),36(?), 37(?),42(?),77(?), 36(B),91(?),92(?)	*s*-BuLi, −78°	2080–2082
$^{-}CH_2CH_2COCH{=}C$	$CO{-}CH_2CH_2SnBu_3$ | $CH_2{-}PO(OPr\text{-}i)_2$	2(B),17(B),42(B), 70(B),77(B),90(B)	BuLi	2083

For footnotes, see Table 6.9.

TABLE 6.5. $^{-}$**C—C—COR**

Equivalence	Source	Electrophiles (Yields)	Reaction Conditions	References
$^{-}CH_2CHMeCOMe$	$ArSO_2CH_2CHMe$-diox[f]	1(C),35(?)	BuLi	2157
2-Cx-one-CH_2^-	Footnote *p*	39(B),62(B),74(B)	$Hg(OAc)_2$, $NaBH_4$	2164
		CO(C)	$Hg(OAc)_2$, $PdCl_3$	1580
Footnote *q*		36(?)	BuLi, −78°	2040
$^{-}CHMeCH_2COMe$	CHMe—OCO—N(Pr-*i*)$_2$ with CHMe bonded to CH=CHMe	32(B),91(?)	BuLi, −78°	2153
$^{-}$CHR—CHR′—COR″	Ph_2PO—CHR—CHR′	32(?),35(?),37(?)	BuLi	2071,2072

For footnotes, see Table 6.9.

TABLE 6.6. $^{-}$**C—C—CO—Ar**

Equivalence	Source	Electrophiles (Yields)	Reaction Conditions	References
$^{-}CH_2CH_2COPh$	CH_2=CHCHPh—NMePh	1(?),30(?),35(?) 36(?),77(B),91(?)	BuLi	2027,2165
	CH_2=CHCHPh—NMeCOPh	2(?),77(?)	BuLi, −78°	2054
Footnote *k*		1(A)	BuLi	2086
•		2(A)	LDA, −78°	2079

For footnotes, see Table 6.9.

TABLE 6.7. $^-$**C—C—COR or** $^-$**C—C—COAr with Additional Functionality**

Equivalence	Source	Electrophiles (Yields)	Reaction Conditions	References
$^-CH_2CH_2COCOOMe$	$MeCH{=}C(COOMe)(NMe{-}Ph)$	1(C),2(B),32(C) 35(C),37(C)	LDA or KDA −78°	1153
$^-CH(CH_2COMe)(COOMe)$	$CH_2{=}CHCHMeOCONR_2$	1(?),23(?),77(?)	LDA	2038
$^-CH(CH_2COMe)(COOR)$	$CH_2COSiMe_3^f$ / CH_2-diox	37(B),41(B),55(C)	BuLi, −60°	2075
$^-CH(CR^2R^3{-}COR^1)(COOMe)$	PhSCHMeCH=CHCOOR Footnote *h*	1(?),2(?)	*t*-BuOK, −78°	2063
	R^1 = Me or Bu-*t*, R^2 = H, or R^1,R^2 = $(CH_2)_4$	1(A-B),2(A-B), 17(A-B),23(B)	LDA, −78°	2067
	Footnote *l*	42(A),85(A),89(B)	$TiCl_4$	2085,2166

$^{-}CH(SMe)(CH_2COCH{=}CH_2)$	$CH_2{=}CHC(OSiEt_3){=}CHCH_2SMe$	1(B),2(B),18(B)	LDA	2082
$^{-}CH(SO_2{-}Ar)(CH{-}COMe)$	$ArSO_2{-}CH{-}CH_2diox^{f}$	1(A),23(A),30(C), 35(A)	BuLi, −70°	2018,2157, 2161
$^{-}CH(CH_2COPh)(COOMe)$	Footnote *h*	1(A)	LDA, −78°	2067
$^{-}CH(CH_2COPh)(SiMe_3)$	$Me_3SiCH_2CH{=}C(Ph)(NMe{-}Ph)$	1(B),2(A),3(A), 9(B),77(B)	BuLi	2027
$^{-}CH(CH_2COPh)(P(O)Ph_2)$	$Ph_2POCH_2CH_2{-}(RO)_2CPh$	32(?),35(?),37(?)	BuLi	2071,2072

For footnotes, see Table 6.9.

TABLE 6.8. $^{-}$**C—C—COOR: R = H, Alkyl, or Aryl**

Equivalence	Source	Electrophiles (Yield)	Reaction Conditions	References
$^{-}CH_2CH_2COOH$	CH_2=$CHCH_2OCONR_2$	32(A-B),35(B),41(B)[b]	BuLi	2041,2153
	CH_2=$CHCH_2OPO(NMe_2)_2$	37(B)	BuLi, −50°	2167
	CH_2=$CHCH_2SiR_3$	25(?),32(?),33(?) 39(?),52(?),62(?)	BuLi,0°	57
	iPr$_3$Si—C≡C—Me	18(?),23(?),30(?) 37(?)	BuLi, −20°	296
	$BrCH_2CH_2COOH$	32(C),35(C),37(C-D)	BuLi, −70° LN	2093
	$ArSOCH_2CH_2COOH$	32(?) diastereosel. 3(?),23(?),37(?),77(?)	LDA, −50°	2168,2169
	$PhSO_2CH_2CH_2COOH$	37(?),77(?)	LDA, −50°	2168
	2-vinyl-1,3-dithiane	1(?),2(?),3(?),7(?) 23(?),24(?),77(?),91(?)	LDA	1266,1270
	MeCH=C(CN)NEt_2	1(?),17(?),23(?), 30(?),39(D),91(?)	BuLi, −60° or LDA, −40°	1299
	CH_2=CHCH(SPh)($SiMe_3$)	32(?),33(?),35(?), 37(?),41(?),42(?)	s-BuLi, −78°	1281
$^{-}CH_2CH_2COOR$	CH_2=$CHCH(OR)_2$	17(?),92(?),93(?)	s-BuLi, −95°	2103
	R′CO—NR—CH_2CH=CH_2	32(A),37(A)[n]	BuLi, −78° $(Et_2N)_3TiCl$	2128

	Footnote *m*	32(A-B),34(B),35(A), 37(?),39(A),43(?), 44(B),52(B), ynone-1,4 (B)	$TiCl_4$ $ZnCl_2$	2097,2098 2171
	cycloPr$^-$ $^+SPh_2$	32(C),37(A)	Eu^{3+}; H_2O_2,HO^-	2090,2091
	$PhSO_2CH_2CH{=}C(CN)(NEt_2)$	2(C)	KOH,Aliquat	2170
	$R_2NCH_2CH_2COOMe$	32(C-D),36(D), 55(B)	Electr. red.	2172
	$Ph_3P{=}CHCH_2COOEt$	35(?)	PPh_3	2088
$^-CHMeCH_2COOH$	$MeCH{=}CHCH(SnBu_3)(OCH_2OMe)$	32(C),35(B),38(C)		2059
	$MeCH{=}CHCH(OSiEt_3)(PO(NMe_2)_2)$	1(B),3(?),23(?), 32(?),35(B),37(?)	LDA, −65°	2100
$^-CHRCH_2COOH$	$Ar_2PO{-}OCH_2CH{=}CHR$	1(C),2(C),3(D), 32(C)	BuLi	2173
	$RCH_2CH{=}C(SR')_2$ (dithiane) R = Me, Ph, H	17(?),18(?),19(?), 23(?),24(D),32(D), 35(C),36(E),37(D) 39(?)	LDA *s*-BuLi, −78°	502, 1265, 1273

TABLE 6.8. (*Continued*)

Equivalence	Source	Electrophiles (Yield)	Reaction Conditions	References
$^{-}CHR'—CH_2COOR$	$R'CH=CHCH(OSiMe_3)PO(OR)_2$	R = Phe: 1(B),2(B), 3(B),9(A),18(A), 23(A),91(C),92(C) R = Pr: 23(B),31(C) R = H: 2(E),23(E)	LDA, −78°	1222, 2099
	$RCH=CHCH(CN)O—CHMe—OEt$	32(?),37(C-D)	LDA, −78° to 0°	2102
	$PhS—CHR—CH_2COOMe$	32(C-E)	Electr. red.	2172
$^{-}CMe_2CH_2COOH$	$Me_2C=CHCH_2—OOC—N(i\text{-}Pr)_2$	32(A)	BuLi, −78°	2153
$^{-}CHR—CHR'COOH$	$RCH_2—CR=C(PO(OEt)_2)NMe_2$	1(?),2(C),3(?), 9(?),10(?),17(?), 18(?),23(?),32(D), 33(E),35(D),91(?),	LDA, −70° *t*-BuLi	2095,2096
	$RCH=CR'CH(OMe)POPh_2$ R,R′ = H, Me, Ph	1(?)[b],32(?),35(?), 77(?)	LDA, −70°	2102

$^{-}CR_2{-}CHR'COOR''$	$R_2C{=}CR'CH(CN)(NMe_2)$	1(A-B),10(?),32(C-D) 33(E),35(C-D),36(?), 37(?),42(?)	LDA, −70°	1299,1300, 1302,2094
$^{-}C{-}C{-}COOH$	$C{=}C{-}CH_2OPONMe_2$	30(A-D),32(C-D), 35(D),36(C),37(B-D), 38(D),42(C),77(B,D)	BuLi, −50°	2092, 2167
	Footnote *o*	32(A),35(B),41(B)	BuLi, −78° (iPrO)$_3$TiCl	2041
$^{-}CH(SO{-}Ar)(CH_2COOH)$	$ArSOCH_2CH_2COOH$	3(?),23(?),33(?)	LDA, −50°	2168
$^{-}CH(SO_2{-}Ar)(CH_2COOH)$	$ArSO_2CH_2CH_2COOH$	32(?),37(?),77(?)	LDA, −50°	2168,2169

For footnotes, see Table 6.9.

TABLE 6.9. $^{-}C{-}C{-}CONR_2$

Equivalence	Source	Electrophiles (Yield)	Reaction Conditions	References
$^{-}CH_2CH_2CONR_2$	$MeC{\equiv}C{-}NR_2$	1(D),2(D),18(D), 20(D),91(D)	*t*-BuLi	2104
$^{-}CHR'{-}CHR''CONHR$	$Bu_3Sn{-}CHR'$ / $RNHCO{-}CHR''$ (bonded CHR'—CHR''); R′,R″ = H or Ph	2(B),3(B),37(B), 41(B),42(B),77(A), 91(B),93(B)	BuLi, −78°	2105,2106

[a] Y = Halogen, OMe, SEt or SBu-*t*.
[b] A mixture of α- and γ-adducts is formed.
[c] R_3Si = Et_3Si or $tBuMe_2Si$.
[d] "Ti" = $ClTi(NEt_2)_3$.
[e] Diox = 2-(1,3-dioxolanyl)-.
[f] Diox = 2-(2-methyl-1,3-dioxolanyl)-.
[g] Cyclopentanecarboxaldehyde 2-anion; R = 1-cyclopentenylmethyl; similar compounds also in the pinane series.
[h] See text, Formula **121.**
[i] See text, Formula **95.**
[j] 1-*t*-Butoxy-1-trimethylsilyloxycyclopropane.

[k]1-Indanone 3-anion, from 1-*N*-pyrrolidinylindene.
[l]See text, Formula **167.**
[m]1-Methoxy-1-trimethylsilyloxycyclopropane.
[n]Enantioselective reaction; R + R′ = urea/ephedrine condensation product.
[o]See text, Formulas **52** and **53.**
[p]1-Trimethylsilyloxy-bicyclo[4.1.0]heptane.
[q]For synthon and source, see Formulas **A** and **B** below.

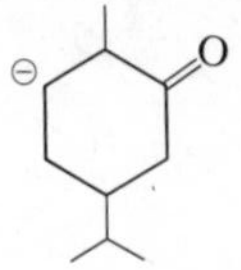

OCON(*i*-Pr)$_2$

A 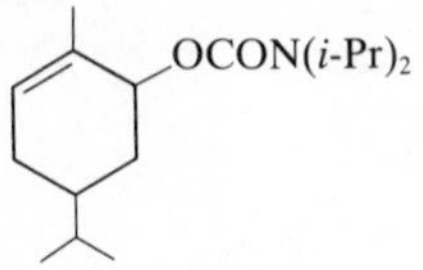 **B**

NEW ADDITIONS

(a) $^-$C—C—CHO: Refs. 2254, 2258, 2259; Ref. 2260 ($^-CH_2$—$CHSiPr_3$—CHO).
(b) $^-$C—C—COR: Refs. 2261–2263; Refs. 2264, 2265 ($MeCOCH_2CH_2^-$ and $MeCOCH_2CH^{2-}$); Ref. 2266 (2-cyclopentenone 2,3-dianion).
(c) $^-$C—C—COOH, esters, amides, nitriles: Refs. 2263, 2267, 2268; Ref. 2269 (also $^-CH_2CH_2CH_2COOEt$).

REFERENCES

1. E. J. Corey, *Pure Appl. Chem.*, **14,** 19 (1967).
2. R. K. Mackie and D. M. Smith, *Guidebook to Organic Synthesis,* Longman, London, 1982, p. 112.
3. R. K. Mackie and D. M. Smith, *Guidebook to Organic Synthesis,* Longman, London, 1982, p. 56, 58, 110, 113.
4. S. Warren, *Organic Synthesis: The Disconnection Approach,* Wiley, New York, 1982, p. 10, 51, 141.
5. J. C. Stowell, *Carbanions in Organic Synthesis,* Wiley, New York, 1979, p. 239.
6. D. Seebach, *Angew. Chem.*, **91,** 259 (1979); *Int. Ed. Engl.,* **18,** 239 (1979).
7. T. Hase and J. K. Koskimies, *Aldrichimica Acta,* **14,** 73 (1981).
8. T. Hase and J. K. Koskimies, *Aldrichimica Acta,* **15,** 35 (1982).
9. R. B. Bates and C. A. Ogle, *Carbanion Chemistry,* Springer-Verlag, Berlin, 1983, p. 75.
10. S. Warren, *Designing Organic Syntheses,* Wiley, New York, 1978.
11. D. Seebach, *Angew. Chem.*, **81,** 690 (1969); *Int. Ed. Engl.,* **8,** 639 (1969).
12. L. Eberson, *Acta Chem. Scand., Ser. B,* **38,** 439 (1984).
13. A. Pross, *Acc. Chem. Res.*, **18,** 212 (1985).
14. P. J. Stang, Z. Rappoport, M. Hanack, and L. R. Subramanian, *Vinyl Cations,* Academic, New York 1979.
15. G. Zweifel and J. A. Miller, *Org. React.*, **32,** 375 (1984).
16. R. H. Shapiro, *Org. React.*, **23,** 405 (1976).
17. J. F. Normant and A. Alexakis, *Synthesis,* 841 (1981).
18. R. M. Adlington and A. G. M. Barrett, *Acc. Chem. Res.*, **16,** 55 (1983).
19. R. F. Heck, *Org. React.*, **27,** 345 (1982).
20. G. H. Posner, *Org. React.*, **22,** 253 (1975).
21. G. H. Posner, *Org. React.*, **19,** 1 (1972).
22. B. M. Trost, *Acc. Chem. Res.*, **13,** 385 (1980).
23. D. C. Billington, *Chem. Soc. Rev.*, **14,** 93 (1985).
24. M. F. Semmelhack, *Org. React.*, **19,** 115 (1972).
25. H. Sakurai, *Pure Appl. Chem.*, **54,** 1 (1982).
26. E. Wenkert and T. W. Ferreira, *Organometallics,* **1,** 1670 (1982).
27. V. Fiandanese, G. Marchese, F. Naso, and L. Ronzini, *J. Chem. Soc., Perkin Trans. 1,* 1115 (1985).
28. M. Marsi and M. Rosenblum, *J. Am. Chem. Soc.*, **106,** 7264 (1984).
29. Y. Tanigawa, K. Nishimura, A. Kawasaki, and S. Murahashi, *Tetrahedron Lett.*, **23,** 5549 (1982).

30. J.-E. Bäckvall, *Pure Appl. Chem.*, **55,** 1669 (1983).
31. H. Takei, H. Sugimura, M. Miura, and H. Okamura, *Chem Lett.*, 1209 (1980).
32. M. Hoshi, Y. Masuda, Y. Nunokawa, and A. Arase, *Chem. Lett.*, 1029 (1984).
33. W. Smadja, *Chem. Rev.*, **83,** 263 (1983).
34. A. Marfat, P. R. McGuirk, and P. Helquist, *J. Org. Chem.*, **44,** 3888 (1979).
35. D. Seebach, K.-H. Geiss, and M. Pohmakotr, *Angew. Chem.*, **88,** 449 (1976); *Int. Ed. Engl.*, **15,** 437 (1976).
36. B. M. Trost and D. M. T. Chan, *J. Am. Chem. Soc.*, **101,** 6429 (1979).
37. B. M. Trost and P. J. Bonk, *J. Am. Chem. Soc.*, **107,** 1778 (1985).
38. B. M. Trost, T. N. Nanninga, and T. Satoh, *J. Am. Chem. Soc.*, **107,** 721 (1985).
39. E. Piers and V. Karunaratne, *J. Org. Chem.*, **48,** 1774 (1983).
40. E. Piers and V. Karunaratne, *J. Chem. Soc., Chem. Commun.*, 935 (1983).
41. E. Piers and V. Karunaratne, *J. Chem. Soc., Chem. Commun.*, 959 (1984).
42. E. Piers and V. Karunaratne, *Can. J. Chem.*, **62,** 629 (1984).
43. J. Ukai, Y. Ikeda, N. Ikeda, and H. Yamamoto, *Tetrahedron Lett.*, **25,** 5173 (1984).
44. B. M. Trost and A. Brandi, *J. Org. Chem.*, **49,** 4811 (1984).
45. B. M. Trost, N. R. Schmuff, and M. J. Miller, *J. Am. Chem. Soc.*, **102,** 5979 (1980).
46. E. Piers and B. W. A. Yeung, *J. Org. Chem.*, **49,** 4569 (1984).
47. E. Piers and B. W. A. Yeung, *J. Org. Chem.*, **49,** 4567 (1984).
48. E. Piers and J. M. Chong, *J. Chem. Soc., Chem. Commun.*, 934 (1983).
49. P. J. Kocienski, *Tetrahedron Lett.*, 2649 (1979).
50. J. J. Eisch, M. Behrooz, and S. K. Dua, *J. Organomet. Chem.*, **285,** 121 (1985).
51. A. Amamria and T. N. Mitchell, *J. Organomet. Chem.*, **210,** C17 (1981).
52. D. Scholz, *Justus Liebigs Ann. Chem.*, 98 (1983).
53. J. Hibino, S. Matsubara, Y. Morizawa, K. Oshima, and H. Nozaki, *Tetrahedron Lett.*, **25,** 2151 (1984).
54. B. M. Trost and B. P. Coppola, *J. Am. Chem. Soc.*, **104,** 6879 (1982).
55. F. T. Bond and R. A. DiPietro, *J. Org. Chem.*, **46,** 1315 (1981).
56. J. B. Hendrickson, G. J. Boudreaux, and P. S. Palumbo, *Tetrahedron Lett.*, **25,** 4617 (1984).
57. R. J. P. Corriu, C. Guerin, and J. M'Boula, *Tetrahedron Lett.*, **22,** 2985 (1981).
58. H. Sano, M. Okawara, and Y. Ueno, *Synthesis,* 933 (1984).
59. R. B. Bates, W. A. Beavers, M. G. Greene, and J. H. Klein, *J. Am. Chem. Soc.*, **96,** 5640 (1974).
60. R. B. Bates, W. A. Beavers, B. Gordon III, and N. S. Mills, *J. Org. Chem.*, **44,** 3800 (1979).
61. P. Ongoka, B. Mauzé, and L. Miginiac, *J. Organomet. Chem.*, **284,** 139 (1985).
62. S. Martin, R. Sauvetre, and J.-F. Normant, *Tetrahedron Lett.*, **23,** 4329 (1982).
63. R. Sauvetre and J.-F. Normant, *Tetrahedron Lett.*, **23,** 4325 (1982).
64. B. Tarnchompoo, Y. Thebtaranonth, S. Utamapanya, and P. Kasemsri, *Chem. Lett.*, 1241 (1981).
65. J. C. Stowell, *Chem. Rev.*, **84,** 409 (1984).
66. J. Hooz, J. G. Calzada, and D. McMaster, *Tetrahedron Lett.*, **26,** 271 (1985).
67. G. S. Mikaelian, A. S. Gybin, W. A. Smit, and R. Caple, *Tetrahedron Lett.*, **26,** 1269 (1985).

68. A. A. Schegolev, W. A. Smit, Y. B. Kalyan, M. Z. Krimer, and R. Caple, *Tetrahedron Lett.,* **23,** 4419 (1982).
69. S. Rajagopalan and G. Zweifel, *Synthesis,* 111 (1984).
70. J. Barluenga, J. R. Fernandez, and M. Yus, *J. Chem. Soc., Chem. Commun.,* 203 (1985).
71. A. M. Caporusso, F. Da Settimo, and L. Lardicci, *Tetrahedron Lett.,* **26,** 5101 (1985).
72. A. Claesson, A. Quader, and C. Sahlberg, *Tetrahedron Lett.,* **24,** 1297 (1983).
73. T. Fujisawa, S. Iida, and T. Sato, *Tetrahedron Lett.,* **25,** 4007 (1984).
74. D. Djahanbini, B. Cazes, and J. Gore, *Tetrahedron Lett.,* **25,** 203 (1984).
75. H. F. Schuster and G. M. Coppola, *Allenes in Organic Synthesis,* Wiley, New York, 1984.
76. J. E. Baldwin, R. M. Adlington, and A. Basak, *J. Chem. Soc., Chem. Commun.,* 1284 (1984).
77. J. Pornet, L. Miginiac, K. Jaworski, and B. Randrianoelina, *Organometallics,* **4,** 333 (1985).
78. C. Huynh and G. Linstrumelle, *J. Chem. Soc., Chem. Commun.,* 1133 (1983).
79. T. Jeffery-Luong and G. Linstrumelle, *Synthesis,* 738 (1982).
80. K. Furuta, M. Ishiguro, R. Haruta, N. Ikeda, and H. Yamamoto, *Bull. Chem. Soc. Jpn.,* **57,** 2768 (1984).
81. G. Zweifel, S. J. Backlund, and T. Leung, *J. Am. Chem. Soc.,* **100,** 5561 (1978).
82. N. R. Pearson, G. Hahn, and G. Zweifel, *J. Org. Chem.,* **47,** 3364 (1982).
83. J. Pornet and B. Randrianoelina, *Tetrahedron Lett.,* **22,** 1327 (1981).
84. X. Creary, *J. Am. Chem. Soc.,* **99,** 7632 (1977).
85. H. Westmijze, H. Kleijn, H. J. T. Bos, and P. Vermeer, *J. Organomet. Chem.,* **199,** 293 (1980).
86. J.-C. Clinet and G. Linstrumelle, *Synthesis,* 875 (1981).
87. J.-C. Clinet and G. Linstrumelle, *Nouv. J. Chim.,* **1,** 373 (1977).
88. G. Linstrumelle and D. Michelot, *J. Chem. Soc., Chem. Commun.,* 561 (1975).
89. D. Michelot, J.-C. Clinet, and G. Linstrumelle, *Synth. Commun.,* **12,** 739 (1982).
90. L. Brandsma and H. D. Verkruijsse, *Synthesis of Acetylenes, Allenes and Cumulenes,* Elsevier, Amsterdam, 1981.
91. A. G. Angoh and D. L. J. Clive, *J. Chem. Soc., Chem. Commun.,* 534 (1984).
92. F. Scott, G. Cahiez, J.-F. Normant, and J. Villieras, *J. Organomet. Chem.,* **144,** 13 (1978).
93. C. Sahlberg, A. Quader, and A. Claesson, *Tetrahedron Lett.,* **24,** 5137 (1983).
94. S. Araki, M. Ohmura, and Y. Butsugan, *Synthesis,* 963 (1985).
95. R. S. Brown, S. C. Eyley, and P. J. Parsons, *J. Chem. Soc., Chem. Commun.,* 438 (1984).
96. D. Djahanbini, B. Cazes, and J. Gore, *Tetrahedron* **40,** 3645 (1984).
97. E. Wenkert, M. H. Leftin, and E. L. Michelotti, *J. Chem. Soc., Chem. Commun.,* 617 (1984).
98. Y. Langlois, N. V. Bac, and Y. Fall, *Tetrahedron Lett.,* **26,** 1009 (1985).
99. T. Hayashi, M. Yanagida, M. Matsuda, and T. Oishi, *Tetrahedron Lett.,* **24,** 2665 (1983).
100. C.-N. Hsiao and H. Shechter, *Tetrahedron Lett.,* **25,** 1219 (1984).
101. R. Bloch, D. Hassan, and X. Mandard, *Tetrahedron Lett.,* **24,** 4691 (1983).
102. A. P. Kozikowski and Y. Kitigawa, *Tetrahedron Lett.,* **23,** 2087 (1982).
103. T. Chou, H.-H. Tso, and L.-J. Chang, *J. Chem. Soc., Chem. Commun.,* 236 (1985).
104. S. Nunumoto and Y. Yamashita, *J. Org. Chem.,* **44,** 4788 (1979).
105. K. Kondo, S. Dobashi, and M. Matsumoto, *Chem. Lett.,* 1077 (1976).

106. S. Halazy and A. Krief, *Tetrahedron Lett.*, **21,** 1997 (1980).
107. P. A. Brown and P. R. Jenkins, *Tetrahedron Lett.*, **23,** 3733 (1982).
108. Y. Ueno, H. Sano, S. Aoki, and M. Okawara, *Tetrahedron Lett.*, **22,** 2675 (1981).
109. K. J. Shea and P. Q. Pham, *Tetrahedron Lett.*, **24,** 1003 (1983).
110. E. Wada, S. Kanemasa, I. Fujiwara, and O. Tsuge, *Bull. Chem. Soc. Jpn.*, **58,** 1942 (1985).
111. B. Cazes, E. Guittet, S. Julia, and O. Ruel, *J. Organomet. Chem.*, **177,** 67 (1979).
112. A. Carpita, F. Bonaccorsi, and R. Rossi, *Tetrahedron Lett.*, **25,** 5193 (1984).
113. Y. N. Bubnov and M. Y. Etinger, *Tetrahedron Lett.*, **26,** 2797 (1985).
114. P. A. A. Klusener, H. H. Hommes, H. D. Verkruijsse, and L. Brandsma, *J. Chem. Soc., Chem. Commun.*, 1677 (1985).
115. Y. Naruta, N. Nagai, Y. Arita, and K. Maruyama, *Chem. Lett.*, 1683 (1983).
116. A. Hosomi, Y. Araki, and H. Sakurai, *J. Org. Chem.*, **48,** 3122 (1983).
117. A. G. Martinez and J. L. M. Contelles, *Synthesis,* 742 (1982).
118. H. Sakurai, A. Hosomi, M. Saito, K. Sasaki, H. Iguchi, J.-I. Sasaki, and Y. Araki, *Tetrahedron,* **39,** 883 (1983).
119. D. Seyferth, J. Pornet, and R. M. Weinstein, *Organometallics,* **1,** 1651 (1982).
120. H. Yasuda and A. Nakamura, *J. Organomet. Chem.*, **285,** 15 (1985).
121. E. Piers and H. E. Morton, *J. Org. Chem.*, **45,** 4263 (1980).
122. H. J. Reich, K. E. Yelm, and I. L. Reich, *J. Org. Chem.*, **49,** 3438 (1984).
123. M. Furber, R. J. K. Taylor, and S. C. Burford, *Tetrahedron Lett.*, **26,** 3285 (1985).
124. T. Hayashi, I. Hori, and T. Oishi, *J. Am. Chem. Soc.*, **105,** 2909 (1983).
125. W. E. Paget, K. Smith, M. G. Hutchings, and G. E. Martin, *J. Chem. Res. (S),* 30 (1983).
126. M. G. Hutchings, W. E. Paget, and K. Smith, *J. Chem. Res. (S),* 31 (1983).
127. T. A. Baer and R. L. Carney, *Tetrahedron Lett.*, 4697 (1976).
128. T. Fujisawa, T. Sato, Y. Gotoh, M. Kawashima, and T. Kawara, *Bull. Chem. Soc. Jpn.*, **55,** 3555 (1982).
129. Y. Abe, M. Sato, H. Goto, R. Sugawara, E. Takahashi, and T. Kato, *Chem. Pharm. Bull.,* **31,** 4346 (1983).
130. B. M. Trost, J. Cossy, and J. Burks, *J. Am. Chem. Soc.*, **105,** 1052 (1983).
131. T. Sato, M. Takeuchi, T. Itoh, M. Kawashima, and T. Fujisawa, *Tetrahedron Lett.*, **22,** 1817 (1981).
132. T. Fujisawa, T. Sato, and T. Itoh, *Chem. Lett.,* 219 (1982).
133. T. Fujisawa, M. Takeuchi, and T. Sato, *Chem. Lett.,* 1521 (1982).
134. T. Sato, M. Kawashima, and T. Fujisawa, *Tetrahedron Lett.*, **22,** 2375 (1981).
135. G. P. Boldrini, D. Savoia, E. Tagliavini, C. Trombini, and A. Umani-Ronchi, *J. Org. Chem.*, **48,** 4108 (1983).
136. J. J. Fitt and H. W. Gschwend, *J. Org. Chem.*, **45,** 4257 (1980).
137. K. Tanaka, H. Yoda, Y. Isobe, and A. Kaji, *Tetrahedron Lett.*, **26,** 1337 (1985).
138. S. E. Drewes and R. F. A. Hoole, *Synth. Commun.*, **15,** 1067 (1985).
139. D. J. Kempf, K. D. Wilson, and P. Beak, *J. Org. Chem.*, **47,** 1610 (1982).
140. M. M. Midland, A. Tramontano, and J. R. Cable, *J. Org. Chem.*, **45,** 28 (1980).
141. R. W. Hoffmann and B. Landmann, *Tetrahedron Lett.*, **24,** 3209 (1983).
142. Y. Naruta, H. Uno, and K. Maruyama, *Chem. Lett.,* 609 (1982).
143. Y. Naruta, H. Uno, and K. Maruyama, *Chem. Lett.,* 961 (1982).
144. P. Albaugh-Robertson and J. A. Katzenellenbogen, *J. Org. Chem.*, **48,** 5288 (1983).
145. I. Fleming, J. Goldwill, and I. Paterson, *Tetrahedron Lett.*, 3209 (1979).

146. K. Itoh, M. Fukui, and Y. Kurachi, *J. Chem. Soc., Chem. Commun.*, 500 (1977).
147. M. Majewski, G. B. Mpango, M. T. Thomas, A. Wu, and V. Snieckus, *J. Org. Chem.*, **46,** 2029 (1981).
148. G. Voss and H. Gerlach, *Helv. Chim. Acta,* **66,** 2294 (1983).
149. T. Fujisawa, T. Itoh, and T. Sato, *Chem. Lett.*, 1901 (1983).
150. E. Dinjus, D. Walther, H. Schütz, and W. Schade, *Z. Chem.,* **23,** 303 (1983).
151. D. Seebach, M. Pohmakotr, C. Schregenberger, B. Weidmann, R. S. Mali, and S. Pohmakotr, *Helv. Chim. Acta,* **65,** 419 (1982).
152. D. Scholz, *Chem. Ber.*, **114,** 909 (1981).
153. N. Joshi, V. R. Mamdapur, and M. S. Chadha, *J. Chem. Soc., Perkin Trans. 1,* 2963 (1983).
154. H. Takahashi, K. Oshima, H. Yamamoto, and H. Nozaki, *J. Am. Chem. Soc.*, **95,** 5803 (1973).
155. D. Scholz, *Justus Liebigs Ann. Chem.*, 264 (1984).
156. G. Rosini, R. Ballini, and M. Petrini, *Synthesis,* 269 (1985).
157. B. M. Trost and J. Vercauteren, *Tetrahedron Lett.*, **26,** 131 (1985).
158. J. Gorzynski Smith, *Synthesis,* 629 (1984).
159. A. Carpita and R. Rossi, *Synthesis,* 469 (1982).
160. R. S. Brown, S. C. Eyley, and P. J. Parsons, *J. Chem. Soc., Chem. Commun.*, 438 (1984).
161. Y. Ishino, K. Wakamoto, and T. Hirashima, *Chem. Lett.*, 765 (1984).
162. J. Tsuji, H. Kataoka, and Y. Kobayashi, *Tetrahedron Lett.*, **22,** 2575 (1981).
163. M. Trost and G. A. Molander, *J. Am. Chem. Soc.*, **103,** 5969 (1981).
164. J.-G. Duboudin, B. Jousseaume, and A. Bonakdar, *C. R. Hebd. Seances Acad. Sci., Ser. C.*, **284,** 351 (1977).
165. T. Fujisawa, Y. Kurita, M. Kawashima, and T. Sato, *Chem. Lett.*, 1641 (1982).
166. E. Wenkert, E. L., Michelotti, C. S. Swindell, and M. Tingoli, *J. Org. Chem.*, **49,** 4894 (1984).
167. E. Wenkert, V. F. Ferreira, E. L. Michelotti, and M. Tingoli, *J. Org. Chem.*, **50,** 719 (1985).
168. H.-J. Liu and I. V. Oppong, *Can. J. Chem.*, **60,** 94 (1982).
169. J. Barluenga, F. J. Fananas, and M. Yus, *J. Org. Chem.*, **44,** 4798 (1979).
170. K. Utimoto, K. Uchida, and H. Nozaki, *Chem. Lett.*, 1493 (1974).
171. J. Barluenga, J. Florez, and M. Yus, *J. Chem. Soc., Perkin Trans. 1,* 3019 (1983).
172. P. G. M. Wutz, P. A. Thompson, and G. R. Callen, *J. Org. Chem.*, **48,** 5398 (1983).
173. E. J. Corey and G. N. Widiger, *J. Org. Chem.*, **40,** 2975 (1975).
174. H. Nishiyama, H. Yokoyama, S. Narimatsu, and K. Itoh, *Tetrahedron Lett.*, **23,** 1267 (1982).
175. J. Barluenga, J. R. Fernandez, and M. Yus, *J. Chem. Soc., Perkin Trans. 1,* 447 (1985).
176. J. Pornet, B. Randrianoelina, and L. Miginiac, *Tetrahedron Lett.*, **25,** 651 (1984).
177. R. H. Wollenberg, K. F. Albizati, and R. Peries, *J. Am. Chem. Soc.*, **99,** 7365 (1977).
178. M. C. Pirrung and J. R. Hwu, *Tetrahedron Lett.*, **24,** 565 (1983).
179. N. Meyer and D. Seebach, *Chem. Ber.*, **113,** 1290 (1980).
180. P. E. Eaton, G. F. Cooper, R. C. Johnson, and R. H. Mueller, *J. Org. Chem.*, **37,** 1947 (1972).
181. R. J. Anderson and C. A. Henrick, *J. Am. Chem. Soc.*, **97,** 4327 (1975).
182. E. Dimitriadis and R. A. Massy-Westropp, *Aust. J. Chem.*, **37,** 619 (1984).

183. T. P. Burns and R. D. Rieke, *J. Org. Chem.,* **48,** 4141 (1983).
184. C. Rücker, *Tetrahedron Lett.,* **25,** 4349 (1984).
185. J. Barluenga, J. Florez, and M. Yus, *Synthesis,* 846 (1985).
186. R. M. Carlson and L. L. White, *Synth. Commun.,* **13,** 237 (1983).
187. R. M. Carlson, *Tetrahedron Lett.,* 111 (1978).
188. E. J. Corey and R. H. Wollenberg, *J. Org. Chem.,* **40,** 2265 (1975).
189. H. Nishiyama, S. Narimatsu, and K. Itoh, *Tetrahedron Lett.,* **22,** 5289 (1981).
190. J. W. Patterson, *Synthesis,* 337 (1985).
191. R. B. Miller and M. I. Al-Hassan, *J. Org. Chem.,* **48,** 4113 (1983).
192. F. Sato, H. Ishikawa, H. Watanabe, T. Miyake, and M. Sato, *J. Chem. Soc., Chem. Commun.,* 718 (1981).
193. T. Cuvigny, M. Julia, and C. Rolando, *J. Chem. Soc., Chem. Commun.,* 8 (1984).
194. D. A. Evans, G. C. Andrews, T. T. Fujimoto, and D. Wells, *Tetrahedron Lett.,* 1389 (1973).
195. D. A. Evans, G. C. Andrews, T. T. Fujimoto, and D. Wells, *Tetrahedron Lett.,* 1385 (1973).
196. V. Calo, L. Lopez, G. Marchese, and G. Pesce, *Synthesis,* 885 (1979).
197. E. Gössinger, *Tetrahedron Lett.,* **21,** 2229 (1980).
198. J. Barluenga, J. R. Fernandez, J. Florez, and M. Yus, *Synthesis,* 736 (1983).
199. Y. Naruta and K. Maruyama, *J. Chem. Soc., Chem. Commun.,* 1264 (1983).
200. A. M. Moiseenkov, E. V. Polunin, and A. V. Semenovsky, *Tetrahedron Lett.,* 4759 (1979).
201. K. Kondo and M. Matsumoto, *Tetrahedron Lett.,* 391 (1976).
202. J. Flahaut and P. Miginiac, *Helv. Chim. Acta,* **61,** 2275 (1978).
203. P. S. Reddy and J. S. Yadav, *Synth. Commun.,* **14,** 327 (1984).
204. A. V. Rama Rao, J. S. Yadav, G. V. M. Sharma, and K. S. Bhide, *Synth. Commun.,* **14,** 321 (1984).
205. W. E. Willy, D. R. McKean, and B. A. Garcia, *Bull. Chem. Soc. Jpn.,* **49,** 1989 (1976).
206. D. Michelot, *Synthesis,* 130 (1983).
207. Y. Le Merrer, A. Duréault, C. Gravier, D. Languin, and J. C. Depezay, *Tetrahedron Lett.,* **26,** 319 (1985).
208. M. Furber and R. J. K. Taylor, *J. Chem. Soc., Chem. Commun.,* 782 (1985).
209. P. T. Lansbury, G. E. Bebernitz, S. C. Maynard, and C. J. Spagnuolo, *Tetrahedron Lett.,* **26,** 169 (1985).
210. E. Vedejs and D. M. Gapinski, *Tetrahedron Lett.,* **22,** 4913 (1981).
211. K. Oshima, H. Takahashi, H. Yamamoto, and H. Nozaki, *J. Am. Chem. Soc.,* **95,** 2693 (1973).
212. C. J. Wiesner and S.-H. Tan, *Chem. and Ind. (London),* 627 (1980).
213. K. Okano, T. Morimoto, and M. Sekiya, *J. Chem. Soc., Chem. Commun.,* 883 (1984).
214. T. Morimoto, T. Takahashi, and M. Sekiya, *J. Chem. Soc., Chem. Commun.,* 794 (1984).
215. H. E. Zaugg, *Synthesis,* 85 (1984).
216. J. Barluenga, A. M. Bayon, and G. Asensio, *J. Chem. Soc., Chem. Commun.,* 427, (1984).
217. L. E. Overman and R. M. Burk, *Tetrahedron Lett.,* **25,** 1635 (1984).
218. M. Wada, Y. Sakurai, and K. Akiba, *Tetrahedron Lett.,* **25,** 1079 (1984).
219. A. R. Katritzky, K. Burgess, and R. C. Patel, *J. Heterocycl. Chem.,* **19,** 741 (1982).
220. A. Woderer, P. Assithianakis, W. Wiesert, D. Speth, and H. Stamm, *Chem. Ber.,* **117,** 3348 (1984).

221. R. A. Holton and R. A. Kjonaas, *J. Am. Chem. Soc.,* **99,** 4177 (1977).
222. J. T. Gupton and C. Colon, *Synth. Commun.,* **14,** 271 (1984).
223. A. Hosomi, Y. Sakata, and H. Sakurai, *Chem. Lett.,* 1117 (1984).
224. J. Barluenga, F. J. Fananas, and M. Yus, *J. Org. Chem.,* **46,** 1281 (1981).
225. J. Barluenga, F. J. Fananas, J. Villamana, and M. Yus, *J. Org. Chem.,* **47,** 1560 (1982).
226. I. Shimizu and J. Tsuji, *Chem. Lett.,* 233 (1984).
227. J. Barluenga, F. J. Fananas, J. Villamana, and M. Yus, *J. Chem. Soc., Perkin Trans. 1,* 2685 (1984).
228. L. Duhamel and J.-M. Poirier, *Bull. Soc. Chim. Fr.,* 297, (1982).
229. F. Z. Basha and J. F. DeBernardis, *Tetrahedron Lett.,* **25,** 5271 (1984).
230. A. Miodownik, J. Kreisberger, M. Nussim, and D. Avnir, *Synth. Commun.,* **11,** 241 (1981).
231. J. J. Fitt and H. W. Gschwend, *J. Org. Chem.,* **46,** 3349 (1981).
232. R. J. P. Corriu, V. Huynh, and J. J. E. Moreau, *Tetrahedron Lett.,* **25,** 1887 (1984).
233. M. Feustel and G. Himbert, *Justus Liebigs Ann. Chem.,* 586 (1984).
234. V. Ratovelomanana and G. Linstrumelle, *Tetrahedron Lett.,* **22,** 315 (1981).
235. V. Ratovelomanana, G. Linstrumelle, and J.-F. Normant, *Tetrahedron Lett.,* **26,** 2575 (1985).
236. A. S. Kende and P. Fludzinski, *Tetrahedron Lett.,* **23,** 2373 (1982).
237. B. Manze, P. Ongoka, and L. Miginiac, *J. Organomet. Chem.,* **264,** 1 (1984).
238. T. Mandai, J. Nokami, T. Yano, Y. Yoshinaga, and J. Otera, *J. Org. Chem.,* **49,** 172 (1984).
239. E. Ehlinger and P. Magnus, *J. Chem. Soc., Chem. Commun.,* 421 (1979).
240. A. Hosomi, M. Ando, and H. Sakurai, *Chem. Lett.,* 1385 (1984).
241. K. Hirai and Y. Kishida, *Tetrahedron Lett.,* 2743 (1972).
242. K. Hirai and Y. Kishida, *Org. Synth.,* **56,** 77 (1977).
243. L. Hegedus and R. J. Perry, *J. Org. Chem.,* **49,** 2570 (1984).
244. F. Barbot, A. Kadib-Elban, and P. Miginiac, *J. Organomet. Chem.,* **255,** 1 (1983).
245. K. Oshima, H. Yamamoto, and H. Nozaki, *J. Am. Chem. Soc.,* **95,** 4446 (1973).
246. W. A. Kinney, G. D. Crouse, and L. A. Paquette, *J. Org. Chem.,* **48,** 4986 (1983).
247. Y. Tanigawa, Y. Fuse, and S.-I. Murahashi, *Tetrahedron Lett.,* **23,** 557 (1982).
248. H. Lehmkuhl, K. Hauschild, and M. Bellenbaum, *Chem. Ber.,* **117,** 383 (1984).
249. T. Hirao, J. Enda, Y. Ohshiro, and T. Agawa, *Tetrahedron Lett.,* **22,** 3079 (1981).
250. J. Kang, W. Cho, and W. K. Lee. *J. Org. Chem.,* **49,** 1838 (1984).
251. L. Birkofer and D. Wundram, *Chem. Ber.,* **115,** 1132 (1982).
252. D. E. Seitz and A. Zapata, *Tetrahedron Lett.,* **21,** 3451 (1980).
253. D. Ager, *J. Org. Chem.,* **49,** 168 (1984).
254. Z. H. Aiube and C. Eaborn, *J. Organomet. Chem.,* **269,** 217 (1984).
255. D. Seyferth, J. L. Lefferts, and R. L. Lambert, *J. Organomet. Chem.,* **142,** 39 (1977).
256. G. L. Larson and O. Rosario, *J. Organomet. Chem.,* **168,** 13 (1979).
257. T. Cohen, J. P. Sherbine, J. R. Matz, R. R. Hutchins, B. M. McHenry and P. R. Willey, *J. Am. Chem. Soc.,* **106,** 3245 (1984).
258. S. Halazy, W. Dumont, and A. Krief, *Tetrahedron Lett.,* **22,** 4737 (1981).
259. L. A. Paquette, G. J. Wells, K. A. Horn, and T.-H. Yan, *Tetrahedron,* **39,** 913 (1983).
260. Y. Yamamoto, Y. Saito, and K. Maruyama, *J. Chem. Soc. Chem. Commun.,* 1326 (1982).
261. F. Sato, Y. Suzuki, and M. Sato, *Tetrahedron Lett.,* **23,** 4589 (1982).
262. Y. Yamamoto, Y. Saito, and K. Maruyama, *Tetrahedron Lett.,* **23,** 4597 (1982).

263. D. J. S. Tsai and D. S. Matteson, *Tetrahedron Lett.,* **22,** 2751 (1981).
264. Y. Yamamoto, H. Yatagai, Y. Saito, and K. Maruyama, *J. Org. Chem.,* **49,** 1096 (1984).
265. J. A. Soderquist and W. W.-H. Leong, *Tetrahedron Lett.,* **24,** 2361 (1983).
266. K. Takaki, M. Yasumura, and K. Negoro, *J. Org. Chem.,* **48,** 54 (1983).
267. R. K. Boeckman, Jr. and K. J. Bruza, *J. Org. Chem.,* **44,** 4781 (1979).
268. K. Karabelas and A. Hallberg, *Tetrahedron Lett.,* **26,** 3131 (1985).
269. M. Obayashi, K. Utimoto, and H. Nozaki, *J. Organomet. Chem.,* **177,** 145 (1979).
270. F. Sato, H. Watanabe, Y. Tanaka, T. Yamaji, and M. Sato, *Tetrahedron Lett.,* **24,** 1041 (1983).
271. F. E. Ziegler and K. Mikami, *Tetrahedron Lett.,* **25,** 131 (1984).
272. F. Sato, Y. Tanaka, and M. Sato, *J. Chem. Soc., Chem. Commun.,* 165 (1983).
273. G. Zweifel, R. E. Murray, and H. P. On, *J. Org. Chem.,* **46,** 1292 (1981).
274. B. B. Snider, M. Karras, and R. S. E. Conn, *J. Am. Chem. Soc.,* **100,** 4624 (1978).
275. F. Sato, H. Kanbara, and Y. Tanaka, *Tetrahedron Lett.,* **25,** 5063 (1984).
276. K. K. Wang, S. S. Nikam, and C. D. Ho, *J. Org. Chem.,* **48,** 5376 (1983).
277. B. M. Trost and D. M. T. Chan, *J. Am. Chem. Soc.,* **104,** 3733 (1982).
278. S. R. Wilson and A. Shedrinsky, *J. Org. Chem.,* **47,** 1983 (1982).
279. S. R. Wilson, A. Shedrinsky, and M. S. Hague, *Tetrahedron,* **39,** 895 (1983).
280. P. R. Jenkins, R. Gut, H. Wetter, and A. Eschenmoser, *Helv. Chim. Acta,* **62,** 1922 (1979).
281. M. Ochiai, T. Ukita, and E. Fujita, *Tetrahedron Lett.,* **24,** 4025 (1983).
282. S. E. Denmark and T. K. Jones, *J. Am. Chem. Soc.,* **104,** 2642 (1982).
283. S. D. Burke, S. M. Smith Strickland, and T. H. Powner, *J. Org. Chem.,* **48,** 454 (1983).
284. R. F. Cunico and F. J. Clayton, *J. Org. Chem.,* **41,** 1480 (1976).
285. C.-N. Hsiao and H. Shechter, *Tetrahedron Lett.,* **24,** 2371 (1983).
286. D. Seyferth and S. C. Vick, *J. Organomet. Chem.,* **144,** 1 (1978).
287. K. Uchida, K. Utimoto, and H. Nozaki, *Tetrahedron,* **33,** 2987 (1977).
288. I. Fleming and F. Roessler, *J. Chem. Soc., Chem. Commun.,* 276 (1980).
289. P. Knochel and J. F. Normant, *Tetrahedron Lett.,* **25,** 4383 (1984).
290. Y. Morizawa, H. Oda, K. Oshima, and H. Nozaki, *Tetrahedron Lett.,* **25,** 1163 (1984).
291. M. Bourgain-Commercon, J. P. Foulon, and J. F. Normant, *Tetrahedron Lett.,* **24,** 5077 (1983).
292. R. Corriu, N. Escudie, and C. Guerin, *J. Organomet. Chem.,* **264,** 207 (1984).
293. E. Ehlinger and P. Magnus, *J. Am. Chem. Soc.,* **102,** 5004 (1980).
294. M. A. Tius, *Tetrahedron Lett.,* **22,** 3335 (1981).
295. T. H. Chan and K. Koumaglo, *J. Organomet. Chem.,* **285,** 109 (1985).
296. E. J. Corey and C. Rücker, *Tetrahedron Lett.,* **23,** 719 (1982).
297. C. Burford, F. Cooke, G. Roy, and P. Magnus, *Tetrahedron,* **39,** 867 (1983).
298. F. Cooke, G. Roy, and P. Magnus, *Organometallics,* **1,** 893 (1982).
299. M. Obayashi, K. Utimoto, and H. Nozaki, *Tetrahedron Lett.,* 1805 (1977).
300. P. Auvray, P. Knochel, and J. F. Normant, *Tetrahedron Lett.,* **26,** 2329 (1985).
301. C. P. Forbes, G. L. Wenteler, and A. Wiechers, *J. Chem. Soc., Perkin Trans. 1,* 2353 (1977).
302. S. A. Bal, A. Marfat, and P. Helquist, *J. Org. Chem.,* **47,** 5045 (1982).
303. S. Patai, Ed., *Chemistry of the Carbonyl Group,* Vol. 1, Wiley-Interscience, London, 1966.
304. J. Zabicky, Ed., *Chemistry of the Carbonyl Group,* Vol. 2, Wiley-Interscience, London, 1970.

305. S. G. Wilkinson, in *Comprehensive Organic Chemistry,* Vol. 1, D. H. R. Barton and W. D. Ollis, Eds., Pergamon, Oxford, 1979, p. 579.
306. R. Brettle, in *Comprehensive Organic Chemistry,* Vol. 1, D. H. R. Barton and W. D. Ollis, Eds., Pergamon, Oxford, 1979, p. 943.
307. A. J. Waring, in *Comprehensive Organic Chemistry,* Vol. 1, D. H. R. Barton and W. D. Ollis, Eds., Pergamon, Oxford, 1979, p. 1017.
308. T. Laird, in *Comprehensive Organic Chemistry,* Vol. 1, D. H. R. Barton and W. D. Ollis, Eds., Pergamon, Oxford, 1979, p. 1105.
309. T. Laird, in *Comprehensive Organic Chemistry,* Vol. 1, D. H. R. Barton and W. D. Ollis, Eds., Pergamon, Oxford, 1979, p. 1161.
310. M. S. Kharasch and O. Reinmuth, *Grignard Reactions of Non-metal Substances,* Prentice-Hall, New York, 1954.
311. H. Normant, *Adv. Org. Chem.,* **2,** 1 (1960).
312. B. J. Wakefield, *The Chemistry of Organolithium Compounds,* Pergamon, Oxford, 1974.
313. B. J. Wakefield, in *Comprehensive Organic Chemistry,* Vol. 3, D. H. R. Barton and W. D. Ollis, Eds., Pergamon, Oxford, 1979, p. 943.
314. W. G. Brown, *Org. React.,* **6,** 469 (1951).
315. G. Zweifel, in *Comprehensive Organic Chemistry,* Vol. 3, D. H. R. Barton and W. D. Ollis, Eds., Pergamon, Oxford, 1979, p. 1013.
316. E. R. Grandbois, S. I. Howard, and J. D. Morrison, in *Asymmetric Synthesis,* Vol. 2, J. D. Morrison, Ed., Academic, New York, 1983, p. 71.
317. A. Pelter and K Smith, in *Comprehensive Organic Chemistry,* Vol. 3, D. H. R. Barton and W. D. Ollis, Eds., Pergamon, Oxford, 1979 p. 695.
318. M. Midland, in *Asymmetric Synthesis,* Vol. 2, J. D. Morrison, Ed., Academic, New York, 1983, p. 45.
319. G. Wittig and G. Geissler, *Justus Liebigs Ann. Chem.,* **580,** 44 (1953).
320. G. Wittig and V. Schöllkopf, *Chem. Ber.,* **87,** 1318 (1954).
321. A. Maercker, *Org. React.,* **14,** 270 (1965).
322. A. W. Johnson, *Ylid Chemistry,* Academic, New York, 1966.
323. M. Schlosser, *Topics Stereochem.,* **5,** 1 (1970).
324. W. S. Wadsworth, Jr., *Org. React.,* **25,** 73 (1977).
325. J. Boutagy and R. Thomas, *Chem. Rev.,* **74,** 87 (1974).
326. I. Gosney and A. G. Rowley, in *Organophosphorus Reagents in Organic Synthesis,* J. I. G. Cadogan, Ed., Academic, London, 1979, p. 17.
327. E. W. Colvin, *Silicon in Organic Synthesis,* Butterworths, London, 1981.
328. E. Lukevics, *Russ. Chem. Rev. (Engl. Transl.),* **46,** 264 (1977).
329. E. Lukevics, Z. V. Belyakova, M. G. Pomerantseva, and M. G. Voronkov, *Organomet. Chem. Rev.,* **5,** 1 (1977).
330. D. Valentine, Jr. and J. W. Scott, *Synthesis,* 329 (1978).
331. R. C. Cookson and S. A. Smith, *J. Chem. Soc., Chem. Commun.,* 145 (1979).
332. P. E. Eaton and R. H. Mueller, *J. Am. Chem. Soc.,* **94,** 1014 (1972).
333. M. F. Semmelhack, J. S. Foos, and S. Katz, *J. Am. Chem. Soc.,* **95,** 7325 (1973).
334. S. D. Burke, C. W. Murtiashaw, M. S. Dike, S. M. S. Strickland, and J. O. Saunders, *J. Org. Chem.,* **46,** 2400 (1981).
335. W. S. Johnson, *Org. React.,* **2,** 114 (1944).
336. C. C. Price, *Org. React.,* **3,** 1 (1946).
337. E. Berliner, *Org. React.,* **5,** 229 (1949).
338. *Friedel Crafts and Related Reactions,* G. A. Olah, Ed., Wiley, New York, Vol. 1, 1963.

339. R. Anliker, M. Müller, M. Perelman, J. Wohlfahrt, and C. Heusser, *Helv. Chim. Acta,* **42,** 1071 (1959).
340. J. K. Groves, *Chem. Soc. Rev.,* **1,** 73 (1972).
341. I. Fleming and A. Pearce, *J. Chem. Soc., Chem. Commun.,* 633 (1975).
342. T. H. Chan, P. W. K. Lau, and W. Mychajlowskij, *Tetrahedron Lett.,* 3317 (1977).
343. B.-W. Au-Yeung and I. Fleming, *J. Chem. Soc., Chem. Commun.,* 79 (1977).
344. M. J. Carter and I. Fleming, *J. Chem. Soc., Chem. Commun.,* 679 (1976).
345. T. H. Chan and I. Fleming, *Synthesis,* 761 (1979).
346. E. J. Corey and S. W. Walinsky, *J. Am. Chem. Soc.,* **94,** 8932 (1972).
347. H. J. Dauben, Jr., L. R. Honnen, and K. M. Harmon, *J. Org. Chem.,* **25,** 1442 (1960).
348. I. Paterson and L. G. Price, *Tetrahedron Lett.,* **22,** 2829 (1981).
349. A. Murai, K. Kato, and T. Masamune, *Tetrahedron Lett.,* **23,** 2887 (1982).
350. A. R. Chamberlin and J. Y. L. Chung, *Tetrahedron Lett.,* **23,** 2619 (1982).
351. C. Westerlund, *Tetrahedron Lett.,* **23,** 4835 (1982).
352. P. G. Gassman and D. R. Amick, *Tetrahedron Lett.,* 3463 (1974).
353. J. H. Rigby, A. Kotnis, and J. Kramer, *Tetrahedron Lett.,* **24,** 2939 (1983).
354. J. Klaveness and K. Undheim, *Acta Chem. Scand., Ser. B,* **37,** 687 (1983).
355. R. S. Brinkmeyer, *Tetrahedron Lett.,* 207 (1979).
356. E. C. Taylor and J. L. LaMattina, *Tetrahedron Lett.,* 2077 (1977).
357. C. G. Kruse, N. L. J. M. Broekhof, A. Wijsman, and A. van der Gen, *Tetrahedron Lett.,* 885 (1977).
358. C. G. Kruse, A. Wijsman, and A. van der Gen, *J. Org. Chem.,* **44,** 1847 (1979).
359. G. Picotin and P. Miginiac, *J. Org. Chem.,* **50,** 1299 (1985).
360. K. Hatanaka, S. Tanimoto, T. Sugimoto, and M. Okano, *Tetrahedron Lett.,* **22,** 3243 (1981).
361. I. Degani and R. Fochi, *J. Chem. Soc., Perkin Trans. 1,* 323 (1976).
362. I. Degani and R. Fochi, *J. Chem. Soc., Perkin Trans. 1,* 1886 (1976).
363. A. Pelter, P. Rupani, and P. Stewart, *J. Chem. Soc., Chem. Commun.,* 164 (1981).
364. I. Fleming, *Chem. Soc. Rev.,* **10,** 83 (1981).
365. I. Fleming and S. K. Patel, *Tetrahedron Lett.,* **22,** 2321 (1981).
366. I. Fleming and D. A. Perry, *Tetrahedron,* **37,** 4027 (1981).
367. D. J. Ager, *Tetrahedron Lett.,* **24,** 419 (1983).
368. L. Field, *Synthesis,* 101 (1972).
369. L. Field, *Synthesis,* 713 (1978).
370. O. W. Lever, Jr., *Tetrahedron,* **32,** 1943 (1976).
371. S. F. Martin, *Synthesis,* 633 (1979).
372. S. F. Martin, *Tetrahedron,* **36,** 419 (1980).
373. B. T. Gröbel and D. Seebach, *Synthesis,* 357 (1977).
374. A. Krief, *Tetrahedron,* **36,** 2531 (1980).
375. G. Wittig and M. Schlosser, *Chem. Ber.,* **94,** 1373 (1961).
376. M. Green, *J. Chem. Soc.,* 1324 (1963).
377. E. J. Corey and M. Chaykovsky, *J. Am. Chem. Soc.,* **87,** 1353 (1965).
378. T. Mukaiyama, S. Fukuyama, and T. Kumamoto, *Tetrahedron Lett.,* 3787 (1968).
379. H. J. Bestmann and J. Angerer, *Tetrahedron Lett.,* 3665 (1969).
380. I. Shahak and J. Almog, *Synthesis,* 170 (1969).

381. I. Shahak and J. Almog, *Synthesis,* 145 (1970).
382. E. J. Corey and J. I. Shulman, *J. Org. Chem.,* **35,** 777 (1970).
383. E. J. Corey and J. I. Shulman, *J. Am. Chem. Soc.,* **92,** 5522 (1970).
384. S. F. Martin and R. Gompper, *J. Org. Chem.,* **39,** 2814 (1974).
385. S. F. Martin, *J. Org. Chem.,* **41,** 3337 (1976).
386. M. Mikolajczyk, S. Grzejszczak, W. Midura, and A. Zatorski, *Synthesis,* 278 (1975).
387. J. I. Grayson and S. Warren, *J. Chem. Soc., Perkin Trans. 1,* 2263 (1977).
388. D. L. Comins, A. F. Jacobine, J. L. Marshall, and M. M. Turnbull, *Synthesis,* 309 (1978).
389. A. G. Cameron and A. T. Hewson, *J. Chem. Soc., Perkin Trans. 1,* 2979 (1983).
390. S. G. Levine, *J. Am. Chem. Soc.,* **80,** 6150 (1958).
391. G. Wittig and E. Knauss, *Angew. Chem.,* **71,** 127 (1959).
392. C. Earnshaw, C. J. Wallis, and S. Warren, *J. Chem. Soc., Chem. Commun.,* 314 (1977).
393. B. M. Trost and T. R. Verhoeven, *J. Am. Chem. Soc.,* **99,** 3867 (1977).
394. N. Petragnani, R. Rodrigues, and J. V. Comasseto, *J. Organomet. Chem.,* **114,** 281 (1976).
395. N. Petragnani, J. V. Comasseto, R. Rodrigues, and T. J. Brocksom, *J. Organomet. Chem.,* **124,** 1 (1977).
396. N. L. Allinger, T. J. Walter, and M. G. Newton, *J. Am. Chem. Soc.,* **96,** 4588 (1974).
397. M. Fétizon, F. J. Kakis, and V. Ignatiadou-Ragoussis, *J. Org. Chem.,* **38,** 1732 (1973).
398. J. P. Marino and J. C. Jaén, *Synth. Commun.,* **13,** 1057 (1983).
399. G. Tennant, in *Comprehensive Organic Chemistry,* Vol. 2, D. H. R. Barton and W. D. Ollis, Eds., Pergamon, Oxford, 1979, p. 383.
400. A. C. Cope, H. L. Holmes, and H. O. House, *Org. React.,* **9,** 107 (1957).
401. H. O. House and W. F. Fisher, Jr., *J. Org. Chem.,* **34,** 3615 (1969).
402. C. M. Starks and R. M. Owens, *J. Am. Chem. Soc.,* **95,** 3613 (1973).
403. J. W. Zubrick, B. I. Dunbar, and H. D. Durst, *Tetrahedron Lett.,* 71 (1975).
404. M. Mikolajczyk, S. Grzejszczak, A. Zatorski, F. Montanari, and M. Cinquini, *Tetrahedron Lett.,* 3757 (1975).
405. D. S. Watt, *J. Org. Chem.,* **39,** 2799 (1974).
406. S. J. Selikson and D. S. Watt, *J. Org. Chem.,* **40,** 267 (1975).
407. K. A. Parker and J. L. Kallmerten, *Tetrahedron Lett.,* 4557 (1977).
408. E. Vedejs and J. E. Telschow, *J. Org. Chem.,* **41,** 740 (1976).
409. S. Arseniyadis, K. S. Kyler, and D. S. Watt, *Org. React.,* **31,** 1 (1984).
410. D. A. Konen, L. S. Silbert, and P. E. Pfeffer, *J. Org. Chem.,* **40,** 3253 (1975).
411. G. K. Koch and J. M. M. Kop, *Tetrahedron Lett.,* 603 (1974).
412. O. Toussaint, P. Capdevielle, and M. Maumy, *Tetrahedron Lett.,* **25,** 3819 (1984).
413. E. Winterfeldt, in *Chemistry of Acetylenes,* H. G. Viehe, Ed., Dekker, New York, 1969.
414. J. H. Saunders, *Org. Synth., Collect. Vol.* **3,** 22 (1955).
415. S. Swaminathan and K. V. Narayanan, *Chem. Rev.,* **71,** 429 (1971).
416. J. S. Mills, H. J. Ringold, and C. Djerassi, *J. Am. Chem. Soc.,* **80,** 6118 (1958).
417. G. Stork and R. Borch, *J. Am. Chem. Soc.,* **86,** 935 (1964).
418. G. Stork and R. Borch, *J. Am. Chem. Soc.,* **86,** 936 (1964).
419. E. J. Corey and R. H. Wollenberg, *J. Am. Chem. Soc.,* **96,** 5581 (1974).
420. G. Pattenden, in *Comprehensive Organic Chemistry,* Vol. 1, D. H. R. Barton and W. D. Ollis, Eds., Pergamon, Oxford, 1979, p. 171.
421. D. Seebach, *Synthesis,* 17 (1969).
422. D. Seebach and E. J. Corey, *J. Org. Chem.,* **40,** 231 (1975).

423. S. E. Browne, S. E. Asher, E. H. Cornwall, J. K. Frisoli, L. J. Harris, E. A. Salot, E. A. Sauter, M. A. Trecoske, and P. S. Veale, Jr., *J. Am. Chem. Soc.*, **106,** 1432 (1984).
424. E. J. Corey and D. Seebach, *Angew. Chem.*, **77,** 1134 (1965); *Int. Ed. Engl.*, **4,** 1075 (1965).
425. D. Seebach, D. Steinmüller, and F. Demuth, *Angew. Chem.*, **80,** 618 (1968); *Int. Ed. Engl.*, **7,** 620 (1968).
426. D. Seebach and E.-M. Wilka, *Synthesis,* 476 (1976).
427. D. Seebach, N. R. Jones, and E. J. Corey, *J. Org. Chem.*, **33,** 300 (1968).
428. B. M. Trost and R. A. Kunz, *J. Am. Chem. Soc.*, **97,** 7152 (1975).
429. D. Tatone, T. C. Dich, R. Nacco, and C. Botteghi, *J. Org. Chem.*, **40,** 2987 (1975).
430. S. L. Schreiber and T. J. Sommer, *Tetrahedron Lett.*, **24,** 4781 (1983).
431. N. H. Andersen, D. A. McCrae, D. B. Grotjahn, S. Y. Gabhe, L. J. Theodore, R. M. Ippolito, and T. K. Sarkar, *Tetrahedron,* **37,** 4069 (1981).
432. E. L. Eliel, A. A. Hartmann, and A. G. Abatjoglou, *J. Am. Chem. Soc.*, **96,** 1807 (1974).
433. E. L. Eliel, *Angew. Chem.*, **84,** 779 (1972); *Int. Ed. Engl.*, **11,** 739 (1972).
434. E. L. Eliel, A. Abatjoglou, and A. A. Hartmann, *J. Am. Chem. Soc.*, **94,** 4786 (1972).
435. A. G. Abatjoglou, E. L. Eliel, and L. F. Kuyper, *J. Am. Chem. Soc.*, **99,** 8262 (1977).
436. E. Juaristi, J. S. Cruz-Sánchez, and F. R. Ramos-Morales, *J. Org. Chem.*, **49,** 4912 (1984).
437. J. B. Jones and R. Grayshan, *Can. J. Chem.*, **50,** 1407 (1972).
438. E. Vedejs and P. L. Fuchs, *J. Org. Chem.*, **36,** 366 (1971).
439. A. I. Meyers and R. S. Brinkmeyer, *Tetrahedron Lett.*, 1749 (1975).
440. T. Nakata, M. Fukui, H. Ohtsuka, and T. Oishi, *Tetrahedron Lett.*, **24,** 2661 (1983).
441. T. Nakata, M. Fukui, H. Ohtsuka, and T. Oishi, *Tetrahedron,* **40,** 2225 (1984).
442. J. B. Jones and R. Graushan, *Can. J. Chem.*, **50,** 810 (1972).
443. E. J. Corey and M. G. Bock, *Tetrahedron Lett.*, 2643 (1975).
444. D. R. Williams and S.-Y. Sit, *J. Am. Chem. Soc.*, **106,** 2949 (1984).
445. H. Paulsen, V. Sinnwell, and P. Stadler, *Chem. Ber.*, **105,** 1978 (1972).
446. A. I. Meyers and R. C. Strickland, *J. Org. Chem.*, **37,** 2579 (1972).
447. J. B. Jones and R. Grayshan, *Can. J. Chem.*, **50,** 1414 (1972).
448. S. R. Wilson and J. Mathew, *Synthesis,* 625 (1980).
449. D. N. Crouse and D. Seebach, *Chem. Ber.*, **101,** 3113 (1968).
450. A. G. Cameron and A. T. Hewson, *Tetrahedron Lett.*, **23,** 561 (1982).
451. I. Kawamoto, S. Muramatsu, and Y. Yura, *Tetrahedron Lett.*, 4223 (1974).
452. P. C. B. Page, M. B. van Niel, and P. H. Williams, *J. Chem. Soc., Chem. Commun.*, 742 (1985).
453. L. Duhamel, P. Duhamel, and N. Mancelle, *Bull. Soc. Chim. Fr.*, 331 (1974).
454. P. Duhamel, L. Duhamel, and N. Mancelle, *Tetrahedron Lett.*, 2991 (1972).
455. S. Wattanasin and F. G. Kathawala, *Tetrahedron Lett.*, **25,** 811 (1984).
456. A. Hassner and A. Kascheres, *Tetrahedron Lett.*, 4623 (1970).
457. A. Hassner, P. Munger, and B. A. Belinka, *Tetrahedron Lett.*, **23,** 699 (1982).
458. M. F. Semmelhack, H. T. Hall, M. Yoshifuji, and G. Clark, *J. Am. Chem. Soc.*, **97,** 1247 (1975).
459. M. F. Semmelhack and G. Clark, *J. Am. Chem. Soc.*, **99,** 1675 (1977).
460. H. G. Raubenheimer, G. J. Kruger, A. v. A. Lombard, L. Linford, and J. C. Viljoen, *Organometallics,* **4,** 275 (1985).
461. D. Seebach, M. Kolb, and B. T. Gröbel, *Chem. Ber.*, **106,** 2277 (1973).

462. K. Rustemeier and E. Breitmaier, *Chem. Ber.*, **115,** 3898 (1982).
463. C. A. Brown and A. Yamaichi, *J. Chem. Soc., Chem. Commun.*, 100 (1979).
464. E. Vedejs and B. Nader, *J. Org. Chem.*, **47,** 3193 (1982).
465. J. Lucchetti, W. Dumont, and A. Krief, *Tetrahedron Lett.*, 2695 (1979).
466. L. Wartski and M. El-Bouz, *Tetrahedron,* **38,** 3285 (1982).
467. M. El-Bouz and L. Wartski, *Tetrahedron Lett.*, **21,** 2897 (1980).
468. J. Lucchetti and A. Krief, *J. Organomet. Chem.*, **194,** C49 (1980).
469. W. Dumont, J. Lucchetti, and A. Krief, *J. Chem. Soc., Chem. Commun.*, 66 (1983).
470. D. Seebach and R. Locker, *Angew. Chem.*, **91,** 1024 (1979); *Int. Ed. Engl.*, **18,** 957 (1979).
471. G. B. Mpango, K. K. Mahalanabis, Z. Mahdavi-Damghani, and V. Snieckus, *Tetrahedron Lett.*, **21,** 4823 (1980).
472. D. Seebach and H. F. Leitz, *Angew. Chem.*, **81,** 1047 (1969); *Int. Ed. Engl.*, **8,** 983 (1969).
473. D. Seebach, V. Ehrig, H. F. Leitz, and R. Henning, *Chem. Ber.*, **108,** 1946 (1975).
474. D. Seebach, H. F. Leitz, and V. Ehrig, *Chem. Ber.*, **108,** 1924 (1975).
475. E. J. Corey, D. Seebach, and R. Freedman, *J. Am. Chem. Soc.*, **89,** 434 (1967).
476. A. G. Brook, J. M. Duff, P. F. Jones, and N. R. Davis, *J. Am. Chem. Soc.*, **89,** 431 (1967).
477. A. G. Brook and H. W. Kucera, *J. Organomet. Chem.*, **87,** 263 (1975).
478. D. Seebach, I. Willert, A. K. Beck, and B. T. Gröbel, *Helv. Chim. Acta,* **61,** 2510 (1978).
479. E. Juaristi, L. Valle, C. Mora-Uzeta, B. A. Valenzuela, P. Joseph-Nathan, and M. F. Fredrich, *J. Org. Chem.*, **47,** 5038 (1982).
480. D. Seebach, K.-H. Geiss, A. K. Beck, B. Graf, and H. Daum, *Chem. Ber.*, **105,** 3280 (1972).
481. K. Arai and M. Oki, *Bull. Chem. Soc. Jpn.*, **49,** 553 (1976).
482. R. A. J. Smith and A. R. Lal, *Aust. J. Chem.*, **32,** 353 (1979).
483. R. J. Hughes, S. Ncube, A. Pelter, K. Smith, E. Negishi, and T. Yoshida, *J. Chem. Soc., Perkin Trans. 1,* 1172 (1977).
484. E. Hunt and B. Lythgoe, *J. Chem. Soc., Chem. Commun.*, 757 (1972).
485. E. Vedejs and J. Eustache, *J. Org. Chem.*, **46,** 3353 (1981).
486. R. D. Balanson, V. M. Kobal, and R. R. Schumaker, *J. Org. Chem.*, **42,** 393 (1977).
487. D. M. Walba and M. D. Wand, *Tetrahedron Lett.*, **23,** 4995 (1982).
488. D. Seebach, E. J. Corey, and A. K. Beck, *Chem. Ber.*, **107,** 367 (1974).
489. D. Seebach, M. Kolb, and B. T. Gröbel, *Tetrahedron Lett.*, 3171 (1974).
490. D. Seebach and A. K. Beck, *Org. Synth.*, **51,** 39 (1971).
491. M. Kolb, in *The Chemistry of Ketenes, Allenes and Related Compounds,* S. Patai, Ed., Wiley, London, 1980, p. 670.
492. T. Harada, Y. Tamaru, and Z. Yoshida, *Tetrahedron Lett.*, 3525 (1979).
493. F. A. Carey and A. S. Court, *J. Org. Chem.*, **37,** 1926 (1972).
494. B. T. Gröbel, R. Bürstinghaus, and D. Seebach, *Synthesis,* 121 (1976).
495. G. A. Russell and L. A. Ochrymowycz, *J. Org. Chem.*, **35,** 764 (1970).
496. J. H. Fried, *U.S. Pat.*, 3,525,751; *Chem. Abstr.*, **73,** 98999n (1970).
497. H. Yoshida, T. Ogata, and S. Inokawa, *Synthesis,* 552 (1976).
498. Y. Nagao, K. Seno, and E. Fujita, *Tetrahedron Lett.*, 4403 (1979).
499. A. Closse and R. Huguenin, *Helv. Chim. Acta,* **57,** 533 (1974).
500. E. J. Corey and A. P. Kozikowski, *Tetrahedron Lett.*, 925 (1975).
501. J. Meijer, P. Vermeer, and L. Brandsma, *Recl. Trav. Chim. Pays-Bas,* **94,** 83 (1975).

502. A. P. Kozikowski and Y.-Y. Chen, *J. Org. Chem.*, **45,** 2236 (1980).
503. D. Seebach, B. T. Gröbel, A. K. Beck, M. Braun, and K.-H. Geiss, *Angew. Chem.*, **84,** 476 (1972); *Int. Ed. Engl.*, **11,** 443 (1972).
504. N. H. Andersen, Y. Yamamoto, and A. D. Denniston, *Tetrahedron Lett.*, 4547 (1975).
505. R. Gompper and W. Reiser, *Tetrahedron Lett.*, 1263 (1976).
506. P. F. Jones, M. F. Lappert, and A. C. Szary, *J. Chem. Soc., Perkin Trans. 1*, 2272 (1973).
507. H. J. Bestmann, R. Engler, H. Hartung, and K. Roth, *Chem. Ber.*, **112,** 28 (1979).
508. M. Mikolajczyk, S. Grzejszczak, A. Zatorski, B. Mlotkowska, H. Gross, and B. Costisella, *Tetrahedron,* **34,** 3081 (1978).
509. T.-J. Lee, W. J. Holtz, and R. L. Smith, *J. Org. Chem.*, **47,** 4750 (1982).
510. F. A. Carey and J. R. Neergaard, *J. Org. Chem.*, **36,** 2731 (1971).
511. D. M. Lemal and E. H. Banitt, *Tetrahedron Lett.*, 245 (1964).
512. E. J. Corey and G. Märkl, *Tetrahedron Lett.*, 3201 (1967).
513. T. Oishi, H. Takechi, and Y. Ban, *Tetrahedron Lett.*, 3757 (1974).
514. F. E. Ziegler, and J.-M. Fang, *J. Org. Chem.*, **46,** 825 (1981).
515. M. G. Saulnier and G. W. Gribble, *Tetrahedron Lett.*, **24,** 3831 (1983).
516. R. Amstutz, D. Seebach, P. Seiler, B. Schweizer, and J. D. Dunitz, *Angew. Chem.*, **92,** 59 (1980); *Int. Ed. Engl.*, **19,** 53 (1980).
517. E. L. Eliel, *Tetrahedron,* **30,** 1503 (1974).
518. D. Seebach, B. W. Erickson, and G. Singh, *J. Org. Chem.*, **31,** 4303 (1966).
519. P. G. Strange, J. Staunton, H. R. Wiltshire, A. R. Battersby, K. R. Hanson, and E. A. Havir, *J. Chem. Soc., Perkin Trans. 1,* 2364 (1972).
520. H. Ikehira, S. Tanimoto, and T. Oida, *J. Chem. Soc., Perkin Trans. 1,* 1223 (1984).
521. D. Seebach and D. Steinmüller, *Angew. Chem.*, **80,** 617 (1968); *Int. Ed. Engl.*, **7,** 619 (1968).
522. E. J. Corey and D. Seebach, *Angew. Chem.*, **77,** 1135 (1965); *Int. Ed. Engl.*, **4,** 1077 (1965).
523. W. D. Woessner, *Synth. Commun.*, **9,** 147 (1979).
524. M. Mori, T. Chuman, M. Kohno, K. Kato, M. Noguchi, H. Nomi, and K. Mori, *Tetrahedron Lett.*, **23,** 667 (1982).
525. D. J. Morgans, Jr. and G. B. Feigelson, *J. Am. Chem. Soc.*, **105,** 5477 (1983).
526. R. A. Ellison and W. D. Woessner, *J. Chem. Soc., Chem. Commun.*, 529 (1972).
527. W. B. Sudweeks and H. S. Broadbent, *J. Org. Chem.*, **40,** 1131 (1975).
528. T. Hylton and V. Boekelheide, *J. Am. Chem. Soc.*, **90,** 6887 (1968).
529. J. C. Sih, *J. Org. Chem.*, **47,** 4311 (1982).
530. J. P. O'Brien, A. I. Rachlin, and S. Teitel, *J. Med. Chem.*, **12,** 1112 (1969).
531. M. Mori, T. Chuman, K. Kato, and K. Mori, *Tetrahedron Lett.*, **23,** 4593 (1982).
532. R. A. Ellison, E. R. Lukenbach, and C. Chiu, *Tetrahedron Lett.*, 499 (1975).
533. G. S. Annapurna and V. H. Deshpande, *Synth. Commun.*, **13,** 1075 (1983).
534. D. Seebach, R. Bürstinghaus, B.-T. Gröbel, and M. Kolb, *Justus Liebigs Ann. Chem.*, 830 (1977).
535. R. M. Carlson and P. M. Helquist, *Tetrahedron Lett.*, 173 (1969).
536. G. H. Posner and D. J. Brunelle, *J. Org. Chem.*, **38,** 2747 (1973).
537. D. B. Grotjahn and N. H. Andersen, *J. Chem. Soc., Chem. Commun.*, 306 (1981).
538. J. A. Marshall and J. L. Belletire, *Tetrahedron Lett.*, 871 (1971).
539. P. M. Weintraub, *J. Heterocycl. Chem.*, **16,** 1081 (1979).

540. B. M. Trost and M. Preckel, *J. Am. Chem. Soc.*, **95,** 7862 (1973).
541. H. J. Reich, P. M. Gold, and F. Chow, *Tetrahedron Lett.*, 4433 (1979).
542. J. Redpath and F. J. Zeelen, *Chem. Soc. Rev.*, **12,** 75 (1983).
543. M. Koreeda and N. Koizumi, *Tetrahedron Lett.*, 1641 (1978).
544. J. R. Schauder and A. Krief, *Tetrahedron Lett.*, **23,** 4389 (1982).
545. W. Sucrow and M. van Nooy, *Justus Liebigs Ann. Chem.*, 1897 (1982).
546. F. Sher, J. L. Isidor, H. R. Taneja, and R. M. Carlson, *Tetrahedron Lett.*, 577 (1973).
547. J. A. Katzenellenbogen and S. B. Bowlus, *J. Org. Chem.*, **38,** 627 (1973).
548. A. M. Sepulchre, A. Gateau-Olesker, G. Lukacs, G. Vass, S. D. Gero, and W. Voelter, *Tetrahedron Lett.*, 3945 (1972).
549. B. Tarnchompoo and Y. Thebtaranonth, *Tetrahedron Lett.*, **25,** 5567 (1984).
550. S. R. Wilson, R. N. Misra, and G. M. Georgiadis, *J. Org. Chem.*, **45,** 2460 (1980).
551. E. Juaristi and E. L. Eliel, *Tetrahedron Lett.*, 543 (1977).
552. T. Yamamori and I. Adachi, *Tetrahedron Lett.*, **21,** 1747 (1980).
553. B. M. Trost, K. Hiroi, and L. N. Jungheim, *J. Org. Chem.*, **45,** 1839 (1980).
554. B. M. Trost and K. Hiroi, *J. Am. Chem. Soc.*, **98,** 4313 (1976).
555. G. A. Poulton and T. D. Cyr, *Can. J. Chem.*, **60,** 2821 (1982).
556. D. Seebach and H. Meyer, *Angew. Chem.*, **86,** 40 (1974); *Int. Ed. Engl.*, **13,** 77 (1974).
557. M. J. Taschner and G. A. Kraus, *J. Org. Chem.*, **43,** 4235 (1978).
558. E. W. Colvin, T. A. Purcell, and R. A. Raphael, *J. Chem. Soc., Chem. Commun.*, 1031 (1972).
559. E. J. Corey, D. H. Hua, B.-C. Pan, and S. P. Seitz, *J. Am. Chem. Soc.*, **104,** 6818 (1982).
560. J. C. Sih, D. R. Graber, S. A. Mizsak, and T. A. Scahill, *J. Org. Chem.*, **47,** 4362 (1982).
561. A. V. R. Rao, G. Venkatswamy, M. Javeed, V. H. Deshpande, and B. R. Rao, *J. Org. Chem.*, **48,** 1552 (1983).
562. E. J. Corey and D. Crouse, *J. Org. Chem.*, **33,** 298 (1968).
563. J. Blumbach, D. A. Hammond, and D. A. Whiting, *Tetrahedron Lett.*, **23,** 3949 (1982).
564. P. C. Ostrowski and V. V. Kane, *Tetrahedron Lett.*, 3549 (1977).
565. F. E. Ziegler and J. A. Schwartz, *J. Org. Chem.*, **43,** 985 (1978).
566. F. E. Ziegler and J. A. Schwartz, *Tetrahedron Lett.*, 4643 (1975).
567. P. S. Tobin, S. K. Basu, R. S. Grosserode, and D. M. S. Wheeler, *J. Org. Chem.*, **45,** 1250 (1980).
568. R. S. Grosserode, P. S. Tobin, and D. M. S. Wheeler, *Synth. Commun.*, **6,** 377 (1976).
569. G. B. Mpango and V. Snieckus, *Tetrahedron Lett.*, **21,** 4827 (1980).
570. S. R. Wilson and R. N. Misra, *J. Org. Chem.*, **43,** 4903 (1978).
571. J. A. Thomas and C. H. Heathcock, *Tetrahedron Lett.*, **21,** 3235 (1980).
572. C. H. Heathcock, M. J. Taschner, T. Rosen, J. A. Thomas, C. R. Hadley, and G. Popj'ak, *Tetrahedron Lett.*, **23,** 4747 (1982).
573. P. Knochel and D. Seebach, *Tetrahedron Lett.*, **23,** 3897 (1982).
574. A. Padwa, M. Dharan, J. Smolanoff, and S. I. Wetmore, Jr., *J. Am. Chem. Soc.*, **95,** 1954 (1973).
575. W. H. Baarschers and T. L. Loh, *Tetrahedron Lett.*, 3483 (1971).
576. R. A. Ellison, W. D. Woessner, and C. C. Williams, *J. Org. Chem.*, **37,** 2757 (1972).
577. W. D. Woessner and P. S. Solera, *Synth. Commun.*, **8,** 279 (1978).
578. A. G. Brook, D. G. Anderson, J. M. Duff, P. F. Jones, and D. M. MacRae, *J. Am. Chem. Soc.*, **90,** 1076 (1968).

579. P. R. Jones and R. West, *J. Am. Chem. Soc.*, **90,** 6978 (1968).
580. E. J. Corey and D. Seebach, *J. Org. Chem.*, **31,** 4097 (1966).
581. I. Stahl and J. Gosselck, *Synthesis,* 561 (1980).
582. J. A. Marshall and A. E. Greene, *J. Org. Chem.*, **36,** 2035 (1971).
583. H. W. Gschwend, *J. Am. Chem. Soc.*, **94,** 8430 (1972).
584. H. Zinner, H. Brandhoff, H. Schmandke, H. Kristen, and R. Haun, *Chem. Ber.*, **92,** 3151 (1959).
585. H. Paulsen and H. Redlich, *Chem. Ber.*, **107,** 2992 (1974).
586. S. Torii, K. Uneyama, and M. Isihara, *J. Org. Chem.*, **39,** 3645 (1974).
587. C. A. Reece, J. O. Rodin, R. G. Brownlee, W. G. Duncan, and R. M. Silverstein, *Tetrahedron,* **24,** 4249 (1968).
588. T. Matsumoto, H. Shirahama, A. Ichihara, H. Shin, and S. Kagawa, *Bull. Chem. Soc. Jpn.*, **45,** 1144 (1972).
589. H. Paulsen and W. Stenzel, *Chem. Ber.*, **107,** 3020 (1974).
590. E. J. Corey, N. H. Andersen, R. M. Carlson, J. Paust, E. Vedejs, I. Vlattas, and R. E. K. Winter, *J. Am. Chem. Soc.*, **90,** 3245 (1968).
591. M. L. Wolfrom, D. I. Weisblat, and A. R. Hanze, *J. Am. Chem. Soc.*, **66,** 2065 (1944).
592. H. Muxfeldt, W.-D. Unterweger, and G. Helmchen, *Synthesis,* 694 (1976).
593. M. M. Campos and H. Hauptmann, *J. Am. Chem. Soc.*, **74,** 2962 (1952).
594. T.-L. Ho, H. C. Ho, and C. M. Wong, *Can. J. Chem.*, **51,** 153 (1973).
595. S. W. Lee and C. Dougherty, *J. Org. Chem.*, **5,** 81 (1940).
596. B. Gauthier, *Ann. Pharm. Fr.*, **12,** 281 (1954); *Chem. Abstr.*, **48,** 13547i (1954).
597. J. B. Chattopadhyaya and A. V. R. Rao, *Tetrahedron Lett.*, 3735 (1973).
598. B. M. Trost and T. N. Salzmann, *J. Org. Chem.*, **40,** 148 (1975).
599. G. R. Newkome and H.-W. Lee, *J. Am. Chem. Soc.*, **105,** 5956 (1983).
600. E. J. Corey and B. W. Erickson, *J. Org. Chem.*, **36,** 3553 (1971).
601. W. Amrein and K. Schaffner, *Helv. Chim. Acta,* **58,** 380 (1975).
602. R. B. Greenwald, D. H. Evans, and J. R. DeMember, *Tetrahedron Lett.*, 3885 (1975).
603. W. F. J. Huurdeman, H. Wynberg, and D. W. Emerson, *Tetrahedron Lett.*, 3449 (1971).
604. P. R. Heaton, J. M. Midgley, and W. B. Whalley, *J. Chem. Soc., Chem. Commun.*, 750 (1971).
605. M. Hojo and R. Masuda, *Synthesis,* 678 (1976).
606. P. C. Bulman-Page, S. V. Ley, J. A. Morton, and D. J. Williams, *J. Chem. Soc., Perkin Trans. 1,* 457 (1981).
607. S. Takano, M. Takahashi, S. Hatakeyama, and K. Ogasawara, *J. Chem. Soc., Chem. Commun.*, 556 (1979).
608. S. Takano, S. Hatakeyama, and K. Ogasawara, *J. Am. Chem. Soc.*, **98,** 3022 (1976).
609. M. Fetizon and M. Jurion, *J. Chem. Soc., Chem. Commun.*, 382 (1972).
610. R. L. Markezich, W. E. Willy, B. E. McCarry, and W. S. Johnson, *J. Am. Chem. Soc.*, **95,** 4414 (1973).
611. H.-L. W. Chang, *Tetrahedron Lett.*, 1989 (1972).
612. T.-L. Ho and C. M. Wong, *Synthesis,* 561 (1972).
613. R. M. Munavu and H. H. Szmant, *Tetrahedron Lett.*, 4543 (1975).
614. E. J. Corey and T. Hase, *Tetrahedron Lett.*, 3267 (1975).
615. I. Stahl, M. Hetschko, and J. Gosselck, *Tetrahedron Lett.*, 4077 (1971).
616. I. Stahl and J. Gosselck, *Tetrahedron,* **29,** 2323 (1973).
617. T. Oishi, K. Kamemoto, and Y. Ban, *Tetrahedron Lett.*, 1085 (1972).

618. T. Oishi, H. Takechi, K. Kamemoto, and Y. Ban, *Tetrahedron Lett.*, 11 (1974).
619. G. Karmas, *Tetrahedron Lett.*, 1093 (1964).
620. R. Kuhn and F. A. Neugebauer, *Chem. Ber.*, **94,** 2629 (1961).
621. H. Nieuwenhuyse and R. Louw, *Tetrahedron Lett.*, 4141 (1971).
622. S. J. Daum and R. L. Clarke, *Tetrahedron Lett.*, 165 (1967).
623. M. A. Abdallah and J. N. Shah, *J. Chem. Soc., Perkin Trans. 1,* 888 (1975).
624. T.-L. Ho, *Synthesis,* 347 (1973).
625. H. C. Ho, T.-L. Ho, and C. M. Wong, *Can. J. Chem.*, **50,** 2718 (1972).
626. M. T. M. El-Wassimy, K. A. Jørgensen, and S.-O. Lawesson, *J. Chem. Soc., Perkin Trans. 1,* 2201 (1983).
627. K. Fuji, K. Ichikawa, and E. Fujita, *Tetrahedron Lett.*, 3561 (1978).
628. T.-L. Ho and C. M. Wong, *Can. J. Chem.*, **50,** 3740 (1972).
629. T.-L. Ho, H. C. Ho, and C. M. Wong, *J. Chem. Soc., Chem. Commun.*, 791 (1972).
630. M. Platen and E. Steckhan, *Tetrahedron Lett.*, **21,** 511 (1980).
631. J. F. Arens, M. Fröling, and A. Fröling, *Recl. Trav. Chim. Pays-Bas,* **78,** 663 (1959).
632. A. Fröling, and J. F. Arens, *Recl. Trav. Chim. Pays-Bas,* **81,** 1009 (1962).
633. J. Hine, R. B. Bayer, and G. G. Hammer, *J. Am. Chem. Soc.*, **84,** 1751 (1962).
634. D. Seebach and A. K. Beck, *J. Am. Chem. Soc.*, **91,** 1540 (1969).
635. G. A. Wildschut, H. J. T. Bos, L. Brandsma, and J. F. Arens, *Monatsh. Chem.*, **98,** 1043 (1967).
636. D. Seebach, *Chem. Ber.*, **105,** 487 (1972).
637. H. Baganz and H.-J. May, *Chem. Ber.*, **99,** 3771 (1966).
638. F. Weygand and H. J. Bestmann, *Chem. Ber.*, **90,** 1230 (1957).
639. A. I. Meyers, T. A. Tait, and D. L. Comins, *Tetrahedron Lett.*, 4657 (1978).
640. S. Masson and A. Thuillier, *Tetrahedron Lett.*, **23,** 4087 (1982).
641. K. Fuji, T. Kawabata, M. Node, and E. Fujita, *Tetrahedron Lett.*, **22,** 875 (1981).
642. R. Kaya and N. R. Beller, *Synthesis,* 814 (1981).
643. R. Kaya and N. R. Beller, *J. Org. Chem.*, **46,** 196 (1981).
644. F. E. Ziegler and C. M. Chan, *J. Org. Chem.*, **43,** 3065 (1978).
645. D. A. Konen, P. E. Pfeffer, and L. S. Silbert, *Tetrahedron,* **32,** 2507 (1976).
646. P. Blatcher and S. Warren, *J. Chem. Soc., Perkin Trans. 1,* 1074 (1979).
647. G. Schill and C. Merkel, *Synthesis,* 387 (1975).
648. T. Cohen, D. Kuhn, and J. R. Falck, *J. Am. Chem. Soc.*, **97,** 4749 (1975).
649. P. Blatcher and S. Warren, *J. Chem. Soc., Perkin Trans. 1,* 1055 (1985).
650. M. F. Semmelhack and J. C. Tomesch, *J. Org. Chem.*, **42,** 2657 (1977).
651. T. Mukaiyama, K. Narasaka, and M. Furusato, *J. Am. Chem. Soc.*, **94,** 8641 (1972).
652. S. Yamamoto, M. Shiono, and T. Mukaiyama, *Chem. Lett.*, 961 (1973).
653. A. Mendoza and D. S. Matteson, *J. Org. Chem.*, **44,** 1352 (1979).
654. T. Cohen, R. E. Gapinski, and R. R. Hutchins, *J. Org. Chem.*, **44,** 3599 (1979).
655. J. N. Denis, S. Desauvage, L. Hevesi, and A. Krief, *Tetrahedron Lett.*, **22,** 4009 (1981).
656. C. Huynh, V. Ratovelomanana, and S. Julia, *Bull. Soc. Chim. Fr.*, 710 (1977).
657. V. B. Jigajinni, R. H. Wightman, and M. M. Campbell, *J. Chem. Res. (S),* 187 (1983).
658. T. Cohen, G. Herman, J. R. Falck, and A. J. Mura, Jr., *J. Org. Chem.*, **40,** 812 (1975).
659. S. R. Wilson, G. M. Georgiadis, H. N. Khatri, and J. E. Bartmess, *J. Am. Chem. Soc.*, **102,** 3577 (1980).
660. S. Tanimoto, S. Jo, and T. Sugimoto, *Synthesis,* 53 (1981).

661. G. A. Olah, Y. D. Vankar, M. Arvanaghi, and G. K. S. Prakash, *Synthesis,* 720 (1979).
662. N. J. Cussans, S. V. Ley, and D. H. R. Barton, *J. Chem. Soc., Perkin Trans. 1,* 1654 (1980).
663. G. A. Olah, S. C. Narang, and G. F. Salem, *Synthesis,* 657 (1980).
664. D. H. R. Barton, N. J. Cussans, and S. V. Ley, *J. Chem. Soc., Chem. Commun.*, 751 (1977).
665. G. A. Olah, S. C. Narang, A. Garcia-Luma, and G. F. Salem, *Synthesis,* 146 (1981).
666. T. T. Takahashi, C. Y. Nakamura, and J. Y. Satoh, *J. Chem. Soc., Chem. Commun.*, 680 (1977).
667. S. Ncube, A. Pelter, K. Smith, P. Blatcher, and S. Warren, *Tetrahedron Lett.*, 2345 (1978).
668. K. Mori, H. Hashimoto, Y. Takenaka, and T. Takigawa, *Synthesis,* 720 (1975).
669. S. Rozen, I. Shahak, and E. D. Bergmann, *Tetrahedron Lett.*, 1837 (1972).
670. I. Hori, T. Hayashi, and H. Midorikawa, *Synthesis,* 705 (1974).
671. T. Nakai and M. Okawara, *Chem. Lett.*, 731 (1974).
672. L. Leger and M. Saquet, *Bull. Soc. Chim. Fr.*, 657 (1975).
673. S. Masson, M. Saquet, and A. Thuillier, *Tetrahedron,* **33,** 2949 (1977).
674. A. Thuillier, *Phosphorus Sulfur,* **23,** 253 (1985).
675. S. Bernstein and L. Dorfman, *J. Am. Chem. Soc.*, **68,** 1152 (1946).
676. D. A. Evans, L. K. Truesdale, K. G. Grimm, and S. L. Nesbitt, *J. Am. Chem. Soc.*, **99,** 5009 (1977).
677. D. A. Evans, K. G. Grimm, and L. K. Truesdale, *J. Am. Chem. Soc.*, **97,** 3229 (1975).
678. M. Tazaki and M. Takagi, *Chem. Lett.*, 767 (1979).
679. D. Seebach and M. Braun, *Angew. Chem.*, **84,** 60 (1972); *Int. Ed. Engl.*, **11,** 49 (1972).
680. M. Braun and D. Seebach, *Chem. Ber.*, **109,** 669 (1976).
681. D. Seebach and R. Bürstinghaus, *Angew. Chem.*, **87,** 37 (1975); *Int. Ed. Engl.*, **14,** 57 (1975).
682. R. Bürstinghaus and D. Seebach, *Chem. Ber.*, **110,** 841 (1977).
683. J. Lucchetti and A. Krief, *Synth. Commun.*, **13,** 1153 (1983).
684. D. J. Ager, *Tetrahedron Lett.*, **21,** 4763 (1980).
685. W. E. Truce and F. E. Roberts, *J. Org. Chem.*, **28,** 961 (1963).
686. P. Blatcher, J. I. Grayson, and S. Warren, *J. Chem. Soc., Chem. Commun.*, 547 (1976).
687. P. Blatcher and S. Warren, *J. Chem. Soc., Chem. Commun.*, 1055 (1976).
688. P. Blatcher, J. I. Grayson, and S. Warren, *J. Chem. Soc., Chem. Commun.*, 657 (1978).
689. P. Blatcher and S. Warren, *Tetrahedron Lett.*, 1247 (1979).
690. P. Brownbridge and S. Warren, *J. Chem. Soc., Chem. Commun.*, 465 (1977).
691. J. Durman, J. Elliott, A. B. McElroy, and S. Warren, *Tetrahedron Lett.*, **24,** 3927 (1983).
692. T. Cohen, D. Ouellette, K. Pushpananda, A. Senaratne, and L.-C. Yu, *Tetrahedron Lett.*, **22,** 3377 (1981).
693. T. Cohen, W. M. Daniewski, and R. B. Weisenfeld, *Tetrahedron Lett.*, 4665 (1978).
694. T. Cohen, R. B. Weisenfeld, and R. E. Gapinski, *J. Org. Chem.*, **44,** 4744 (1979).
695. T. Cohen and J. R. Matz, *Synth. Commun.*, **10,** 311 (1980).
696. J.-F. Biellmann, H. d'Orchymont, and J.-L. Schmitt, *J. Am. Chem. Soc.*, **101,** 3283 (1979).
697. T. Cohen, R. J. Ruffner, D. W. Shull, E. R. Fogel, and J. R. Falck, *Org. Synth.*, **59,** 202 (1980).
698. K. Mori and H. Watanabe, *Tetrahedron,* **40,** 299 (1984).
699. K. Mori, T. Uematsu, H. Watanabe, K. Yanagi, and M. Minobe, *Tetrahedron Lett.*, **25,** 3875 (1984).

700. S. Ncube, A. Pelter, and K. Smith, *Tetrahedron Lett.*, 1893 (1979).

701. C. A. Brown, R. D. Miller, C. M. Lindsay, and K. Smith, *Tetrahedron Lett.*, **25,** 991 (1984).

702. C. M. Lindsay, K. Smith, C. A. Brown, and K. Betterton-Cruz, *Tetrahedron Lett.*, **25,** 995 (1984).

703. K. Ogura and G. Tsuchihashi, *Tetrahedron Lett.*, 3151 (1971).

704. R. Kuhn, W. Baschang-Bister, and W. Dafeldecker, *Justus Liebigs Ann. Chem.*, **641,** 160 (1961).

705. K. Ogura, N. Katoh, and G. Tsuchihashi, *Bull. Chem. Soc. Jpn.*, **51,** 889 (1978).

706. G. Schill and P. R. Jones *Synthesis,* 117 (1974).

707. K. Ogura and G. Tsuchihashi, *Tetrahedron Lett.*, 2681 (1972).

708. K. Ogura, S. Furukawa, and G. Tsuchihashi, *Chem. Lett.*, 659 (1974).

709. G. R. Newkome, J. M. Robinson, and J. D. Sauer, *J. Chem. Soc., Chem. Commun.*, 410 (1974).

710. K. Ogura, M. Yamashita, and G. Tsuchihashi, *Tetrahedron Lett.*, 1303 (1978).

711. K. Ogura, and G. Tsuchihashi, *J. Am. Chem. Soc.*, **96,** 1960 (1974).

712. K. Ogura, Y. Ito, and G. Tsuchihashi, *Synthesis,* 736 (1980).

713. K. Ogura, J. Watanabe, K. Takahashi, and H. Iida, *J. Org. Chem.*, **47,** 5404 (1982).

714. K. Ogura, M. Yamashita, and G. Tsuchihashi, *Synthesis,* 385 (1975).

715. K. Ogura, S. Furukawa, and G. Tsuchihashi, *Bull. Chem. Soc. Jpn.*, **48,** 2219 (1975).

716. K. Ogura, N. Katoh, I. Yoshimura, and G. Tsuchihashi, *Tetrahedron Lett.*, 375 (1978).

717. J. L. Herrmann, G. R. Kieczykowski, R. F. Romanet, P. J. Wepplo, and R. H. Schlessinger, *Tetrahedron Lett.*, 4711 (1973).

718. J. L. Herrmann, G. R. Kieczykowski, R. F. Romanet, and R. H. Schlessinger, *Tetrahedron Lett.*, 4715 (1973).

719. B. Cazes, C. Huynh, S. Julia, V. Ratovelomanana, and O. Ruel, *J. Chem. Res. (S),* 68 (1978).

720. R. F. Romanet and R. H. Schlessinger, *J. Am. Chem. Soc.*, **96,** 3701 (1974).

721. M. Hojo, R. Masuda, T. Saeki, K. Fujimori, and S. Tsutsumi, *Tetrahedron Lett.*, 3883 (1977).

722. A. Deljac, Z. Stefanac, and K. Balenovic, *Tetrahedron Suppl.*, **8,** 33 (1966).

723. J. E. Richman, J. L. Herrmann, and R. H. Schlessinger, *Tetrahedron Lett.*, 3267 (1973).

724. J. L. Herrmann, J. E. Richman, P. J. Wepplo, and R. H. Schlessinger, *Tetrahedron Lett.*, 4707 (1973).

725. J. L. Herrmann, J. E. Richman, and R. H. Schlessinger, *Tetrahedron Lett.*, 3271 (1973).

726. R. M. Carlson and P. M. Helquist, *J. Org. Chem.*, **33,** 2596 (1968).

727. F. A. Carey, O. D. Dailey, Jr., and O. Hernandez, *J. Org. Chem.*, **41,** 3979 (1976).

728. R. F. Bryan, F. A. Carey, O. D. Dailey, Jr., R. J. Maher, and R. W. Miller, *J. Org. Chem.*, **43,** 90 (1978).

729. F. A. Carey, O. D. Dailey, Jr., O. Hernandez, and J. R. Tucker, *J. Org. Chem.*, **41,** 3975 (1976).

730. L. Colombo, C. Gennari, C. Scolastico, G. Guanti, and E. Narisano, *J. Chem. Soc., Chem. Commun.*, 591 (1979).

731. L. Colombo, C. Gennari, G. Resnati, and C. Scolastico, *Synthesis,* 74 (1981).

732. L. Colombo, C. Gennari, G. Resnati, and C. Scolastico, *J. Chem. Soc., Perkin Trans. 1,* 1284 (1981).

733. G. Guanti, E. Narisano, L. Banfi, and C. Scolastico, *Tetrahedron Lett.*, **24,** 817 (1983).

734. K. Ogura, M. Fujita, T. Inaba, K. Takahashi, and H. Iida, *Tetrahedron Lett.*, **24,** 503 (1983).

735. L. Colombo, C. Gennari, C. Scolastico, G. Guanti, and E. Narisano, *J. Chem. Soc., Perkin Trans. 1,* 1278 (1981).

736. K. Ogura, K. Ohtsuki, M. Nakamura, N. Yahata, K. Takahashi, and H. Iida, *Tetrahedron Lett.,* **26,** 2455 (1985).

737. K. Ogura, M. Yamashita, M. Suzuki, and G. Tsuchihashi, *Tetrahedron Lett.*, 3653 (1974).

738. K. Ogura, M. Yamashita, S. Furukawa, M. Suzuki, and G. Tsuchihashi, *Tetrahedron Lett.*, 2767 (1975).

739. K. Ogura, M. Yamashita, and G. Tsuchihashi, *Tetrahedron Lett.,* 759 (1976).

740. K. Ogura, M. Suzuki, J. Watanabe, M. Yamashita, H. Iida, and G. Tsuchihashi, *Chem. Lett.,* 813 (1982).

741. H. Friebolin, H. G. Schmid, S. Kabuss, and W. Faisst, *Org. Magn. Reson.,* **1,** 147 (1969).

742. K. Pihlaja and P. Pasanen, *Acta Chem. Scand.,* **24,** 2257 (1970).

743. P. Pasanen and K. Pihlaja, *Acta Chem. Scand.,* **25,** 1908 (1971).

744. K. Fuji, M. Ueda, and E. Fujita, *J. Chem. Soc., Chem. Commun.,* 814 (1977).

745. K. Fuji, M. Ueda, K. Sumi, K. Kajiwara, E. Fujita, T. Iwashita, and I. Miura, *J. Org. Chem.,* **50,** 657 (1985).

746. E. L. Eliel, J. K. Koskimies, and B. Lohri, *J. Am. Chem. Soc.,* **100,** 1614 (1978).

747. E. L. Eliel and S. Morris-Natschke, *J. Am. Chem. Soc.,* **106,** 2937 (1984).

748. J. E. Lynch and E. L. Eliel, *J. Am. Chem. Soc.,* **106,** 2943 (1984).

749. E. L. Eliel and W. J. Frazee, *J. Org. Chem.,* **44,** 3598 (1979).

750. J. de Lattre, *Bull. Soc. Chim. Belg.,* **26,** 323 (1912).

751. S. Oae, T. Masuda, K. Tsujihara, and N. Furukawa, *Bull. Chem. Soc. Jpn.,* **45,** 3586 (1972).

752. A. de Groot and B. J. M. Jansen, *Tetrahedron Lett.,* **22,** 887 (1981).

753. V. H. Rawal, M. Akiba, and M. P. Cava, *Synth. Commun.,* **14,** 1129 (1984).

754. B. M. Trost and C. H. Miller, *J. Am. Chem. Soc.,* **97,** 7182 (1975).

755. T. Mandai, M. Takeshita, K. Mori, M. Kawada, and J. Otera, *Chem. Lett.,* 1909 (1983).

756. H. Hackett and T. Livinghouse, *Tetrahedron Lett.,* **25,** 3539 (1984).

757. T. Mandai, M. Yamaguchi, Y. Nakayama, J. Otera, and M. Kawada, *Tetrahedron Lett.,* **26,** 2675 (1985).

758. T. Mandai, K. Hara, T. Nakajima, M. Kawada, and J. Otera, *Tetrahedron Lett.,* **24,** 4993 (1983).

759. H. J. E. Loewenthal, in *Protective Groups in Organic Chemistry,* J. F. W. McOmie, Ed., Plenum, London, 1973, p. 323.

760. D. W. Emerson and H. Wynberg, *Tetrahedron Lett.,* 3445 (1971).

761. T. Mandai, H. Irei, M. Kawada, and J. Otera, *Tetrahedron Lett.,* **25,** 2371 (1984).

762. K. Fuji, M. Ueda, and E. Fujita, *J. Chem. Soc., Chem. Commun.,* 49 (1983).

763. K. Fuji, M. Ueda, K. Sumi, and E. Fujita, *J. Org. Chem.,* **50,** 662 (1985).

764. H. Gilman and F. J. Webb, *J. Am. Chem. Soc.,* **62,** 987 (1940).

765. D. J. Ager, *J. Chem. Soc., Perkin Trans. 1,* 1131 (1983).

766. P. J. Kocienski, *Tetrahedron Lett.,* **21,** 1559 (1980).

767. F. A. Carey and A. S. Court, *J. Org. Chem.,* **37,** 939 (1972).

768. D. J. Ager and R. C. Cookson, *Tetrahedron Lett.,* **21,** 1677 (1980).

769. F. A. Carey and O. Hernandez, *J. Org. Chem.,* **38,** 2670 (1973).

770. D. J. Ager, *Tetrahedron Lett.,* **22,** 2803 (1981).

771. D. J. Ager, *Chem. Soc. Rev.,* **11,** 493 (1982).
772. B. T. Gröbel and D. Seebach, *Chem. Ber.,* **110,** 852 (1977).
773. D. J. Peterson, *J. Org. Chem.,* **33,** 780 (1968).
774. D. J. Ager, *Synthesis,* 384 (1984).
775. K. H. Geiss, D. Seebach, and B. Seuring, *Chem. Ber.,* **110,** 1833 (1977).
776. T. A. Hase and L. Lahtinen, *J. Organomet. Chem.,* **240,** 9 (1982).
777. P. J. Kocienski, *J. Chem. Soc., Chem. Commun.,* 1096 (1980).
778. D. J. Ager, *Tetrahedron Lett.,* **22,** 2923 (1981).
779. D. J. Ager, *Tetrahedron Lett.,* **22,** 587 (1981).
780. F. Ogura, T. Otsubo, and N. Ohira, *Synthesis,* 1006 (1983).
781. D. J. Ager, *Tetrahedron Lett.,* **21,** 4759 (1980).
782. D. J. Ager, *Tetrahedron Lett.,* **24,** 95 (1983).
783. D. J. Ager, *J. Chem. Soc., Chem. Commun.,* 486 (1984).
784. G. W. Gokel, H. M. Gerdes, D. E. Miles, J. M. Hufnal, and G. A. Zerby, *Tetrahedron Lett.,* 3375 (1979).
785. K. Tanaka, S. Matsui, and A. Kaji, *Bull. Chem. Soc. Jpn.,* **53,** 3619 (1980).
786. P. A. Wade, H. R. Hinney, N. V. Amin, P. D. Vail, S. D. Morrow, S. A. Hardinger, and M. S. Saft, *J. Org. Chem.,* **46,** 765 (1981).
787. F. de Reinach-Hirtzbach and T. Durst, *Tetrahedron Lett.,* 3677 (1976).
788. T. Durst, K.-C. Tin, F. de Reinach-Hirtzbach, J. M. Decesare, and M. D. Ryan, *Can. J. Chem.,* **57,** 258 (1979).
789. E. J. Corey and D. Boger, *Tetrahedron Lett.,* 9 (1978).
790. E. J. Corey and D. Boger, *Tetrahedron Lett.,* 13 (1978).
791. D. L. Boger, *J. Org. Chem.,* **43,** 2296 (1978).
792. H. Chikashita and K. Itoh, *Heterocycles,* **23,** 295 (1985).
793. H. Gilman and J. A. Beel, *J. Am. Chem. Soc.,* **71,** 2328 (1949).
794. A. Ricci, M. Fiorenza, M. A. Grifagni, and G. Bartolini, *Tetrahedron Lett.,* **23,** 5079 (1982).
795. H. G. Viehe, R. Merényi, L. Stella, and Z. Janousek, *Angew. Chem.,* **91,** 982 (1979); *Int. Ed. Engl.,* **18,** 917 (1979).
796. M. Makosza, E. Bialecka, and M. Ludwikow, *Tetrahedron Lett.,* 2391 (1972).
797. M. Yamashita, J. Onozuka, G. Tsuchihashi, and K. Ogura, *Tetrahedron Lett.,* **24,** 79 (1983).
798. T. Kitahara and K. Mori, *J. Org. Chem.,* **49,** 3281 (1984).
799. D. Morgans, Jr. and G. B. Feigelson, *J. Org. Chem.,* **47,** 1131 (1982).
800. N.-Y. Wang, S. Su, and L. Tsai, *Tetrahedron Lett.,* 1121 (1979).
801. D. N. Brattesani and C. H. Heathcock, *Tetrahedron Lett.,* 2279 (1974).
802. Y. Masuyama, Y. Ueno, and M. Okawara, *Tetrahedron Lett.,* 2967 (1976).
803. A. M. van Leusen, R. J. Bouma, and O. Possel, *Tetrahedron Lett.,* 3487 (1975).
804. A. V. R. Rao, J. S. Yadav, and G. S. Annapurna, *Synth. Commun.,* **13,** 331 (1983).
805. H. Sasaki and T. Kitagawa, *Chem. Pharm. Bull.,* **31,** 2868 (1983).
806. P. Yadagiri and J. S. Yadav, *Synth. Commun.,* **13,** 1067 (1983).
807. M. S. Frazza and B. W. Roberts, *Tetrahedron Lett.,* **22,** 4193 (1981).
808. O. Possel and A. M. van Leusen, *Tetrahedron Lett.,* 4229 (1977).
809. U. Schöllkopf and E. Blume, *Tetrahedron Lett.,* 629 (1973).
810. O. H. Oldenziel and A. M. van Leusen, *Tetrahedron Lett.,* 163 (1974).

811. O. H. Oldenziel and A. M. van Leusen, *Tetrahedron Lett.*, 167 (1974).
812. D. van Leusen and A. M. van Leusen, *Tetrahedron Lett.*, 4233 (1977).
813. S. Halazy and P. Magnus, *Tetrahedron Lett.*, **25,** 1421 (1984).
814. B. M. Trost and T. N. Salzmann, *J. Am. Chem. Soc.*, **95,** 6840 (1973).
815. P. A. Grieco and C.-L. J. Wang, *J. Chem. Soc., Chem. Commun.*, 714 (1975).
816. B. M. Trost and Y. Tamaru, *Tetrahedron Lett.*, 3797 (1975).
817. S. Yamagiwa, N. Hoshi, H. Sato, H. Kosugi, and H. Uda, *J. Chem. Soc., Perkin Trans. 1,* 214 (1978).
818. B. M. Trost, M. J. Crimmin, and D. Butler, *J. Org. Chem.*, **43,** 4549 (1978).
819. B. M. Trost and Y. Tamaru, *J. Am. Chem. Soc.*, **97,** 3528 (1975).
820. J. Nokami, M. Kawada, R. Okawara, S. Torii, and H. Tanaka, *Tetrahedron Lett.*, 1045 (1979).
821. E. Block, *Reactions of Organosulfur Compounds,* Academic, New York, 1978.
822. S. Iriuchijima, K. Maniwa, and G. Tsuchihashi, *J. Am. Chem. Soc.*, **96,** 4280 (1974).
823. S. Iriuchijima, K. Maniwa, and G. Tsuchihashi, *J. Am. Chem. Soc.*, **97,** 596 (1975).
824. G. Tsuchihashi, S. Iriuchijima, and M. Ishibashi, *Tetrahedron Lett.*, 4605 (1972).
825. G. A. Russell and L. A. Ochrymowycz, *J. Org. Chem.*, **34,** 3618 (1969).
826. T. L. Moore, *J. Org. Chem.*, **32,** 2786 (1967).
827. P. D. Magnus, *Tetrahedron,* **33,** 2019 (1977).
828. P. J. Kocienski, *Chem. Ind. (London),* 548 (1981).
829. R. D. Little and S. O. Myong, *Tetrahedron Lett.*, **21,** 3339 (1980).
830. J. R. Hwu, *J. Org. Chem.*, **48,** 4432 (1983).
831. H. Kotake, K. Inomata, H. Kinoshita, Y. Sakamoto, and Y. Kaneto, *Bull. Chem. Soc. Jpn.*, **53,** 3027 (1980).
832. J.-B. Baudin, M. Julia, and C. Rolando, *Tetrahedron Lett.*, **26,** 2333 (1985).
833. P. G. Gassman and H. R. Drewes, *J. Am. Chem. Soc.*, **96,** 3002 (1974).
834. P. Bakuzis, M. L. F. Bakuzis, C. C. Fortes, and R. Santos, *J. Org. Chem.*, **41,** 2769 (1976).
835. V. Reutrakul and P. Poochaivatananon, *Tetrahedron Lett.*, **24,** 531 (1983).
836. S. Julia, C. Huynh, and D. Michelot, *Tetrahedron Lett.*, 3587 (1972).
837. T. Nakai and K. Mikami, *Chem. Lett.*, 1243 (1978).
838. J. P. Marino and R. C. Landick, *Tetrahedron Lett.*, 4531 (1975).
839. K. Sachdev and H. S. Sachdev, *Tetrahedron Lett.*, 4223 (1976).
840. J. D. White, M. Kang, and B. G. Sheldon, *Tetrahedron Lett.*, **24,** 4539 (1983).
841. D. Seebach and N. Peleties, *Angew. Chem.*, **81,** 465 (1969); *Int. Ed. Engl.*, **8,** 450 (1969).
842. D. van Ende, W. Dumont, and A. Krief, *J. Organomet. Chem.*, **149,** C10 (1978).
843. H. J. Reich and S. K. Shah, *J. Org. Chem.*, **42,** 1773 (1977).
844. D. van Ende, A. Cravador, and A. Krief, *J. Organomet. Chem.*, **177,** 1 (1979).
845. D. Seebach, *Angew. Chem.*, **79,** 468 (1967); *Int. Ed. Engl.*, **6,** 442 (1967).
846. A. Burton, L. Hevesi, W. Dumont, A. Cravador, and A. Krief, *Synthesis,* 877 (1979).
847. G. Stork and S. R. Dowd, *J. Am. Chem. Soc.*, **85,** 2178 (1963).
848. H. O. House, W. C. Liang, and P. D. Weeks, *J. Org. Chem.*, **39,** 3102 (1974).
849. J.-F. Le Borgne, *J. Organomet. Chem.*, **122,** 123 (1976).
850. J.-F. Le Borgne, *J. Organomet. Chem.*, **122,** 129 (1976).
851. J.-F. Le Borgne, *J. Organomet. Chem.*, **122,** 139 (1976).
852. D. A. McCrae and L. Dolby, *J. Org. Chem.*, **42,** 1607 (1976).
853. G. Stork, A. Y. W. Leong, and A. M. Touzin, *J. Org. Chem.*, **41,** 3491 (1976).

854. F. G. Cowherd, M.-C. Doria, E. Galeazzi, and J. M. Muchowski, *Can. J. Chem.*, **55,** 2919 (1977).

855. R. A. Abramovitch, S. S. Mathur, D. W. Saunders, and D. P. Vanderpool, *Tetrahedron Lett.*, **21,** 705 (1980).

856. G. Doleschall, *Tetrahedron Lett.*, 1889 (1975).

857. H.-W. Wanzlick and E. Schikora, *Chem. Ber.*, **94,** 2389 (1961).

858. H.-W. Wanzlick and H.-J. Kleiner, *Chem. Ber.*, **96,** 3024 (1963).

859. H.-W. Wanzlick, *Angew. Chem.*, **74,** 129 (1962); *Int. Ed. Engl.*, **1,** 75 (1962).

860. E. C. Taylor and A. McKillop, *Adv. Org. Chem.*, **7,** 1 (1970).

861. J. D. Albright, *Tetrahedron*, **39,** 3207 (1983).

862. G. Stork, R. M. Jacobson, and R. Levitz, *Tetrahedron Lett.*, 771 (1979).

863. C. R. Hauser, H. M. Taylor, and T. G. Ledford, *J. Am. Chem. Soc.*, **82,** 1786 (1960).

864. J. W. Stanley, J. G. Beasley, and I. W. Mathison, *J. Org. Chem.*, **37,** 3746 (1972).

865. G. F. Morris and C. R. Hauser, *J. Org. Chem.*, **27,** 465 (1962).

866. D. J. Bennett, G. W. Kirby, and V. A. Moss, *J. Chem. Soc., Chem. Commun.*, 218 (1967).

867. H. Ahlbrecht, W. Raab, and C. Vonderheid, *Synthesis*, 127 (1979).

868. E. B. Sanders, H. V. Secor, and J. I. Seeman, *J. Org. Chem.*, **41,** 2658 (1976).

869. E. B. Sanders, H. V. Secor, and J. I. Seeman, *J. Org. Chem.*, **43,** 324 (1978).

870. K. Takahashi, M. Matsuzaki, K. Ogura, and H. Iida, *J. Org. Chem.*, **48,** 1909 (1983).

871. H. M. Taylor and C. R. Hauser, *J. Am. Chem. Soc.*, **82,** 1960 (1960).

872. E. Hebert, N. Maigrot, and Z. Welvart, *Tetrahedron Lett.*, **24,** 4683 (1983).

873. M. Makosza, B. Serafinowa, and T. Boleslawska, *Rocz. Chem.*, **42,** 817 (1968); *Chem. Abstr.*, **69,** 106174 (1968).

874. C. R. Hauser and G. F. Morris, *J. Org. Chem.*, **26,** 4740 (1961).

875. S. F. Dyke, E. P. Tiley, A. W. C. White, and D. P. Gale, *Tetrahedron*, **31,** 1219 (1975).

876. S. F. Dyke, A. C. Ellis, R. G. Kinsman, and A. W. C. White, *Tetrahedron*, **30,** 1193 (1974).

877. F. J. McEvoy and J. D. Albright, *J. Org. Chem.*, **44,** 4597 (1979).

878. J. Dockx, *Synthesis*, 441 (1973).

879. J. D. Albright, F. J. McEvoy, and D. B. Moran, *J. Heterocycl. Chem.*, **15,** 881 (1978).

880. E. Leete, M. R. Chedekel, and G. B. Bodem, *J. Org. Chem.*, **37,** 4465 (1972).

881. V. Reutrakul, S. Nimgirawath, S. Panichanun, and P. Ratananukul, *Chem. Lett.*, 399 (1979).

882. E. Leete and G. B. Bodem, *J. Am. Chem. Soc.*, **98,** 6321 (1976).

883. E. Leete, *J. Org. Chem.*, **41,** 3438 (1976).

884. M. Zervos and L. Wartski, *Tetrahedron Lett.*, **25,** 4641 (1984).

885. T. Wakamatsu, S. Hobara, and Y. Ban, *Heterocycles*, **19,** 1395 (1982).

886. G. Stork, A. A. Ozorio, and A. Y. W. Leong, *Tetrahedron Lett.*, 5175 (1978).

887. D. Enders and H. Lotter, *Tetrahedron Lett.*, **23,** 639 (1982).

888. H. Ahlbrecht and K. Pfaff, *Synthesis*, 897 (1978).

889. V. Reutrakul, P. Ratananukul, and S. Nimgirawath, *Chem. Lett.*, 71 (1980).

890. H. Schick, F. Theil, H. Jablokoff, and S. Schwarz, *Z. Chem.*, **21,** 68 (1981).

891. H. Ahlbrecht and H.-M. Kompter, *Synthesis*, 645 (1983).

892. K. Takahashi, K. Shibasaki, K. Ogura, and H. Iida, *J. Org. Chem.*, **48,** 3566 (1983).

893. D. Enders, H. Lotter, N. Maigrot, J.-P. Mazaleyrat, and Z. Welvart, *Nouv. J. Chim.*, **8,** 747 (1984).

894. L. N. Mander and J. V. Turner, *J. Org. Chem.,* **38,** 2915 (1973).
895. T. A. Bryson and W. E. Pye, *J. Org. Chem.,* **42,** 3215 (1977).
896. G. Büchi and H. Wüest, *J. Am. Chem. Soc.,* **96,** 7573 (1974).
897. L. Stella and A. Amrollah-Madjdabadi, *Synth. Commun.,* **14,** 1141 (1984).
898. L. Stella, *Tetrahedron Lett.,* **25,** 3457 (1984).
899. L. N. Mander and L. T. Palmer, *Aust. J. Chem.,* **32,** 823 (1979).
900. L. N. Mander, J. V. Turner, and B. G. Coombe, *Aust. J. Chem.,* **27,** 1985 (1974).
901. R. G. Coombes, in *Comprehensive Organic Chemistry,* Vol. 2, D. H. R. Barton and W. D. Ollis, Eds., Pergamon, Oxford, 1979, p. 303.
902. H. H. Baer and L. Urbas, in *The Chemistry of the Nitro and Nitroso Groups,* H. Feuer, Ed., Wiley, New York, 1970, Part 2, Ch. 3, p. 75.
903. N. Kornblum *Org. React.,* **12,** 101 (1962).
904. F. Lehr, J. Gonnermann, and D. Seebach, *Helv. Chim. Acta,* **62,** 2258 (1979).
905. D. St. C. Black, *Tetrahedron Lett.,* 1331 (1972).
906. A. T. Hewson and D. T. MacPherson, *Tetrahedron Lett.,* **24,** 647 (1983).
907. W. T. Monte, M. M. Baizer, and R. D. Little, *J. Org. Chem.,* **48,** 803 (1983).
908. J. E. McMurry and J. Melton, *J. Am. Chem. Soc.,* **93,** 5309 (1971).
909. J. Nokami, T. Sonoda, and S. Wakabayashi, *Synthesis,* 763 (1983).
910. S. Ranganathan, D. Ranganathan, and A. K. Mehrotra, *J. Am. Chem. Soc.,* **96,** 5261 (1974).
911. R. M. Jacobson, *Tetrahedron Lett.,* 3215 (1974).
912. N. Kornblum and R. A. Brown, *J. Am. Chem. Soc.,* **87,** 1742 (1965).
913. W. E. Noland, *Chem. Rev.,* **55,** 137 (1955).
914. T. Severin and D. König, *Chem. Ber.,* **107,** 1499 (1974).
915. H. Lerche, D. König, and T. Severin, *Chem. Ber.,* **107,** 1509 (1974).
916. J. R. Williams, L. R. Unger, and R. H. Moore, *J. Org. Chem.,* **43,** 1271 (1978).
917. A. H. Pagano and H. Shechter, *J. Org. Chem.,* **35,** 295 (1970).
918. N. Kornblum and P. A. Wade, *J. Org. Chem.,* **38,** 1418 (1973).
919. H. Shechter and F. T. Williams, *J. Org. Chem.,* **27,** 3699 (1962).
920. J. R. Hanson, *Synthesis,* 1 (1974).
921. J. R. Hanson and T. D. Organ, *J. Chem. Soc. C,* 1182 (1970).
922. J. E. McMurry, J. Melton, and H. Padgett, *J. Org. Chem.,* **39,** 259 (1974).
923. J. E. McMurry, *Acc. Chem. Res.,* **7,** 281 (1974).
924. T.-L. Ho and C. M. Wong, *Synthesis,* 196 (1974).
925. J. E. McMurry and J. Melton, *J. Org. Chem.,* **38,** 4367 (1973).
926. J. E. Arrowsmith, M. J. Cook, and D. J. Hardstone, *J. Chem. Soc., Perkin Trans. 1,* 2364 (1979).
927. R. M. Adlington, J. E. Baldwin, J. C. Bottaro, and M. W. D. Perry, *J. Chem. Soc., Chem. Commun.,* 1040 (1983).
928. J. E. Baldwin, R. M. Adlington, J. C. Bottaro, A. U. Jain, J. N. Kolhe, M. W. D. Perry, and I. M. Newington, *J. Chem. Soc., Chem. Commun.,* 1095 (1984).
929. D. Hoppe, *Angew. Chem.* **86,** 878 (1974); *Int. Ed. Engl.,* **13,** 789 (1974).
930. H. M. Walborsky and G. E. Niznik, *J. Am. Chem. Soc.,* **91,** 7778 (1969).
931. G. E. Niznik, W. H. Morrison, and H. M. Walborsky, *J. Org. Chem.,* **39,** 600 (1974).
932. Y. Yamamoto, K. Kondo, and I. Moritani, *J. Org. Chem.,* **40,** 3644 (1975).
933. Y. Yamamoto, K. Kondo, and I. Moritani, *Tetrahedron Lett.,* 793 (1974).

934. Y. Yamamoto, K. Kondo, and I. Moritani, *Tetrahedron Lett.*, 2689 (1975).
935. R. Chênevert, R. Plante, and N. Voyer, *Synth. Commun.*, **13,** 403 (1983).
936. A. T. Au, *Synth. Commun.*, **14,** 743, 749 (1984).
937. W. Nagata and M. Yoshioka, *Tetrahedron Lett.*, 1913 (1966).
938. G. Stork and L. Maldonado, *J. Am. Chem. Soc.*, **93,** 5286 (1971).
939. G. Stork, J. C. Depezay, and J. d'Angelo, *Tetrahedron Lett.*, 389 (1975).
940. A. Kalir and D. Balderman, *Synthesis,* 358 (1973).
941. G. A. Garcia, H. Muñoz, and J. Tamariz, *Synth. Commun.*, **13,** 569 (1983).
942. M.-C. Roux-Schmitt, N. Seuron, and J. Seyden-Penne, *Synthesis,* 494 (1983).
943. D. M. Bailey, C. G. DeGrazia, D. Wood, J. Siggins, H. R. Harding, G. O. Potts, and T. W. Skulan, *J. Med. Chem.*, **17,** 702 (1974).
944. M. Hamana, T. Endo, and S. Saeki, *Tetrahedron Lett.*, 903 (1975).
945. E. Aufderhaar, J. E. Baldwin, D. H. R. Barton, D. J. Faulkner, and M. Slaytor, *J. Chem. Soc. C,* 2175 (1971).
946. F. Zymalkowski and W. Schauer, *Arch. Pharm.*, **290,** 218 (1957).
947. T. Endo, S. Saeki, and M. Hamana, *Heterocycles,* **3,** 19 (1975).
948. D. A. Evans, L. K. Truesdale, and G. L. Carroll, *J. Chem. Soc., Chem. Commun.*, 55 (1973).
949. D. A. Evans and L. K. Truesdale, *Tetrahedron Lett.*, 4929 (1973).
950. D. A. Evans, J. M. Hoffman, and L. K. Truesdale, *J. Am. Chem. Soc.*, **95,** 5822 (1973).
951. D. A. Evans, G. L. Carroll, and L. K. Truesdale, *J. Org. Chem.*, **39,** 914 (1974).
952. W. Lidy and W. Sundermeyer, *Chem. Ber.*, **106,** 587 (1973).
953. K. Deuchert, U. Hertenstein, and S. Hünig, *Synthesis,* 777 (1973).
954. K. Deuchert, U. Hertenstein, S. Hünig, and G. Wehner, *Chem. Ber.*, **112,** 2045 (1979).
955. S. Hünig and G. Wehner, *Synthesis,* 180 (1975).
956. W. E. Parham and C. S. Roosevelt, *Tetrahedron Lett.*, 923 (1971).
957. W. E. Parham and C. S. Roosevelt, *J. Org. Chem.*, **37,** 1975 (1972).
958. I. Fleming, J. Iqbal, and E. P. Krebs, *Tetrahedron,* **39,** 841 (1983).
959. G. Boche, F. Bosold, and M. Niessner, *Tetrahedron Lett.*, **23,** 3255 (1982).
960. S. Hünig and G. Wehner, *Chem. Ber.*, **112,** 2062 (1979).
961. S. Hünig and G. Wehner, *Synthesis,* 391 (1975).
962. S. Hünig and G. Wehner, *Chem. Ber.*, **113,** 302 (1980).
963. S. Hünig and G. Wehner, *Chem. Ber.*, **113,** 324 (1980).
964. G. A. Kraus, H. Cho, S. Crowley, B. Roth, H. Sugimoto, and S. Prugh, *J. Org. Chem.*, **48,** 3439 (1983).
965. S. Hünig and M. Öller, *Chem. Ber.*, **114,** 959 (1981).
966. F. Duboudin, P. Cazeau, O. Babot, and F. Moulines, *Tetrahedron Lett.*, **24,** 4335 (1983).
967. T. Takahashi, T. Nagashima, and J. Tsuji, *Tetrahedron Lett.*, **22,** 1359 (1981).
968. T. Takahashi, H. Ikeda, and J. Tsuji, *Tetrahedron Lett.*, **22,** 1363 (1981).
969. T. Takahashi, T. Nagashima, H. Ikeda, and J. Tsuji, *Tetrahedron Lett.*, **23,** 4361 (1982).
970. T. Takahashi, H. Nemoto, and J. Tsuji, *Tetrahedron Lett.*, **24,** 2005 (1983).
971. T. Takahashi, H. Nemoto, J. Tsuji, and I. Miura, *Tetrahedron Lett.*, **24,** 3485 (1983).
972. T. Takahashi, K. Kitamura, H. Nemoto, J. Tsuji, and I. Miura, *Tetrahedron Lett.*, **24,** 3489 (1983).
973. T. Takahashi, K. Kitamura, and J. Tsuji, *Tetrahedron Lett.*, **24,** 4695 (1983).
974. T. Takahashi, I. Minami, and J. Tsuji, *Tetrahedron Lett.*, **22,** 2651 (1981).

975. G. L. A. Maldonado, *Rev. Soc. Quim. Mex.*, **16,** 200 (1972); *Chem. Abstr.*, **78,** 15437 (1973).
976. M. B. Floyd, *U.S. Pat.*, 4,076,732; *Chem. Abstr.*, **89,** 108341 (1978).
977. J. Ficini, J. d'Angelo, and J. Noiré, *J. Am. Chem. Soc.*, **96,** 1213 (1974).
978. B. Cazes and S. Julia, *Synth. Commun.*, **7,** 113 (1977).
979. A. Casares and L. A. Maldonado, *Synth. Commun.*, **6,** 11 (1976).
980. G. Stork and L. Maldonado, *J. Am. Chem. Soc.*, **96,** 5272 (1974).
981. N. Seuron, L. Wartski, and J. Seyden-Penne, *Tetrahedron Lett.*, **22,** 2175 (1981).
982. E. Guittet and S. Julia, *Tetrahedron Lett.*, 1155 (1978).
983. B. Cazes and S. Julia, *Tetrahedron,* **35,** 2655 (1979).
984. B. Cazes and S. Julia, *Bull. Soc. Chim. Fr.*, 925 (1977).
985. B. Cazes and S. Julia, *Bull. Soc. Chim. Fr.*, 931 (1977).
986. W. S. Ide and J. S. Buck, *Org. React.*, **4,** 269 (1948).
987. K. B. Wiberg, *J. Am. Chem. Soc.*, **76,** 5371 (1954).
988. J. P. Kuebrich, R. L. Schowen, M.-S. Wang, and M. E. Lupes, *J. Am. Chem. Soc.*, **93,** 1214 (1971).
989. J. P. Kuebrich and R. L. Schowen, *J. Am. Chem. Soc.*, **93,** 1220 (1971).
990. S. M. McElvain, *Org. React.*, **4,** 256 (1948).
991. H. Stetter and M. Schreckenberg, *Tetrahedron Lett.*, 1461 (1973).
992. H. Stetter and H. Kuhlmann, *Tetrahedron,* **33,** 353 (1977).
993. H. Stetter and M. Schreckenberg, *U.S. Pat.*, 4,014,889; *Chem. Abstr.*, **87,** 39116 (1977).
994. H. Stetter and M. Schreckenberg, *Chem. Ber.*, **107,** 2453 (1974).
995. H. Stetter and M. Schreckenberg, *Tetrahedron Lett.*, 1461 (1973).
996. H. Stetter, M. Schreckenberg, and K. Wiemann, *Chem. Ber.*, **109,** 541 (1976).
997. H. Stetter and B. Rajh, *Chem. Ber.*, **109,** 534 (1976).
998. H. Stetter and M. Schreckenberg, *Chem. Ber.*, **107,** 210 (1974).
999. H. Stetter and M. Schreckenberg, *Angew. Chem.*, **85,** 89 (1973); *Int. Ed. Engl.*, **12,** 81 (1973).
1000. H. Stetter, *Angew. Chem.*, **88,** 695 (1976); *Int. Ed. Engl.*, **15,** 639 (1976).
1001. H. Stetter and H. T. Leinen, *Chem. Ber.*, **116,** 254 (1983).
1002. H. Stetter and J. Krasselt, *J. Heterocycl. Chem.*, **14,** 573 (1977).
1003. L. Novák, B. Majoros, and S. Szantay, *Heterocycles,* **12,** 369 (1979).
1004. H. Stetter and K. Kuhlmann, *Angew. Chem.*, **86,** 589 (1974); *Int. Ed. Engl.*, **13,** 539 (1974).
1005. H. Stetter and K. Kuhlmann, *Tetrahedron Lett.*, 4505 (1974).
1006. H. Stetter and H. Kuhlmann, *Synthesis,* 379 (1975).
1007. T.-L. Ho and S.-H. Liu, *Synth. Commun.*, **13,** 1125 (1983).
1008. H. Stetter, R. Y. Raemsch, and H. Kuhlmann, *Synthesis,* 733 (1976).
1009. H. Stetter and G. Daembkes, *Synthesis,* 309 (1980).
1010. H. Stetter and R. Y. Raemsch, *Synthesis,* 477 (1981).
1011. J. C. Sheehan and T. Hara, *J. Org. Chem.*, **39,** 1196 (1974).
1012. W. Tagaki, Y. Tamura, and Y. Yano, *Bull. Chem. Soc. Jpn.*, **53,** 478 (1980).
1013. L. Novák, G. Baán, J. Marosfalvi, and S. Szántay, *Tetrahedron Lett.*, 487 (1978).
1014. H. Stetter, G. Hilboll, and H. Kuhlmann, *Chem. Ber.*, **112,** 84 (1979).
1015. H. Stetter and W. Schlenker, *Tetrahedron Lett.*, **21,** 3479 (1980).
1016. H. Stetter, W. Basse, and J. Nienhaus, *Chem. Ber.*, **113,** 690 (1980).

1017. L. Novák, G. Baán, J. Marosfalvi, and C. Szántay, *Chem. Ber.*, **113,** 2939 (1980).
1018. H. Stetter and W. Haese, *Chem. Ber.*, **117,** 682 (1984).
1019. T. Matsumoto, M. Ohishi, and S. Inoue, *J. Org. Chem.*, **50,** 603 (1985).
1020. B. M. Trost, *Chem. Soc. Rev.*, **11,** 141 (1982).
1021. R. C. Cookson and R. M. Lane, *J. Chem. Soc., Chem. Commun.*, 804 (1976).
1022. H. Stetter and H. Kuhlmann, *Chem. Ber.*, **109,** 2890 (1976).
1023. H. Stetter, W. Basse, and K. Wiemann, *Chem. Ber.*, **111,** 431 (1978).
1024. B. M. Trost, C. D. Shuey, F. DiNinno, and S. S. McElvain, *J. Am. Chem. Soc.*, **101,** 1284 (1979).
1025. H. Stetter and G. Lorenz, *Chem. Ber.*, **118,** 1115 (1985).
1026. R. Lohmar and W. Steglich, *Angew. Chem.*, **90,** 493 (1978); *Int. Ed. Engl.*, **17,** 450 (1978).
1027. B. Kübel, P. Gruber, R. Hurnaus, and W. Steglich, *Chem. Ber.*, **112,** 128 (1979).
1028. R. Lohmar and W. Steglich, *Chem. Ber.*, **113,** 3706 (1980).
1029. W. Steglich and P. Gruber, *Angew. Chem.*, **83,** 727 (1971); *Int. Ed. Engl.*, **10,** 655 (1971).
1030. W. Steglich, P. Gruber, G. Höfle, and W. Konig, *Angew. Chem.*, **83,** 725 (1971); *Int. Ed. Engl.*, **10,** 653 (1971).
1031. H. Wegmann, G. Schultz, and W. Steglich, *Justus Leibigs Ann. Chem.*, 1736 (1980).
1032. G. Schultz and W. Steglich, *Chem. Ber.*, **113,** 787 (1980).
1033. R. A. Olofson and D. M. Zimmerman, *J. Am. Chem. Soc.*, **89,** 5057 (1967).
1034. T. Cohen and I. W. Song, *J. Am. Chem. Soc.*, **87,** 3780 (1965).
1035. J. C. Craig and L. R. Kray, *J. Org. Chem.*, **33,** 871 (1968).
1036. Y. Watanabe, Y. Tsuji, and R. Takeuchi, *Bull. Chem. Soc. Jpn.*, **56,** 1428 (1983).
1037. J. P. Quintard, B. Elissondo, and M. Pereyre, *J. Organomet. Chem.*, **212,** C31 (1981).
1038. J. P. Quintard, B. Elissondo, and M. Pereyre, *J. Org. Chem.*, **48,** 1559 (1983).
1039. A. I. Meyers and A. L. Campbell, *Tetrahedron Lett.*, 4155 (1979).
1040. H. H. Wasserman and B. H. Lipshutz, *Tetrahedron Lett.*, 4611 (1975).
1041. W. Adam, O. Cueto, and V. Ehrig, *J. Org. Chem.*, **41,** 370 (1976).
1042. E. J. Corey, R. H. Wollenberg, and D. R. Williams, *Tetrahedron Lett.*, 2243 (1977).
1043. B. M. Trost and Y. Tamaru, *J. Am. Chem. Soc.*, **99,** 3101 (1977).
1044. P. E. Pfeffer and L. S. Silbert, *Tetrahedron Lett.*, 699 (1970).
1045. J. Nokami, T. Yamamoto, M. Kawada, M. Izumi, N. Ochi, and R. Okawara, *Tetrahedron Lett.*, 1047 (1979).
1046. T. Wakamatsu, K. Akasaka, and Y. Ban, *Tetrahedron Lett.*, 3879 (1974).
1047. T. Wakamatsu, K. Akasaka, and Y. Ban, *Tetrahedron Lett.*, 3883 (1974).
1048. A. G. Brook, T. J. D. Vandersar, and W. Limburg, *Can. J. Chem.*, **56,** 2758 (1978).
1049. A. Degl'Innocenti, S. Pike, D. R. M. Walton, G. Seconi, A. Ricci, and M. Fiorenza, *J. Chem. Soc., Chem. Commun.*, 1201 (1980).
1050. D. Schinzer and C. H. Heathcock, *Tetrahedron Lett.*, 1881 (1981).
1051. A. Ricci, A. Degl'Innocenti, S. Chimichi, M. Fiorenza, G. Rossini, and H. J. Bestmann, *J. Org. Chem.*, **50,** 130 (1985).
1052. J. E. Baldwin, G. A. Höfle, and O. W. Lever, *J. Am. Chem. Soc.*, **96,** 7125 (1974).
1053. A. B. Levy and S. J. Schwartz, *Tetrahedron Lett.*, 2201 (1976).
1054. C. G. Chavdarian and C. H. Heathcock, *J. Am. Chem. Soc.*, **97,** 3822 (1975).
1055. J. E. Baldwin, O. W. Lever, Jr., and N. R. Tzodikov, *J. Org. Chem.*, **41,** 2312 (1976).
1056. U. Schöllkopf and P. Hänssle, *Justus Liebigs Ann. Chem.*, **763,** 208 (1972).
1057. E. M. Dexheimer and L. Spialter, *J. Organomet. Chem.*, **107,** 229 (1976).

1058. R. K. Boeckman, K. J. Bruza, J. E. Baldwin, and O. W. Lever, *J. Chem. Soc., Chem. Commun.*, 519 (1975).
1059. M. J. Eis, J. E. Wrobel, and B. Ganem, *J. Am. Chem. Soc.*, **106,** 3693 (1984).
1060. R. Muthukrishnan and M. Schlosser, *Helv. Chim. Acta,* **59,** 13 (1976).
1061. C. E. Russell and L. S. Hegedus, *J. Am. Chem. Soc.*, **105,** 943 (1983).
1062. E.-I. Negishi and F.-T. Luo, *J. Org. Chem.*, **48,** 1560 (1983).
1063. J. Hartmann, M. Stahle, and M. Schlosser, *Synthesis,* 888 (1974).
1064. R. Paul and S. Tchelitcheff, *C. R. Hebd. Seances Acad. Sci.*, **235,** 1226 (1952).
1065. M. Schlosser, *Angew. Chem.*, **76,** 124 (1964); *Int. Ed. Engl.*, **3,** 287 (1964).
1066. R. K. Boeckman and K. J. Bruza, *Tetrahedron Lett.*, 4187 (1977).
1067. R. K. Boeckman and K. J. Bruza, *Tetrahedron,* **37,** 3997 (1981).
1068. A. R. Rossi, B. D. Remillard, and S. J. Gould, *Tetrahedron Lett.*, 4357 (1978).
1069. T. Cohen and M. Bhupathy, *Tetrahedron Lett.*, **24,** 4163 (1983).
1070. P. Kocienski and C. Yeates, *Tetrahedron Lett.*, **24,** 3905 (1983).
1071. W. E. Parham and R. F. Motter, *J. Am. Chem. Soc.*, **81,** 2146 (1959).
1072. W. E. Parham, M. A. Kalnins, and D. R. Theissen, *J. Org. Chem.*, **27,** 2698 (1962).
1073. K. Oshima, K. Shimoji, H. Takahashi, H. Yamamoto, and H. Nozaki, *J. Am. Chem. Soc.*, **95,** 2694 (1973).
1074. R. C. Cookson and P. J. Parsons, *J. Chem. Soc., Chem. Commun.*, 821 (1978).
1075. T. Takeda, H. Furukawa, and T. Fujiwara, *Chem. Lett.*, 593 (1982).
1076. R. C. Cookson and P. J. Parsons, *J. Chem. Soc., Chem. Commun.*, 990 (1976).
1077. P. J. Parsons, Ph.D. Thesis, University of Southhampton, 1978.
1078. B. Harirchian and P. Magnus, *J. Chem. Soc., Chem. Commun.*, 522 (1977).
1079. B. T. Gröbel and D. Seebach, *Chem. Ber.*, **110,** 867 (1977).
1080. E. Schaumann and W. Walter, *Chem. Ber.*, **107,** 3562 (1974).
1081. T. Cohen and R. B. Weisenfeld, *J. Org. Chem.*, **44,** 3601 (1979).
1082. H. Westmijze, J. Meijer, and P. Vermeer, *Tetrahedron Lett.*, 2923 (1975).
1083. E. Erdik, *Tetrahedron,* **40,** 641 (1984).
1084. A. A. Oswald, K. Griesbaum, B. E. Hudson, and J. M. Bregman, *J. Am. Chem. Soc.*, **86,** 2877 (1964).
1085. W. E. Truce, J. A. Simms, and M. M. Boudakian, *J. Am. Chem. Soc.*, **78,** 695 (1956).
1086. T. Cohen and Z. Kosarych, *Tetrahedron Lett.*, **21,** 3955 (1980).
1087. T. Cohen, A. J. Mura, Jr., D. W. Shull, E. R. Fogel, R. J. Ruffner, and J. R. Falck, *J. Org. Chem.*, **41,** 3218 (1976).
1088. B. M. Trost and A. C. Lavoie, *J. Am. Chem. Soc.*, **105,** 5075 (1983).
1089. M. Kakimoto, T. Yamamoto, and M. Okawara, *Tetrahedron Lett.*, 623 (1979).
1090. H. C. Brown and M. H. Rei, *J. Am. Chem. Soc.*, **91,** 5646 (1969).
1091. I. Shahak and J. Almog, *Synthesis,* 170 (1969).
1092. K. Watt and E. J. Corey *Tetrahedron Lett.*, 4651 (1972).
1093. F. A. Carey and A. S. Court, *J. Org. Chem.*, **37,** 4474 (1972).
1094. T. Mukaiyama, M. Shiono, and T. Sato, *Chem. Lett.*, 37, (1974).
1095. M. Mikolajczyk, S. Grzejszczak, and K. Korbacz, *Tetrahedron Lett.*, 3097 (1981).
1096. D. S. Matteson and K. H. Arne, *Organometallics,* **1,** 280 (1982).
1097. Y. Ikeda, H. Furuta, N. Meguriya, N. Ikeda, and H. Yamamoto, *J. Am. Chem. Soc.*, **104,** 7663 (1982).
1098. B. M. Trost, K. Hiroi, and S. Kurozumi, *J. Am. Chem. Soc.*, **97,** 438 (1975).

1099. S. Kano, T. Yokomatsu, T. Ono, S. Hibino, and S. Shibuya, *J. Chem. Soc., Chem. Commun.,* 414 (1978).
1100. T. Nakai and T. Mimura, *Tetrahedron Lett.,* 531 (1979).
1101. B. M. Trost, K. Hiroi, and N. Holy, *J. Am. Chem. Soc.,* **97,** 5873 (1975).
1102. H. Oda, M. Sato, Y. Morizawa, K. Oshima, and H. Nozaki, *Tetrahedron Lett.,* **24,** 2877 (1983).
1103. R. R. Schmidt and B. Schmid, *Tetrahedron Lett.,* 3583 (1977).
1104. G. Rosenkranz, S. Kaufmann, and J. Romo, *J. Am. Chem. Soc.,* **71,** 3689 (1949).
1105. J. Romo, M. Romero, C. Djerassi, and G. Rosenkrantz, *J. Am. Chem. Soc.,* **73,** 1528 (1951).
1106. M. Iyoda, M. Morigaki, and M. Nakagawa, *Tetrahedron Lett.,* 3677 (1974).
1107. A. J. Mura, Jr., G. Majetich, P. A. Grieco, and T. Cohen, *Tetrahedron Lett.,* 4437 (1975).
1108. T. Mukaiyama, K. Kamio, S. Kobayashi, and H. Takei, *Bull. Chem. Soc. Jpn.,* **45,** 3723 (1972).
1109. H. J. Bestmann and J. Angerer, *Justus Liebigs Ann. Chem.,* 2085 (1974).
1110. G. Solladie and G. Moine, *J. Am. Chem. Soc.,* **106,** 6097 (1984).
1111. I. Vlattas, L. D. Vecchia, and A. O. Lee, *J. Am. Chem. Soc.,* **98,** 2008 (1976).
1112. I. Fleming, in *Comprehensive Organic Chemistry,* Vol. 3, D. H. R. Barton and W. D. Ollis, Eds., Pergamon, Oxford, 1979, p. 541.
1113. R. Yamaguchi, H. Kawasaki, and M. Kawanisi, *Synth. Commun.,* **12,** 1027 (1982).
1114. R. B. Miller, M. I. Al-Hassan, and G. McGarvey, *Synth. Commun.,* **13,** 969 (1983).
1115. B. T. Gröbel and D. Seebach, *Angew. Chem.,* **86,** 102 (1974); *Int. Ed. Engl.,* **13,** 83 (1974).
1116. A. G. Brook, J. M. Duff, and W. F. Reynolds, *J. Organomet. Chem.,* **121,** 293 (1976).
1117. G. Zweifel and W. Lewis, *J. Org. Chem.,* **43,** 2739 (1978).
1118. C. Huynh and G. Linstrumelle, *Tetrahedron Lett.,* 1073 (1979).
1119. T. H. Chan, W. Mychajlowskij, B. S. Ong, and D. N. Harpp, *J. Organomet. Chem.,* **107,** C1 (1976).
1120. T. H. Chan, B. S. Ong, and W. Mychajlowskij, *Tetrahedron Lett.,* 3253 (1976).
1121. W. Mychajlowskij and T. H. Chan, *Tetrahedron Lett.,* 4439 (1976).
1122. T. H. Chan and B. S. Ong, *J. Org. Chem.,* **43,** 2994 (1978).
1123. R. Amouroux and T. H. Chan, *Tetrahedron Lett.,* 4453 (1978).
1124. T. H. Chan, W. Mychajlowskij, B. S. Ong, and D. N. Harpp, *J. Org. Chem.,* **43,** 1526 (1978).
1125. L. A. Paquette, K. A. Horn, and G. J. Wells, *Tetrahedron Lett.,* **23,** 259 (1982).
1126. R. K. Boeckman and K. J. Bruza, *Tetrahedron Lett.,* 3365 (1974).
1127. R. K. Boeckman and M. Ramaiah, *J. Org. Chem.,* **42,** 1581 (1977).
1128. A. G. Brook and J. M. Duff, *Can. J. Chem.,* **51,** 2024 (1973).
1129. I. Matsuda, *Chem. Lett.,* 773 (1978).
1130. R. B. Miller and T. Reichenbach, *Tetrahedron Lett.,* 543 (1974).
1131. K. Uchida, K. Utimoto, and H. Nozaki, *J. Org. Chem.,* **41,** 2941 (1976).
1132. J. J. Eisch and G. A. Damasevitz, *J. Org. Chem.,* **41,** 2214 (1976).
1133. K. Uchida, K. Utimoto, and H. Nozaki, *J. Org. Chem.,* **41,** 2215 (1976).
1134. W. Dumont, D. Van Ende, and A. Krief, *Tetrahedron Lett.,* 485 (1979).
1135. H. Sakurai, K.-L. Nishiwaki, and M. Kira, *Tetrahedron Lett.,* 4193 (1973).
1136. D. Ayalon-Chass, E. Ehlinger, and P. Mangus, *J. Chem. Soc., Chem. Commun.,* 772 (1977).

1137. K. Yamamoto, M. Ohta, and J. Tsuji, *Chem. Lett.*, 713 (1979).
1138. G. Stork and E. Colvin, *J. Am. Chem. Soc.*, **93,** 2080 (1971).
1139. J. J. Eisch and J. E. Galle, *J. Org. Chem.*, **41,** 2615 (1976).
1140. G. Stork and M. E. Jung, *J. Am. Chem. Soc.*, **96,** 3682 (1974).
1141. G. M. Robbins and G. H. Whitham, *J. Chem. Soc., Chem. Commun.*, 697 (1976).
1142. A. P. Davis, G. J. Hughes, P. R. Lowndes, C. M. Robbins, E. J. Thomas and G. H. Whitham, *J. Chem. Soc., Perkin Trans. 1,* 1934 (1981).
1143. P. F. Hudrlik, J. O. Arcoleo, R. H. Schwartz, R. N. Misra, and R. J. Rona, *Tetrahedron Lett.*, 591 (1977).
1144. G. Nagendrappa, *Tetrahedron,* **38,** 2429 (1982).
1145. K. Tamao, M. Kumada, and K. Maeda, *Tetrahedron Lett.*, **25,** 321 (1984).
1146. M. Sevrin, J. N. Denis, and A. Krief, *Angew. Chem.*, **90,** 550 (1978); *Int. Ed. Engl.*, **17,** 526 (1978).
1147. H. J. Reich, W. W. Willis, and P. D. Clark, *J. Org. Chem.*, **46,** 2775 (1981).
1148. S. Raucher and G. A. Koolpe, *J. Org. Chem.*, **43,** 3794 (1978).
1149. J. N. Denis and A. Krief, *Tetrahedron Lett.*, **23,** 3411 (1982).
1150. J. V. Comasseto, *J. Organomet. Chem.*, **253,** 131 (1983).
1151. R. R. Schmidt, J. Talbiersky, and P. Russegger, *Tetrahedron Lett.*, 4273 (1979).
1152. R. R. Schmidt and H. Speer, *Tetrahedron Lett.*, **22,** 4259 (1981).
1153. H. Ahlbrecht and H. Simon, *Synthesis,* 58 (1983).
1154. H. Ahlbrecht and H. Simon, *Synthesis,* 61 (1983).
1155. J. E. Baldwin and J. C. Bottaro, *J. Chem. Soc., Chem. Commun.*, 1121 (1981).
1156. M. P. Cooke, *J. Am. Chem. Soc.*, **92,** 6080 (1970).
1157. G. Cainelli, F. Manescalchi, A. Umani-Ronchi, and M. Panunzio, *J. Org. Chem.*, **43,** 1598 (1978).
1158. J. P. Collman, S. R. Winter, and R. G. Komoto, *J. Am. Chem. Soc.*, **95,** 249 (1973).
1159. W. O. Siegl and J. P. Collman, *J. Am. Chem. Soc.*, **94,** 2516 (1972).
1160. H. Masada, M. Mizuno, S. Suga, Y. Watanabe, and Y. Takegami, *Bull. Chem. Soc. Jpn.*, **43,** 3824 (1970).
1161. Y. Watanabe, T. Mitsudo, M. Tanaka, K. Yamamoto, T. Okajima, and Y. Takegami, *Bull. Chem. Soc. Jpn.*, **44,** 2569 (1971).
1162. Y. Takegami, Y. Watanabe, T. Mitsudo, and H. Masada, *Bull. Chem. Soc. Jpn.*, **42,** 202 (1969).
1163. V. P. Baillargeon and J. K. Stille, *J. Am. Chem. Soc.*, **105,** 7175 (1983).
1164. M. Ryang, S. Kwang-Myeong, Y. Sawa, and S. Tsutsumi, *J. Organomet. Chem.*, **5,** 305 (1966).
1165. J. Chandrasekhar, J. G. Andrade, and P. v. Ragué Schleyer, *J. Am. Chem. Soc.*, **103,** 5612 (1981).
1166. L. S. Trzupek, T. L. Newirth, E. G. Kelly, N. E. Sbarbati, and G. M. Whitesides, *J. Am. Chem. Soc.*, **95,** 8118 (1973).
1167. N. S. Nudelman and A. A. Vitale, *J. Organomet. Chem.*, **241,** 143 (1983).
1168. P. Jutzi and F.-W. Schröder, *J. Organomet. Chem.*, **24,** C43 (1970).
1169. P. Jutzi and F.-W. Schröder, *J. Organomet. Chem.*, **24,** 1 (1970).
1170. D. Seyferth and R. M. Weinstein, *J. Am. Chem. Soc.*, **104,** 5534 (1982).
1171. R. M. Weinstein, W.-L. Wang, and D. Seyferth, *J. Org. Chem.*, **48,** 3367 (1983).
1172. D. Seyferth, R. M. Weinstein, and W.-L. Wang, *J. Org. Chem.*, **48,** 1144 (1983).

1173. D. Seyferth, R. M. Weinstein, W.-L. Wang, and R. C. Hui, *Tetrahedron Lett.*, **24,** 4907 (1983).
1174. D. Seyferth and R. C. Hui, *Tetrahedron Lett.*, **25,** 2623 (1984).
1175. D. Seyferth and R. C. Hui, *Tetrahedron Lett.*, **25,** 5251 (1984).
1176. W. J. J. M. Sprangers, A. P. van Swieten, and R. Louw, *Tetrahedron Lett.*, 3377 (1974).
1177. D. Seyferth and R. J. Spohn, *J. Am. Chem. Soc.*, **91,** 3037 (1969).
1178. J. P. Collman, P. Faruham, and G. Dolcetti, *J. Am. Chem. Soc.*, **94,** 1788 (1972).
1179. J. P. Collman and N. W. Hoffman, *J. Am. Chem. Soc.*, **95,** 2689 (1973).
1180. Y. Sawa, M. Ryang, and S. Tsutsumi, *J. Org. Chem.*, **35,** 4183 (1970).
1181. J. P. Collman, J. N. Cawse, and J. I. Brauman, *J. Am. Chem. Soc.*, **94,** 5905 (1972).
1182. Y. Sawa, M. Ryang, and S. Tsutsumi, *Tetrahedron Lett.*, 5189 (1969).
1183. E. O. Fischer and V. Kiener, *J. Organomet. Chem.*, **23,** 215 (1970).
1184. J. P. Collman, *Acc. Chem. Res.*, **8,** 342 (1975).
1185. G. Tanguy, B. Weinberger, and H. des Abbayes, *Tetrahedron Lett.*, **25,** 5529 (1984).
1186. M. P. Cooke, Jr. and R. M. Parlman, *J. Am. Chem. Soc.*, **97,** 6863 (1975).
1187. J. Y. Mérour, J. L. Roustan, C. Charrier, J. Collin, and J. Benaim, *J. Organomet. Chem.*, **51,** C24 (1973).
1188. E. J. Corey and L. S. Hegedus, *J. Am. Chem. Soc.*, **91,** 4926 (1969).
1189. M. F. Semmelhack, L. Keller, T. Sato, and E. Spiess, *J. Org. Chem.*, **47,** 4382 (1982).
1190. Y. Sawa, I. Hashimoto, M. Ryang, and S. Tsutsumi, *J. Org. Chem.*, **33,** 2159 (1968).
1191. S. Inaba and R. D. Rieke, *Tetrahedron Lett.*, **24,** 2451 (1983).
1192. S. Inaba and R. D. Rieke, *Chem. Lett.*, 25 (1984).
1193. L. S. Hegedus and Y. Inoue, *J. Am. Chem. Soc.*, **104,** 4917 (1982).
1194. J. H. Merrifield, J. P. Godschalx, and J. K. Stille, *Organometallics*, **3,** 1108 (1984).
1195. G. T. Crisp, W. J. Scott, and J. K. Stille, *J. Am. Chem. Soc.*, **106,** 7500 (1984).
1196. E. Negishi, V. Bagheri, S. Chatterjee, F.-T. Luo, J. A. Miller, and A. T. Stoll, *Tetrahedron Lett.*, **24,** 5181 (1983).
1197. E. Anders and T. Gassner, *Angew. Chem.*, **95,** 635 (1983); *Int. Ed. Engl.*, **22,** 619 (1983).
1198. M. W. Rathke and H. Yu, *J. Org. Chem.*, **37,** 1732 (1972).
1199. B. Giese and U. Erfort, *Chem. Ber.*, **116,** 1240 (1983).
1200. I. I. Lapkin, G. Y. Anvarova, and T. Y. Povarnitsina, *J. Gen. Chem. USSR*, **36,** 1952 (1966).
1201. G. Wittig, L. Gonsior, and H. Vogel, *Justus Liebigs Ann. Chem.*, **688,** 1 (1965).
1202. H. C. Brown, *Acc. Chem. Res.*, **2,** 65 (1969).
1203. K. Chikamatsu, T. Otsubo, F. Ogura, and H. Yamaguchi, *Chem. Lett.*, 1081 (1982).
1204. D. S. Matteson and R. J. Moody, *J. Am. Chem. Soc.*, **99,** 3196 (1977).
1205. H. J. Bestmann and E. Kranz,, *Chem. Ber.*, **102,** 1802 (1969).
1206. W. Waszkuc, T. Janecki, and R. Bodalski, *Synthesis*, 1025 (1984).
1207. G. Köbrich, *Angew. Chem.*, **84,** 557 (1972); *Int. Ed. Engl.*, **11,** 473 (1972).
1208. G. Köbrich and W. Werner, *Tetrahedron Lett.*, 2181 (1969).
1209. P. Blumbergs, M. P. La Montagne, and J. I. Stevens, *J. Org. Chem.*, **37,** 1248 (1972).
1210. H. Taguchi, S. Tanaka, H. Yamamoto, and H. Nozaki, *Tetrahedron Lett.*, 2465 (1973).
1211. G. Köbrich, K. Flory, and W. Drischel, *Angew. Chem.*, **76,** 536 (1964); *Int. Ed. Engl.*, **3,** 513 (1964).
1212. G. Büchi, D. Minster, and J. C. F. Young, *J. Am. Chem. Soc.*, **93,** 4319 (1971).

1213. T. Shono, H. Ohmizu, and N. Kise, *Tetrahedron Lett.,* **23,** 4801 (1982).
1214. A. Pelter, B. Singaram, and J. W. Wilson, *Tetrahedron Lett.,* **24,** 635 (1983).
1215. M. V. Garad, A. Pelter, B. Singaram, and J. W. Wilson, *Tetrahedron Lett.,* **24,** 637 (1983).
1216. P. Magnus, *Aldrichimica Acta,* **13,** 43 (1980).
1217. C. Burford, F. Cooke, E. Ehlinger, and P. Magnus, *J. Am. Chem. Soc.,* **99,** 4536 (1977).
1218. F. Cooke and P. Magnus, *J. Chem. Soc., Chem. Commun.,* 513 (1977).
1219. P. Magnus and G. Roy, *J. Chem. Soc., Chem. Commun.,* 297 (1978).
1220. P. Magnus and G. Roy, *J. Chem. Soc., Chem. Commun.,* 822 (1979).
1221. P. Beak, J. Yamamoto, and C. J. Upton, *J. Org. Chem.,* **40,** 3052 (1975).
1222. M. Sekine, M. Nakajima, A. Kume, A. Hashizume, and T. Hata, *Bull. Chem. Soc. Jpn,* **55,** 224 (1982).
1223. M. Sekine, M. Nakajima, A. Kume, and T. Hata, *Tetrahedron Lett.,* 4475 (1979).
1224. T. Hata, A. Hashizume, M. Nakajima, and M. Sekine,*Tetrahedron Lett.,* 363 (1978).
1225. R. E. Koenigkramer and H. Zimmer, *Tetrahedron Lett.,* **21,** 1017 (1980).
1226. R. E. Koenigkramer and H. Zimmer, *J. Org. Chem.,* **45,** 3994 (1980).
1227. J. Binder and E. Zbiral, *Tetrahedron Lett.,* **25,** 4213 (1984).
1228. J. Villieras, C. Bacquet, and J.-F. Normant, *Bull. Soc. Chim. Fr.,* 1797 (1975).
1229. Y. Ueno and M. Okawara, *Synthesis,* 268 (1975).
1230. H. G. Heine, *Justus Liebigs Ann. Chem.,* **735,** 56, (1970).
1231. T. Hase, *Synthesis,* 36 (1980).
1232. I. Kawenoki, D. Maurel, and J. Kossanyi, *Bull. Soc. Chim. Fr.,* 385 (1982).
1233. R. Scheffold and R. Orlinski, *J. Am. Chem. Soc.,* **105,** 7200 (1983).
1234. K. J. H. Kruithof, R. F. Schmitz, and G. W. Klumpp, *J. Chem. Soc., Chem. Commun.,* 239 (1983).
1235. K. J. H. Kruithof, R. F. Schmitz, and G. W. Klumpp, *Tetrahedron Lett.,* **39,** 3073 (1983).
1236. H. J. Reich and S. K. Shah, *J. Am. Chem. Soc.,* **99,** 263 (1977).
1237. H. J. Reich, S. K. Shah, P. M. Gold, and R. E. Olson, *J. Am. Chem. Soc.,* **103,** 3112 (1981).
1238. H. J. Cristau, H. Christol, and D. Bottaro, *Synthesis,* 826 (1978).
1239. D. J. Pasto, *Tetrahedron,* **40,** 2805 (1984).
1240. S. Hoff, L. Brandsma, and J. F. Arens, *Recl. Trav. Chim. Pays-Bas,* **87,** 916 (1968).
1241. S. Hoff, L. Brandsma, and J. F. Arens, *Recl. Trav. Chim. Pays-Bas,* **87,** 1179 (1968).
1242. A. Kucerovy, K. Neuenschwander, and S. M. Weinreb, *Synth. Commun.,* **13,** 875 (1983).
1243. D. Gange and P. Magnus, *J. Am. Chem. Soc.,* **100,** 7746 (1978).
1244. Y. Leroux and C. Roman, *Tetrahedron Lett.,* 2585 (1973).
1245. Y. Leroux and R. Mantione, *Tetrahedron Lett.,* 591 (1971).
1246. R. Mantione and Y. Leroux, *Tetrahedron Lett.,* 593 (1971).
1247. F. Derguini and G. Linstrumelle, *Tetrahedron Lett.,* **25,** 5763 (1984).
1248. J.-C. Clinet and G. Linstrumelle, *Tetrahedron Lett.,* **21,** 3987 (1980).
1249. R. Mantione and A. Alves, *Tetrahedron Lett.,* 2483 (1969).
1250. R. Mantione, A. Alves, P. P. Montijn, G. A. Wildschut, H. J. T. Bos, and L. Brandsma, *Recl. Trav. Chim. Pays-Bas,* **89,** 97 (1970).
1251. L. Brandsma, C. Jonker, and M. Berg, *Recl. Trav. Chim. Pays-Bas,* **84,** 560 (1965).
1252. L. Brandsma, H. E. Wijers, and J. F. Arens, *Recl. Trav. Chim. Pays-Bas,* **82,** 1040 (1963).
1253. J. H. van Boom, L. Brandsma, J. F. Arens, *Recl. Trav. Chim. Pays-Bas,* **85,** 580 (1966).
1254. D. I. Gasking, and G. H. Whitham, *J. Chem. Soc., Perkin Trans. 1,* 409 (1985).

1255. R. C. Cookson and P. J. Parsons, *J. Chem. Soc., Chem. Commun.*, 822 (1978).
1256. I. Cutting and P. J. Parsons, *J. Chem. Soc., Chem. Commun.*, 1209 (1983).
1257. I. Cutting and P. J. Parsons, *Tetrahedron Lett.*, **24,** 4463 (1983).
1258. R. M. Carlson, R. W. Jones, and A. S. Hatcher, *Tetrahedron Lett.*, 1741 (1975).
1259. R. M. Carlson and J. L. Isidor, *Tetrahedron Lett.*, 4819 (1973).
1260. J. H. S. Weiland and J. F. Arens, *Recl. Trav. Chim. Pays-Bas,* **79,** 1293 (1960).
1261. S. I. Pennanen, *Synth. Commun.*, **12,** 209 (1982).
1262. B. Corbel, J.-P. Paugam, M. Dreux, and P. Savignac, *Tetrahedron Lett.*, 835 (1976).
1263. T. Nakai, T. Mimura, and A. Ari-Izumi, *Tetrahedron Lett.*, 2425 (1977).
1264. D. Seebach and M. Kolb, *Justus Liebigs Ann. Chem.*, 811 (1977).
1265. F. E. Zeigler and C. C. Tam, *J. Org. Chem.*, **44,** 3428 (1979).
1266. W. S. Murphy and S. Wattanasin, *Tetrahedron Lett.*, 1827 (1979).
1267. D. Coffen, T. E. McEntee, and D. R. Williams, *J. Chem. Soc., Chem. Commun.*, 913 (1970).
1268. Y. Köksal, P. Raddatz, and E. Winterfeldt, *Justus Leibigs Ann. Chem.*, 450 (1984).
1269. Y. Köksal, P. Raddatz, and E. Winterfeldt, *Justus Liebigs Ann. Chem.*, 462 (1984).
1270. W. S. Murphy and S. Wattanasin, *J. Chem. Soc., Perkin Trans. 1,* 2678 (1980).
1271. F. E. Ziegler and C. C. Tam, *Tetrahedron Lett.*, 4717 (1979).
1272. F. E. Ziegler, U. R. Chakraborty, and R. T. Wester, *Tetrahedron Lett.*, **23,** 3237 (1982).
1273. F. E. Ziegler, J.-M. Fang, and C. C. Tam, *J. Am. Chem. Soc.*, **104,** 7174 (1982).
1274. J. C. Saddler and P. L. Fuchs, *J. Am. Chem. Soc.*, **103,** 2112 (1981).
1275. N. H. Andersen, A. D. Denniston, and D. A. McRae,, *J. Org. Chem.*, **47,** 1145 (1982).
1276. D. Seebach, M. Kolb, and B. T. Gröbel, *Angew. Chem.*, **85,** 42 (1973); *Int. Ed. Engl.*, **12,** 69 (1973).
1277. G. Just, P. Potvin, and G. H. Hakimelahi, *Can. J. Chem.*, **58,** 2780 (1980).
1278. J. F. Biellmann and J. B. Ducep, *Tetrahedron Lett.*, 5629 (1968).
1279. K. Oshima, H. Yamamoto, and H. Nozaki, *Bull. Chem. Soc. Jpn.*, **48,** 1567 (1975).
1280. T. Nakai, H. Shiono, and M. Okawara, *Tetrahedron Lett.*, 3625 (1974).
1281. K. S. Kyler and D. S. Watt, *J. Org. Chem.*, **46,** 5182 (1981).
1282. T. Mandai, T. Moriyama, Y. Nakayama, K. Sugino,, M. Kawada, and J. Otera, *Tetrahedron Lett.*, **25,** 5913 (1984).
1283. T. Mandai, H. Arase, J. Otera, and M. Kawada, *Tetrahedron Lett.*, **26,** 2677 (1985).
1284. T. Mandai, M. Takeshita, M. Kawada, and J. Otera, *Chem. Lett.*, 1259 (1984).
1285. E. J. Corey and R. Noyori, *Tetrahedron Lett.*, 311 (1970).
1286. E. J. Corey, B. W. Erickson, and R. Noyori, *J. Am. Chem. Soc.*, **93,** 1724 (1971).
1287. H. J. Reich, M. C. Clark, and W. W. Willis, Jr., *J. Org. Chem.*, **47,** 1618 (1982).
1288. U. Hertenstein, S. Hünig, and M. Öller, *Synthesis,* 416 (1976).
1289. R. M. Jacobson and G. P. Lahm, *J. Org. Chem.*, **44,** 462 (1979).
1290. U. Hertenstein, S. Hünig, and M. Öller, *Chem. Ber.*, **113,** 3783 (1980).
1291. R. M. Jacobson, G. P. Lahm, and J. W. Clader, *J. Org. Chem.*, **45,** 395 (1980).
1292. S. Hünig and M. Öller, *Chem. Ber.*, **113,** 3803 (1980).
1293. U. Hertenstein, S. Hünig, H. Reichelt, and R. Schaller, *Chem. Ber.*, **115,** 261 (1982).
1294. U. Hertenstein and S. Hünig, *Angew. Chem.*, **87,** 195 (1975); *Int. Ed. Engl.*, **14,** 179 (1975).
1295. F. E. Ziegler, R. V. Nelson, and T. Wang, *Tetrahedron Lett.*, **21,** 2125 (1980).
1296. F. T. Bond, C.-Y. Ho, and O. McConnell, *J. Org. Chem.*, **41,** 1416 (1976).

1297. F. L. Malanco and L. A. Maldonado, *Synth. Commun.*, **6,** 515 (1976).
1298. J. A. Noguez and L. A. Maldonado, *Synth. Commun.*, **6,** 39 (1976).
1299. B. Lesur, J. Toye, M. Chantrenne, and L. Ghosez, *Tetrahedron Lett.*, 2835 (1979).
1300. H. Ahlbrecht and C. Vonderheid, *Synthesis,* 512 (1975).
1301. K. Takahashi, A. Honma, K. Ogura, and H. Iida, *Chem. Lett.*, 1263 (1982).
1302. R. M. Jacobson and J. W. Clader, *Tetrahedron Lett.*, **21,** 1205 (1980).
1303. P. Tuchinda, V. Prapansiri, W. Naengchomnong, and V. Reutrakul, *Chem. Lett.*, 1427 (1984).
1304. D. Seyferth, G. J. Murphy, and R. A. Woodruff, *J. Am. Chem. Soc.*, **96,** 5011 (1974).
1305. L. Duhamel, J. Chauvin, and A. Messier, *Tetrahedron Lett.*, **21,** 4171 (1980).
1306. Y. Nagao, K. Kaneko, and E. Fujita, *Tetrahedron Lett.*, 4115 (1978).
1307. A. G. Cameron, A. T. Hewson, and M. I. Osammor, *Tetrahedron Lett.*, **25,** 2267 (1984).
1308. R. J. Bryant and E. McDonald, *Tetrahedron Lett.*, 3841 (1975).
1309. H. Stetter and K. H. Mohrmann, *Synthesis,* 129 (1981).
1310. R. Nouri-Bimorghi, *Bull. Soc. Chim. Fr.*, 1876 (1975).
1311. I. Stahl, R. Manske, and J. Gosselck, *Chem. Ber.*, **113,** 800 (1980).
1312. R. E. Ireland and J. A. Marshall, *J. Org. Chem.*, **27,** 1615 (1962).
1313. J. P. Marino and L. C. Katterman, *J. Chem. Soc., Chem. Commun.*, 946 (1979).
1314. E. J. Corey and R. H. K. Chen, *Tetrahedron Lett.*, 3817 (1973).
1315. T. Shahak and Y. Sasson,*Tetrahedron Lett.*, 4207 (1973).
1316. R. K. Dieter and Y. Jenkitkasemwong, *Tetrahedron Lett.*, **23,** 3747 (1982).
1317. R. K. Dieter, J. R. Fishpaugh, and L. A. Silks, *Tetrahedron Lett.*, **23,** 3751 (1982).
1318. R. B. Gammill, D. M. Sobieray, and P. M. Gold, *J. Org. Chem.*, **46,** 3555 (1981).
1319. S. Masson and A. Thuillier, *Tetrahedron Lett.*, **21,** 4085 (1980).
1320. R. R. Schmidt and J. Talbiersky, *Angew. Chem.*, **88,** 193 (1976); *Int. Ed. Engl.*, **15,** 171 (1976).
1321. V. Ramanathan and R. Levine, *J. Org. Chem.*, **27,** 1216 (1962).
1322. G. Büchi and H. Wüest, *J. Org. Chem.*, **31,** 977 (1966).
1323. H. Gilman and F. Breuer, *J. Am. Chem. Soc.*, **56,** 1123 (1934).
1324. H. Zak and U. Schmidt, *Angew. Chem.*, **87,** 454 (1975); *Int. Ed. Engl.*, **14,** 432 (1975).
1325. J. Gombos, E. Haslinger, H. Zak, and U. Schmidt, *Tetrahedron Lett.*, 3391 (1975).
1326. J. Gombos, E. Haslinger, H. Zak, and U. Schmidt, *Monatsh. Chem.*, **106,** 219 (1975).
1327. M. J. Arco, M. H. Trammell, and J. D. White, *J. Org. Chem.*, **41,** 2075 (1976).
1328. C. H. Heathcock, L. G. Gulich, and T. Dehlinger, *J. Heterocycl. Chem.*, **6,** 141 (1969).
1329. E. Niwa and M. Miyake, *Chem. Ber.*, **103,** 997 (1970).
1330. L. Lepage and Y. Lepage, *Synthesis,* 1018 (1983).
1331. R. Ribéreau and G. Quéguiner, *Tetrahedron,* **39,** 3593 (1983).
1332. D. J. Chadwick, M. V. McKnight, and R. Ngochindo, *J. Chem. Soc., Perkin Trans. 1,* 1343 (1982).
1333. T. Masamune, M. Ono, and H. Matsue, *Bull. Chem. Soc. Jpn.*, **48,** 491 (1975).
1334. D. J. Ager, *Tetrahedron Lett.*, **24,** 5441 (1983).
1335. G. Piancatelli, A. Scettri, and M. D'Auria, *Tetrahedron,* **36,** 661 (1980).
1336. P. D. Williams and E. LeGoff, *J. Org. Chem.*, **46,** 4143 (1981).
1337. J. K. MacLeod, G. Bott, and J. Cable, *Aust. J. Chem.*, **30,** 2561 (1977).
1338. J. Jurczak and S. Pikul, *Tetrahedron Lett.*, **26,** 3039 (1985).
1339. I. Kuwajima, I. Azegami, and E. Nakamura, *Chem. Lett.*, 1431 (1978).

1340. E. L. Eliel and A. A. Hartmann, *J. Org. Chem.*, **37,** 505 (1972).
1341. J. H. Hoare and P. Yates, *J. Org. Chem.*, **48,** 3333 (1983).
1342. M. Lissel, *Synth. Commun.*, **11,** 343 (1981).
1343. M. Lissel, *Justus Liebigs Ann. Chem.*, 1589 (1982)
1344. Y. Köksal, V. Osterthun, and E. Winterfeldt, *Justus Liebigs Ann. Chem.*, 1300 (1979).
1345. M. Braun and M. Esdar, *Chem. Ber.*, **114,** 2924 (1981).
1346. J. L. Belletire, D. R. Walley, and S. L. Fremont, *Tetrahedron Lett.*, **25,** 5729 (1984).
1347. M. Farina, M. C. Maestro, M. R. Martin, M. V. Martin, and F. Sanchez, *J. Chem. Res. (S),* 44 (1984).
1348. E. Cossement, R. Binamé, and L. Ghosez, *Tetrahedron Lett.*, 997 (1974).
1349. H. N. Khatri and H. M. Walborsky, *J. Org. Chem.*, **43,** 734 (1978).
1350. S. M. Hannick and Y. Kishi, *J. Org. Chem.*, **48,** 3833 (1983).
1351. G. S. Bates, *J. Chem. Soc., Chem. Commun.*, 161 (1979).
1352. G. S. Bates and S. Ramaswamy, *Can. J. Chem.*, **58,** 716 (1980).
1353. J. L. Herrmann, J. E. Richman, R. H. Schlessinger, *Tetrahedron Lett.*, 2599 (1973).
1354. R. E. Damon and R. H. Schlessinger, *Tetrahedron Lett.*, 4551 (1975).
1355. R. J. Gregge, J. L. Herrmann, J. E. Richman, R. F. Romanet, and R. H. Schlessinger, *Tetrahedron Lett.*, 2595 (1973).
1356. J. González, F. Sánchez, and T. Torres, *Synthesis,* 911 (1983).
1357. G. Massiot, T. Mulamba, and J. Lévy, *Bull. Soc. Chim. Fr.*, 241 (1982).
1358. R. J. Gregge, J. L. Herrmann, and R. H. Schlessinger, *Tetrahedron Lett.*, 2603 (1973).
1359. H. Paulsen, W. Koebernick, and H. Koebernick, *Tetrahedron Lett.*, 2297 (1976).
1360. F. Huet, M. Pellet, and J. M. Conia, *Synthesis,* 33 (1979).
1361. H. Neef and U. Eder, *Tetrahedron Lett.*, 2825 (1977).
1362. M. T. Reetz, H. Heimbach, and K. Schwellnus, *Tetrahedron Lett.*, **25,** 511 (1984).
1363. S. L. Hartzell and M. W. Rathke, *Tetrahedron Lett.*, 2737 (1976).
1364. K. Tanaka, T. Nakai, and N. Ishikawa, *Tetrahedron Lett.*, 4809 (1978).
1365. J. Villieras, J. R. Disnar, P. Perriot, and J. F. Normant, *Synthesis,* 524 (1975).
1366. B. Rague, J.-A. Fehrentz, R. Guegan, Y. Chapleur, and B. Castro, *Bull. Soc. Chim. Fr.*, 230 (1983).
1367. R. M. Carlson, A. R. Oyler, and J. R. Peterson, *J. Org. Chem.*, **40,** 1610 (1975).
1368. R. M. Carlson and A. R. Oyler, *Tetrahedron Lett.*, 2615 (1974).
1369. K. Isobe, M. Fuse, H. Kosugi, and H. Hagiwara, and H. Uda, *Chem. Lett.*, 785 (1979).
1370. Y. Takahashi, K. Isobe, H. Hagiwara, H. Kosugi, and H. Uda, *J. Chem. Soc., Chem. Commun.*, 714 (1981).
1371. M. Utaka, H. Kuriki, T. Sakai, and A. Takeda, *Chem. Lett.*, 911 (1983).
1372. S. J. Gould and B. D. Remillard, *Tetrahedron Lett.*, 4353 (1978).
1373. G. J. McGarvey and J. S. Bajwa, *J. Org. Chem.*, **49,** 4092 (1984).
1374. R. Amouroux, *Heterocycles,* **22,** 1489 (1984).
1375. M. Makosza and T. Goetzen, *Rocz. Chem.*, **46,** 1239 (1972); *Chem. Abstr.*, **77,** 164004v (1972).
1376. H. Stetter, K. H. Mohrmann, and W. Schlenker, *Chem. Ber.*, **114,** 581 (1981).
1377. C. B. BiEkogha, O. Ruel, and S. A. Julia, *Tetrahedron Lett.*, **24,** 4825 (1983).
1378. M. Braun and W. Hild, *Angew. Chem.*, **96,** 701 (1984); *Int. Ed. Engl.*, **23,** 723 (1984).
1379. F. Derguini and G. Linstrumelle, *Tetrahedron Lett.*, **25,** 5763 (1984).
1380. M. Pohmakotr and D. Seebach, *Tetrahedron Lett.*, 2271 (1979).

1381. F. Ozawa, T. Sugimoto, Y. Yuasa, M. Santra, T. Yamamoto, and A. Yamamoto, *Organometallics,* **3,** 683 (1984).

1382. J.-M. Vatele, *Tetrahedron Lett.,* **25,** 5997 (1984).

1383. K. C. Brinkman and J. A. Gladysz, *Organometallics,* **3,** 147 (1984).

1384. E. W. Colvin, in *Carboxylic Acids* in *Comprehensive Organic Chemistry,* Vol. 2, D. H. R. Barton and W. D. Ollis, Eds., Pergamon, Oxford, 1979, p. 591.

1385. S. Patai, Ed., *The Chemistry of Carboxylic Acids and Esters,* Wiley, London, 1969.

1386. D. Seebach and M. Kolb, *Chem. Ind. (London),* 687 (1974).

1387. D. Seebach, *Synthesis,* 17 (1969).

1388. D. Seebach, *Angew. Chem.,* **91,** 427 (1979); *Int. Ed. Engl.,* **18,** 399 (1979).

1389. D. Vorländer, *Ber. Dtsch. Chem. Ges.,* **52,** 263 (1919).

1390. A. Lapworth, *J. Chem. Soc.,* **121,** 416 (1922).

1391. W. O. Kermack and R. Robinson, *J. Chem. Soc.,* **121,** 427 (1922).

1392. J. A. Pople and M. Gordon, *J. Am. Chem. Soc.,* **89,** 4253 (1967).

1393. S. W. Benson, *Angew. Chem.,* **90,** 868 (1978); *Int. Ed. Engl.,* **17,** 812 (1978).

1394. E. Hückel, *Theoretische Grundlagen der Organischen Chemie,* Vol. 2, 8th ed. Geest and Portig, Leipzig, 1957.

1395. H. Staudinger, *Ber. Dtsch. Chem. Ges.,* **41,** 2217 (1908).

1396. F. C. Schaefer, in *The Chemistry of the Cyano Group,* Z. Rappoport, Ed., Interscience, New York, 1970, p. 239.

1397. G. Tennant, *Imines, Nitrones, Nitriles, and Isocyanates,* in *Comprehensive Organic Chemistry,* Vol. 2, D. H. R. Barton and W. D. Ollis, Eds., Pergamon, Oxford, 1979, p. 528.

1398. G. Tennant, *Imines, Nitrones, Nitriles, and Isocyanates,* in *Comprehensive Organic Chemistry,* Vol. 2, D. H. R. Barton and W. D. Ollis, Eds., Pergamon, Oxford, 1979, p. 568.

1399. P. A. Smith, *Open Chain Nitrogen Compounds,* Vol. 1, Benjamin, New York, 1965, p. 223.

1400. T. Saeguso and Y. Ito, in *Isonitrile Chemistry,* I. Ugi, Ed., Academic, New York, 1971, p. 65.

1401. J. U. Nef, *Justus Liebigs Ann. Chem.,* **270,** 267 (1892).

1402. P. Kurtz, *Methoden der organischen Chemie,* (*Houben-Weyl*), Vol. 8, G. Thieme, Stuttgart, 1952, p. 247.

1403. I. Ugi, Ed., *Isonitrile Chemistry,* Academic, New York, 1971.

1404. M. P. Periasamy and H. M. Walborsky, *Org. Prep. Proced. Int.,* **11,** 293 (1979).

1405. I. Ugi, *Angew. Chem.,* **94,** 826 (1982); *Int. Ed. Engl.,* **21,** 810 (1982).

1406. J. U. Nef, *Justus Liebigs Ann. Chem.,* **280,** 291 (1894).

1407. M. Passerini, *Gazz. Chim. Ital.,* **57,** 452 (1927).

1408. M. Okano, Y. Ito, T. Shono, R. Oda, *Bull. Chem. Soc. Jpn.,* **36,** 1314 (1963).

1409. F. Millich, *Chem. Rev.,* **72,** 101 (1972).

1410. D. Marquading, G. Gockel, P. Hoffman, and I. Ugi, in *Isonitrile Chemistry,* I. Ugi, Ed., Academic, New York, 1971, p. 133.

1411. P. Hoffmann, D. Marquading, H. Kliiman, and I. Ugi, in *The Chemistry of the Cyano Group,* Z. Rappaport, Ed., Interscience, New York, 1970, p. 853.

1412. W. Reeve, *Synthesis,* 131 (1971).

1413. P. Galimberti and A. Defranceshi, *Gazz. Chim. Ital.,* **77,** 431 (1947).

1414. H. Gilman and G. R. Wilder, *J. Am. Chem. Soc.,* **77,** 6644 (1955).

1415. J. Jocicz, *Zhur. Russ. Fis.-Kihm. Obshch.,* **29,** 92 (1897); *Chem. Zentr.,* **68,** (1), 1013 (1897).

1416. C. Weizman, M. Sulzbacher, and E. Bergmann, *J. Am. Chem. Soc.*, **70,** 1153 (1948).

1417. E. D. Bergmann, D. Ginsburg, and D. Lavie, *J. Am. Chem. Soc.*, **72,** 5012 (1950).

1418. P. Hébert, *Bull. Soc. Chim. Fr.*, **27,** 45 (1920).

1419. P. F. Hudrlik and A. M. Hudrlik, "Applications of Acetylenes in Organic Chemistry," in *The Chemistry of the Carbon-Carbon Triple Bond,* S. Patai, Ed., Wiley, London, 1978, p. 199.

1420. M. Mailfert, *C. R. Hebd. Seances Acad. Sci.*, **94,** 1186 (1882).

1421. C. Harries, *Ber. Dtsch. Chem. Ges.*, **40,** 4905 (1907).

1422. E. Molinari, *Ber. Dtsch. Chem. Ges.*, **41,** 585 (1908).

1423. E. Molinari, *Ber. Dtsch. Chem. Ges.*, **41,** 2782 (1908).

1424. *Ozonation in Organic Chemistry, Vol. 2: Nonolefinic Compounds,* P. S. Bailey, Ed., Academic, New York, 1982, p. 3.

1425. H. Gopal and A. J. Gordon, *Tetrahedron Lett.*, 2941 (1971).

1426. F. Gerhart, private communication, Merrell Dow Research Institute, Strasbourg Center, France, 1984.

1427. W. Nagata, and M. Yoshioka, *Org. React.*, **25,** 255 (1977)

1428. W. Nagata, M. Yoshioka, and S. Hirai, *Tetrahedron Lett.*, 461 (1962).

1429. W. Nagata, T. Terasawa, and T. Aoki, *Tetrahedron Lett.*, 865 (1963).

1430. W. Nagata, T. Terasawa, and T. Aoki, *Tetrahedron Lett.*, 869 (1963).

1431. W. Nagata, M. Yoshioka, and M. Murakami, *J. Am. Chem. Soc.*, **94,** 4654 (1972).

1432. W. Nagata, M. Yoshioka, and S. Hirai, *J. Am. Chem. Soc.*, **94,** 4635 (1972).

1433. W. Nagata, M. Yoshioka, and M. Murakami, *J. Am. Chem. Soc.*, **94,** 4644 (1972).

1434. W. Nagata, M. Yoshioka, and T. Terasawa, *J. Am. Chem. Soc.*, **94,** 4672 (1972).

1435. W. Nagata, S. Hirai, H. Itazaki, and K. Takeda, *J. Org. Chem.*, **26,** 2413 (1961).

1436. W. Nagata, T. Sugasawa, M. Narisada, T. Wakabayashi, and Y. Hayase, *J. Am. Chem. Soc.*, **85,** 2342 (1963).

1437. W. Nagata, M. Yoshioka, and T. Okumura, *Tetrahedron Lett.*, 847 (1966).

1438. G. Dittus, *Methoden der organischen Chemie* (*Houben-Weyl*), Vol. 6/3, G. Thieme, Stuttgart, 1965, p. 450.

1439. H. O. House and W. F. Fischer, Jr., *J. Org. Chem.*, **34,** 3626 (1969).

1440. E. J. Corey and L. S. Hegedus, *J. Am. Chem. Soc.*, **91,** 1233 (1969).

1441. S. Cacchi, L. Caglioti, and G. Paolucci, *Synthesis,* 120 (1975).

1442. S. Cacchi, L. Caglioti, and G. Paolucci, *Chem. Ind.* (*London*), 213 (1972).

1443. M. C. Chaco and N. Rabjohn, *J. Org. Chem.*, **27,** 2765 (1962).

1444. F. E. Ziegler and P. A. Wender, *J. Am. Chem. Soc.*, **93,** 4318 (1971).

1445. W. P. Weber, *Silicon Reagents for Organic Synthesis,* Springer Verlag, Berlin, 1983, p. 6.

1446. P. G. Gassman and J. J. Talley, *Tetrahedron Lett.*, 3773 (1978).

1447. M. T. Reetz and H. Müller-Starke, *Tetrahedron Lett.*, **25,** 3301 (1984).

1448. W. Kantlehner, E. Haug, W. Frick, P. Speh, and H. J. Bräuner, *Synthesis,* 358 (1984).

1449. T. Shioiri, *Ann. Rep. Pharm. Nagoya City Univ.*, **25,** 1 (1977); *J. Synth. Org. Chem. Jpn.*, **37,** 856 (1979).

1450. S. Harusawa, Y. Hamada, and T. Shioiri, *Synthesis,* 716 (1979).

1451. S. Harusawa, Y. Hamada, and T. Shioiri, *Tetrahedron Lett.*, 4663 (1979).

1452. T. Ishide, M. Inoue, S. Harusawa, Y. Hamada, and T. Shioiri, *Acta Cryst. Sect. B,* **37,** 1881 (1981).

1453. S. Harusawa, R. Yoneda, T. Kurihara, Y. Hamada, and T. Shioiri, *Chem Pharm. Bull.*, **31,** 2932 (1983).

1454. B. C. Challis and J. A. Challis, in *Amides and Related Compounds* in *Comprehensive Organic Chemistry,* Vol. 2, D. H. R. Barton and W. D. Ollis, Eds., Pergamon, Oxford, 1979, p. 964.

1455. K. Ogura and Y. Yamashita, *Tetrahedron Lett.,* 357 (1978).

1456. U. Schöllkopf and H. Bechhaus, *Angew. Chem.,* **15,** 296 (1976); *Int. Ed. Engl.,* **15,** 293 (1976).

1457. L. Cassar, G. P. Chiusoli, and F. Guerrieri, *Synthesis,* 509 (1973).

1458. E. W. Colvin, in *Carboxylic Acids* in *Comprehensive Organic Chemistry,* Vol. 2, D. H. R. Barton and W. D. Ollis, Eds., Pergamon, Oxford, 1979, p. 602.

1459. J. Tsuji, *Organic Synthesis by Means of Transition Metal Complexes,* Springer Verlag, Berlin, 1975.

1460. A. P. Kozikowski and H. F. Wetter, *Synthesis,* 561 (1976).

1461. C. W. Bird, *Transition Metal Intermediates in Organic Synthesis,* Logos, London, 1967.

1462. C. W. Bird, *Chem. Rev.,* **62,** 283 (1962).

1463. R. F. Heck, *Adv. Catal.,* **26,** 323 (1977).

1464. I. Ugi and C. Steinbrückner, *Angew. Chem.,* **72,** 267 (1960).

1465. I. Ugi and U. Fetzer, *Chem. Ber.,* **94,** 1116 (1961).

1466. I. Ugi and K. Rosendahl, *Chem. Ber.,* **94,** 2233 (1961).

1467. I. Ugi and C. Steinbrückner, *Chem. Ber.,* **94,** 2802 (1961).

1468. I. Ugi, *Angew. Chem.,* **1,** 9 (1962); *Int. Ed. Engl.,* **1,** 8 (1962).

1469. J. W. McFarland, *J. Org. Chem.,* **28,** 2179 (1963).

1470. I. Hagedorn and U. Eholzer, *Chem. Ber.,* **98,** 936 (1965).

1471. J. U. Nef, *Justus Liebigs Ann. Chem.,* **287,** 265 (1895).

1472. E. Jungermann and F. W. Smith, *J. Am. Oil Chem. Soc.,* **36,** 388 (1959).

1473. I. Ugi and F. Bodesheim, *Chem. Ber.,* **94,** 2797 (1961).

1474. I. Ugi and C. Steinbrückner, *Chem. Ber.,* **94,** 734 (1961).

1475. M. Schiess and D. Seebach, *Helv. Chim. Acta,* **66,** 1618 (1983).

1476. D. Seebach, B. Weidman, and L. Widler, in *Modern Synthetic Methods,* R. Scheffold, Ed., Wiley, New York, 1983, p. 217.

1477. T. Mukaiyama, K. Watanabe, and M. Shiono, *Chem. Lett.,* 1523 (1974).

1478. A. Schoenberg and R. F. Heck, *J. Org. Chem.,* **39,** 3327 (1974).

1479. H. A. Dieck, R. M. Laine, and R. F. Heck, *J. Org. Chem.,* **40,** 2819 (1975).

1480. M. Yamashita, Y. Watanabe, T. Mitsudo, and Y. Tagekami, *Tetrahedron Lett.,* 1585 (1976).

1481. M. Yamashita, K. Mizushima, Y. Watanabe, T. Mitsudo, and Y. Takegami, *J. Chem. Soc., Chem. Commun.,* 670 (1976).

1482. Y. Watanabe, K. Taniguchi, M. Suga, T. Mitsudo, and Y. Takegami, *Bull. Chem. Soc. Jpn.,* **52,** 1869 (1979).

1483. V. Rautenstrauch and M. Joyeux, *Angew. Chem.,* **91,** 73 (1979); *Int. Ed. Engl.,* **18,** 85 (1979).

1484. T. Tsuda, M. Miwa, and T. Saegusa, *J. Org. Chem.,* **44,** 3734 (1979).

1485. V. Rautenstrauch and F. Delay, *Angew. Chem.,* **92,** 764 (1980); *Int. Ed. Engl.,* **19,** 726 (1980).

1486. H. Takahashi and J. Tsuji, *J. Organomet. Chem.,* **10,** 511 (1967).

1487. S. Murahashi and S. Horiie, *J. Am. Chem. Soc.,* **78,** 4816 (1956).

1488. S. Horiie and S. Murahashi *Bull. Chem. Soc. Jpn.,* **33,** 88 (1960).

1489. V. Rautenstrauch and M. Joyeux, *Angew. Chem.,* **91,** 72 (1979), *Int. Ed. Engl.,* **18,** 83 (1979).

1490. M. R. Kilbourn, P. A. Jerabek, and M. J. Welch, *J. Chem. Soc., Chem. Commun.*, 861 (1983).
1491. N. S. Nudelman and D. Perez, *J. Org. Chem.*, **48,** 133 (1983).
1492. P. Longi, R. Montagna, and R. Mazzocchi, *Chim. Ind. (Milan)*, **47,** 480 (1965).
1493. S. Fukuoka, M. Ryang, and S. Tsutsumi, *J. Org. Chem.*, **36,** 2721 (1971).
1494. S. Fukuoka, M. Ryang, and S. Tsutsumi, *J. Org. Chem.*, **33,** 2973 (1968).
1495. K. Takahashi, K. Shibasaki, K. Ogura, and H. H. Iida, *Chem. Lett.*, 859 (1983).
1496. A. S. Fletcher, K. Smith, and K. Swaminathan, *J. Chem. Soc., Perkin Trans. 1,* 1881 (1977).
1497. U. Zehavi, *J. Org. Chem.*, **42,** 2821 (1977).
1498. M. Igarashi and H. Midorikawa, *J. Org. Chem.*, **28,** 3088 (1963).
1499. G. Barger and A. J. Ewins, *J. Chem. Soc.*, **97,** 284 (1910).
1500. E. Campaigne, G. Skowronski, and B. R. Rogers, *Synth. Commun.*, **3,** 325 (1973).
1501. A. Wohl and L. H. Lips, *Ber. Dtsch. Chem. Ges.*, **40,** 2312 (1907).
1502. J. Schreiber, *C. R. Hebd. Seances Acad. Sci.*, **242,** 139 (1956).
1503. R. Adams, H. B. Bramlet, and F. H. Tendick, *J. Am. Chem. Soc.*, **42,** 2369 (1920).
1504. U. Schöllkopf and F. Gerhart, *Angew. Chem.*, **6,** 819 (1967); *Int. Ed. Engl.*, **6,** 805 (1967).
1505. B. Bánhidai and U. Schöllkopf, *Angew. Chem.*, **12,** 861 (1973); *Int. Ed. Engl.*, **12,** 836 (1973).
1506. R. R. Fraser and P. R. Hubert, *Can. J. Chem.*, **52,** 185 (1974).
1507. K. Smith and K. Swaminathan, *J. Chem. Soc., Chem. Commun.*, 387 (1976).
1508. A. S. Fletcher, W. E. Paget, K. Smith, K. Swaminathan, J. H. Beynon, R. P. Morgan, M. Bozorgzadeg, and M. J. Haley, *J. Chem. Soc., Chem. Commun.*, 347, (1979).
1509. J. Souppe, J.-L. Namy, and H. B. Kagan, *Tetrahedron Lett.*, **25,** 2869 (1984).
1510. W. W. Prichard, *J. Am. Chem. Soc.*, **78,** 6137 (1956).
1511. G. P. Chiusoli and L. Cassar, *Angew. Chem.*, **6,** 177 (1967); *Int. Ed. Engl.*, **6,** 124 (1967).
1512. L. Cassar and M. Foà, *J. Organomet. Chem.*, **51,** 381 (1973).
1513. H. Dieterle and W. Eschenbach, D. R. Pat. 537,610 (1927).
1514. A. T. Larson (E. I. DuPont), U. S. Pat. 1,993,555 (1932).
1515. H. Bliss and R. W. Southworth, U. S. Pat. 2,565,461 (1949).
1516. H. I. Leiben (E. I. DuPont), U. S. Pat. 2,640,071 (1953); U. S. Pat. 2,691,671 (1954); U. S. Pat. 2,734,912 (1955); U. S. Pat. 2,773,090 (1956).
1517. K. Yamamoto and K. Sato, *Bull. Chem. Soc. Jpn.*, **27,** 389 (1954).
1518. G. C. Tustin and R. T. Hembre, *J. Org. Chem.*, **49,** 1761 (1984).
1519. P. M. Henry, *Tetrahedron Lett.*, 2285 (1968).
1520. L. Cassar, M. Foà, and A. Gardano, *J. Organomet. Chem.*, **121,** C55 (1976).
1521. B. Loubinoux, B. Fixari, J. J. Brunet, and P. Caubere, *J. Organomet. Chem.*, **105,** C22 (1976).
1522. H. Alper and H. Des Abbayes, *J. Organomet. Chem.*, **134,** C11 (1977).
1523. L. Cassar and M. Foà, *J. Organomet. Chem.*, **134,** C15 (1977).
1524. J. J. Brunet, C. Sidot, B. Loubinoux, and P. Caubere, *J. Org. Chem.*, **44,** 2199 (1979).
1525. J. J. Brunet, C. Sidot, and P. Caubere, *Tetrahedron Lett.*, **22,** 1013 (1981).
1526. J. J. Brunet, C. Sidot, and P. Caubere, *J. Org. Chem.*, **48,** 1166 (1983).
1527. G. Tanguy, B. Weinberger, and H. Des Abbayes *Tetrahedron Lett.*, **24,** 4005 (1983).
1528. U. Schöllkopf, F. Gerhart, I. Hoppe, R. Harms, K. Hantke, K.-H. Scheunemann, E. Eilers, and F. Blume *Justus Liebigs Ann. Chem.*, 183 (1976).

1529. R. P. Kurkjy and E. V. Brown, *J. Am. Chem. Soc.,* **74,** 6260 (1952).

1530. H. C. Beyerman, P. H.. Berben, and J. S. Bontekoe, *Recl. Trav. Chim. Pays-Bas,* **73,** 325 (1954).

1531. J. Metzger and B. Koether, *Bull. Soc. Chim. Fr.,* 702 (1953).

1532. J. Metzger and B. Koether, *Bull. Soc. Chim. Fr.,* 708 (1953).

1533. J. Beraud and J. Metzger, *Bull. Soc. Chim. Fr.,* 2072 (1962).

1534. J. V. Metzger, *Thiazoles and their Benzo Derivatives,* in *Comprehensive Heterocyclic Chemistry,* Vol. 6, A. R. Katritzky and C. W. Rees, Eds., Pergamon, Oxford,, 1984, p. 235.

1535. N. J. Curtis and R. S. Brown, *J. Org. Chem.,* **45,** 4038 (1980).

1536. R. E. Damon and R. H. Schlessinger, *Tetrahedron Lett.,* 1561 (1976).

1537. J. Gabriel and D. Seebach, *Helv. Chim. Acta,* **67,** 1070 (1984).

1538. M. Mikolajczyk, S. Grzejszczak, A. Zatorski, and B. Mlotkowska, *Tetrahedron Lett.,* 2731 (1976).

1539. G. S. Bates and S. Ramaswamy, *Can. J. Chem.,* **61,** 2006 (1983).

1540. K. Hartke and J. Quanto, *Arch. Pharm.,* **313,** 1029 (1980).

1541. S. Hayashi, M. Furukawa, Y. Fujino, N. Nakao, and S. Inoue, *Chem. Pharm. Bull.,* **19,** 1557 (1971).

1542. G. S. Bates and S. Ramaswamy, *J. Chem. Soc., Chem. Commun.,* 904 (1980).

1543. P. F. Juby, W. R. Goodwin, T. W. Hudyma, and R. A. Partyka, *J. Med. Chem.,* **15,** 1297 (1972).

1544. E. J. Corey and D. J. Beames, *J. Am. Chem. Soc.,* **95,** 5829 (1973).

1545. E. Rothstein, *J. Chem. Soc.,* 1553 (1940).

1546. H. C. Volger and J. F. Arens, *Recl. Trav. Chim. Pays-Bas,* **76,** 847 (1957).

1547. D. S. Tarbell and D. P. Harnish, *Chem. Rev.,* **49,** 1 (1951).

1548. F. Weygand and G. Hilgetag, *Organische chemische Experimentierkunst,* Barth Verlag, Leipzig, 1964, p. 472.

1549. C. R. Johnson, J. R. Shanklin, and R. A. Kirchhoff, *J. Am. Chem. Soc.,* **95,** 6462 (1973).

1550. U. Sehnbert, *Synthesis,* 364 (1978).

1551. T. N. Salzman, R. W. Ratcliffe, B. G. Christensen, and F. A. Bouffard, *J. Am. Chem. Soc.,* **102,** 6161 (1980).

1552. E. L. Compere, Jr., *J. Org. Chem.,* **33,** 2565 (1968).

1553. A. Merz, *Synthesis,* 724 (1974).

1554. P. Kuhl, M. Mühlstaedt, and J. Graefe, *Synthesis,* 825 (1976).

1555. E. L. Compere, Jr. and D. A. Weinstein, *Synthesis,* 852 (1977).

1556. D. Seebach and N. Peleties, *Chem. Ber.,* **105,** 511 (1972).

1557. H. Gross and B. Costisella, *Angew. Chem.,* **80,** 364 (1968); *Int. Ed. Engl.,* **7,** 391 (1968).

1558. N. Kreutzkamp, *Pharm. Ztg.,* **105,** 429 (1960).

1559. G. A. Russell and G. J. Mikol, *J. Am. Chem. Soc.,* **88,** 5498 (1966).

1560. J. Tsuji, M. Morikawa, and J. Kiji, *Tetrahedron Lett.,* 1437 (1963).

1561. J. Tsuji, J. Kiji, and M. Morikawa, *Tetrahedron Lett.,* 1811 (1963).

1562. J. Tsuji, M. Morikawa, and N. Iwamoto, *J. Am. Chem. Soc.,* **86,** 2095 (1964).

1563. J. Tsuji, J. Kiji, S. Imamura, and M. Morikawa, *J. Am. Chem. Soc.,* **86,** 4350 (1964).

1564. J. Tsuji, S. Imamura, and J. Kiji, *J. Am. Chem. Soc.,* **86,** 4491 (1964).

1565. J. Tsuji, M. Morikawa, and J. Kiji, *J. Am. Chem. Soc.,* **86,** 4851 (1964).

1566. J. Tsuji, J. Kiji, and S. Hosaka, *Tetrahedron Lett.,* 605 (1964).

1567. J. Tsuji and S. Hosaka, *J. Am. Chem. Soc.,* **87,** 4075 (1965).

1568. J. Tsuji, S. Hosaka, J. Kiji, and T. Susuki, *Bull. Chem. Soc. Jpn.,* **39,** 141 (1966).
1569. J. Tsuji and S. Imamura, *Bull. Chem. Soc. Jpn.,* **40,** 197 (1967).
1570. J. Tsuji and T. Nogi, *J. Am. Chem. Soc.,* **88,** 1289 (1966).
1571. J. Tsuji and T. Nogi, *J. Org. Chem.,* **31,** 2641 (1966).
1572. J. Tsuji and T. Susuki, *Tetrahedron Lett.,* 3027 (1965).
1573. T. Susuki and J. Tsuji, *Bull. Chem. Soc. Jpn.,* **41,** 1954 (1968).
1574. J. Tsuji and T. Nogi, *Tetrahedron Lett.,* 1801 (1966).
1575. K. Bittler, N. v. Kutepov, D. Neubauer, and H. Reis, *Angew. Chem.,* **7,** 352 (1968); *Int. Ed. Engl.,* **7,** 329 (1968).
1576. G. P. Chiusoli, C. Venturello, and S. Merzoni, *Chem. Ind. (London),* **29,** 977 (1968).
1577. W. T. Dent, R. Long, and G. H. Whitfield, *J. Chem. Soc.,* 1588 (1964).
1578. A. Schoenberg, I. Bartoletti, and R. F. Heck, *J. Org. Chem.,* **39,** 3318 (1974).
1579. J. Haggin, *Chem. Eng. News,* Feb. 4., p. 25 (1985).
1580. I. Ryu, K. Matsumoto, M. Ando, S. Murai, and N. Sonoda, *Tetrahedron Lett.,* **21,** 4283 (1980).
1581. R. F. Heck, *J. Am. Chem. Soc.,* **85,** 2014 (1963).
1582. N. L. Bauld, *Tetrahedron Lett.,* 1841 (1963).
1583. M. Nakayama and T. Mizoroki, *Bull. Chem. Soc. Jpn.,* **44,** 508 (1971).
1584. G. P. Chiusoli, *Angew. Chem.,* **72,** 74 (1960).
1585. G. P. Chiusoli, *Chim. Ind. (Milan),* **41,** 503 (1959).
1586. G. P. Chiusoli and S. Merzoni, *Chim. Ind. (Milan),* **43,** 255 (1961).
1587. G. P. Chiusoli, G. Bottaccio, and A. Cameroni, *Chim. Ind. (Milan),* **44,** 131 (1962).
1588. G. P. Chiusoli and S. Merzoni, *Chim. Ind. (Milan),* **45,** 6 (1963).
1589. G. P. Chiusoli, S. Merzoni, and G. Mondelli, *Chim. Ind. (Milan),* **46,** 743 (1963).
1590. G. P. Chiusoli and G. Bottaccio, *Chim. Ind. (Milan),* **47,** 165 (1965).
1591. G. P. Chiusoli, *Chim. Ind. (Milan),* **41,** 506 (1959).
1592. G. P. Chiusoli, *Chim. Ind. (Milan),* **41,** 513 (1959).
1593. G. P. Chiusoli, *Chim. Ind. (Milan),* **41,** 762 (1959).
1594. Y. Takegami, Y. Watanabe, H. Masada, and I. Kanaya, *Bull. Chem. Soc. Jpn.,* **40,** 1456 (1967).
1595. M. Yamashita, K. Mizushima, Y. Watanabe, T. Mitsudo, and Y. Takegami, *Chem. Lett.,* 1355 (1977).
1596. W. Reeve and C. W. Woods, *J. Am. Chem. Soc.,* **82,** 4062 (1960).
1597. W. Reeve and E. L. Compere, Jr., *J. Am. Chem. Soc.,* **83,** 2755 (1961).
1598. A.-R. B. Manas and R. A. J. Smith, *J. Chem. Soc., Chem. Commun.,* 216 (1975).
1599. W. D. Woessner, *Chem. Lett.,* 43 (1976).
1600. D. Scholz, *Synth. Commun.,* **12,** 527 (1982).
1601. S. Knapp, A. F. Trope, M. S. Theodore, N. Hirata, and J. J. Barchi, *J. Org. Chem.,* **49,** 608 (1984).
1602. J. E. Hengeveld, V. Grief, J. Tadanico, C.-M. Lee, D. Riley, and P. A. Lartey, *Tetrahedron Lett.,* **25,** 4075 (1984).
1603. A. R. Chamberlin, H. D. Nguyen, J. Y. L. Chung, *J. Org. Chem.,* **49,** 1682 (1984).
1604. J. E. Thompson, *J. Org. Chem.,* **32,** 3947 (1967).
1605. K. Ogura, S. Furukawa, and G. Tsuchihashi, *Bull. Chem. Soc. Jpn.,* **48,** 2219 (1975).
1606. K. Ogura, N. Yahata, K. Hashizume, K. Tsuyama, K. Takahashi, and H. Iida, *Chem. Lett.,* 767 (1983).
1607. K. Ogura and G. Tsuchihashi, *Tetrahedron Lett.,* 1383 (1972).

1608. K. Ogura, N. Katoh, I. Yoshimura, and G. Tsuchihashi, *Tetrahedron Lett.,* 375 (1978).
1609. K. Ogura, J. Watanabe, and H. Iida, *Tetrahedron Lett.,* **22,** 4499 (1981).
1610. G. Rousseau, P. Le Perchec, and J. M. Conia, *Synthesis,* 67 (1978).
1611. G. Rousseau, P. Le Perchec, and J. M. Conia, *Tetrahedron Lett.,* 45 (1977).
1612. V. Reutrakul, Y. Srikirin, and S. Panichanun, *Chem. Lett.,* 879 (1982).
1613. Y. Ito, H. Kato, and T. Saegusa, *J. Org. Chem.,* **47,** 741 (1982).
1614. C. d. Neves and H. Jose, *Rev. Port. Farm.,* **22,** 85 (1972); *Chem. Abstr.,* **77,** 163912 (1972).
1615. S. Patai, Ed., *The Chemistry of Carboxylic Acids and Esters,* Wiley, London, 1969, p. 724.
1616. L. Caglioti, *Tetrahedron,* **22,** 487 (1966).
1617. L. Caglioti and G. Magi, *Tetrahedron,* **19,** 1127 (1963).
1618. L. Caglioti and P. Grasseli, *Chim. Ind. (Milan),* **46,** 1492 (1964).
1619. L. Caglioti and P. Grasseli, *Chem. Ind. (London),* 153 (1964).
1620. A. Pinner and F. Klein, *Ber. Dtsch. Chem. Ges.,* **10,** 1889 (1877).
1621. R. A. J. Smith and G. S. Keng, *Tetrahedron Lett.,* 675 (1978).
1622. R. E. Damon, T. Luo, and R. H. Schlessinger, *Tetrahedron Lett.,* 2749 (1976).
1623. D. Seebach and R. Bürstinghaus, *Synthesis,* 461 (1975).
1624. K. Ogura, I. Yoshimura, N. Katoh, and G. Tsuchihashi, *Chem. Lett.,* 803 (1975).
1625. K. M. More and J. Wemple, *J. Org. Chem.,* **43,** 2713 (1978).
1626. A. G. Brook and D. G. Anderson, *Can. J. Chem.,* **46,** 2115 (1968).
1627. A. G. Brook, *Acc. Chem. Res.,* **7,** 77 (1974).
1628. E. Vedejs and M. Mullins, *Tetrahedron Lett.,* 2017 (1975).
1629. B. J. Banks, A. G. M. Barrett, and M. A. Russell, *J. Chem. Soc., Chem. Commun.,* 670 (1984).
1630. H. Ahlbrecht and K. Pfaff, *Synthesis,* 413 (1980).
1631. H. Sugihara, R. Tanikaga, K. Tanaka, and A. Kaji, *Bull. Chem. Soc. Jpn.,* **51,** 655 (1978).
1632. K. Steliou and M.-A. Poupart, *J. Org. Chem.,* **50,** 4971 (1985).
1633. J. d'Angelo, *Tetrahedron,* **32,** 2979 (1976).
1634. E. Buncel and T. Durst, Eds., *Comprehensive Carbanion Chemistry,* Part B, Elsevier, Amsterdam, 1984.
1635. L. Duhamel, F. Tombret, and J.-M. Poirier, *Org. Prep. Proced. Int.,* **17,** 99 (1985).
1636. T. Yamamoto and M. Okawara, *Chem. Lett.,* 581 (1975).
1637. H. Bredereck, F. Effenberger, and R. Gleiter, *Angew. Chem.,* **77,** 964 (1965); *Int. Ed. Engl.,* **4,** 951 (1965).
1638. U. Schöllkopf, R. Schröder, and E. Blume, *Justus Liebigs Ann. Chem.,* **766,** 130 (1972).
1639. Y. Kita, O. Tamura, H. Yasuda, F. Itoh, and Y. Tamura, *Chem. Pharm, Bull.,* **33,** 4235 (1985).
1640. D. A. Evans and D. C. Andrews, *Acc. Chem. Res.,* **7,** 147 (1974).
1641. D. Seebach and D. Enders, *Angew. Chem.,* **87,** 1 (1975); *Int. Ed. Engl.,* **14,** 15 (1975).
1642. P. Beak and L. G. Carter, *J. Org. Chem.,* **46,** 2363 (1981).
1643. A. Pelter, L. Williams, and J. W. Wilson, *Tetrahedron Lett.,* **24,** 627 (1983).
1644. T. M. Dolak and T. A. Bryson, *Tetrahedron Lett.,* 1961 (1977).
1645. E. J. Corey and T. M. Eckrich, *Tetrahedron Lett.,* **24,** 3165 (1983).
1646. R. Lehmann and M. Schlosser, *Tetrahedron Lett.,* **25,** 745 (1984).
1647. H. Siegel, *Top. Curr. Chem.,* **106,** 55 (1982).

1648. J. Villieras and M. Rambaud, *Synthesis,* 644 (1980).
1649. W. C. Still, *J. Am. Chem. Soc.,* **100,** 1481 (1978).
1650. E. J. Corey and T. M. Eckrich, *Tetrahedron Lett.,* **24,** 3163 (1983).
1651. D. J. Peterson, *Organometal. Chem. Rev., Sect. A,* **7,** 295 (1972).
1652. T. Kauffmann, *Top. Curr. Chem.,* **92,** 109 (1980).
1653. D. Seebach and N. Peleties, *Angew. Chem.,* **81,** 465 (1969); *Int. Ed. Engl.,* **8,** 450 (1969).
1654. L. F. Cason and H. G. Brooks, *J. Am. Chem. Soc.,* **74,** 4582 (1952).
1655. D. J. Peterson, *J. Org. Chem.,* **31,** 950 (1966).
1656. T. Kauffmann, H. Ahlers, H. J. Tilhard, and A. Woltermann, *Angew. Chem.,* **89,** 760 (1977); *Int. Ed. Engl.,* **16,** 710 (1977).
1657. D. Seyferth and M. A. Weiner, *J. Am. Chem. Soc.,* **84,** 361 (1962).
1658. S. Wolfe, *Acc. Chem. Res.,* **5,** 102 (1972).
1659. P. Beak and D. B. Reitz, *Chem. Rev.,* **78,** 275 (1978).
1660. A. Streitwieser, Jr. and S. P. Ewing, *J. Am. Chem. Soc.,* **97,** 190 (1975).
1661. A. Streitwieser, Jr. and J. E. Williams, Jr., *J. Am. Chem. Soc.,* **97,** 191 (1975).
1662. F. Bernardi, I. G. Csizmadia, A. Mangini, H. B. Schlegel, M.-H. Whangbo, and S. Wolfe, *J. Am. Chem. Soc.,* **97,** 2209 (1975).
1663. N. D. Epiotis, R. L. Yates, F. Bernardi, and S. Wolfe, *J. Am. Chem. Soc.,* **98,** 5435 (1976).
1664. J.-M. Lehn and G. Wipff, *J. Am. Chem. Soc.,* **98,** 7498 (1976).
1665. F. G. Bordwell, J. E. Bares, J. E. Bartmess, G. E. Drucker, J. Gerhold, G. J. McCollum, M. Van Der Puy, N. R. Vanier, and W. S. Matthews, *J. Org. Chem.,* **42,** 326 (1977).
1666. P. R. Farina and H. Tieckelmann, *J. Org. Chem.,* **38,** 4259 (1973).
1667. B. Renger, H. Hügel, W. Wykypiel, and D. Seebach, *Chem. Ber.,* **111,** 2630 (1978).
1668. U. Schöllkopf, R. Jentsch, K. Madawinata, and R. Harms, *Justus Liebigs Ann. Chem.,* 2105 (1976).
1669. P. Beak, W. J. Zajdel, *J. Am. Chem. Soc.,* **106,** 1010 (1984).
1670. A. I. Meyers, L. N. Fuentes, and Y. Kubota, *Tetrahedron,* **40,** 1361 (1984).
1671. M. Al-Aseer, P. Beak, D. Hay, D. J. Kempf, S. Mills, and S. G. Smith, *J. Am. Chem. Soc.,* **105,** 2080 (1983).
1672. A. I. Meyers, W. F. Rieker, and L. M. Fuentes, *J. Am. Chem. Soc.,* **105,** 2082 (1983).
1673. J. T. B. H. Jastrzebski, G. van Koten, M. Konijn, and C. H. Stam, *J. Am. Chem. Soc.,* **104,** 5490 (1982).
1674. D. L. J. Clive, *Tetrahedron,* **34,** 1049 (1978).
1675. K. Harada, "Additions to the Azomethine Group," in *The Chemistry of the Carbon-Nitrogen Double Bond,* S. Patai, Ed., Interscience, London, 1970, Chapter 6, p. 255.
1676. A. Pictet and T. Spengler, *Ber. Dtsch. Chem. Ges.,* **44,,** 2030 (1911).
1677. F. F. Blicke, *Org. React.,* **1,** 303 (1942).
1678. E. D. Bergmann, *Chem. Rev.,* **53,** 309 (1953).
1679. G. M. Robinson and R. Robinson, *J. Chem. Soc.,* **123,** 532 (1923).
1680. J. E. Saavedra, *J. Org. Chem.,* **50,** 2271 (1985).
1681. D. J. Peterson, *J. Organomet. Chem.,* **21,** P63 (1970).
1682. J. E. Saavedra, *J. Org. Chem.,* **48,** 2388 (1983).
1683. K. Eiter, K.-F. Hebenbrock, and H.-J. Kabbe, *Justus Liebigs Ann. Chem.,* **765,** 55 (1972).
1684. R. Kupper and C. J. Michejda, *J. Org. Chem.,* **45,** 2919 (1980).
1685. J. E. Saavedra, *J. Org. Chem.,* **46,** 2610 (1981).

1686. J. E. Saavedra, Abstract, 16th Annual Middle Atlantic Regional Meeting, ACS, Newark, DE, (1982), Abs. No. ORGN 250.

1687. U. Schöllkopf and F. Gerhart, *Angew. Chem.*, **80,** 842 (1968); *Int. Ed. Engl.*, **7,** 805 (1968).

1688. W. A. Böll, F. Gerhart, A. Nürrenbach, and U. Schöllkopf, *Angew. Chem.*, **82,** 482 (1970); *Int. Ed. Engl.*, **9,** 458 (1970).

1689. U. Schöllkopf and P. Böhme, *Angew. Chem.*, **83,** 490 (1971); *Int. Ed. Engl.*, **10,** 491 (1971).

1690. U. Schöllkopf, K.-W. Henneke, K. Madawinata, and R. Harms, *Justus Liebigs Ann. Chem.*, 40, (1977).

1691. U. Schöllkopf, *Angew. Chem.*, **89,** 351 (1977); *Int. Ed. Engl.*, **16,** 339 (1977).

1692. T. Kauffmann, H. Berg, E. Köppelmann, and D. Kuhlmann, *Chem. Ber.*, **110,** 2659 (1977).

1693. P. Hullot and T. Cuvigny, *Bull. Soc. Chim. Fr.*, 2989 (1973).

1694. D. Hoppe, *Angew. Chem.*, **87,** 449 (1975); *Int. Ed. Engl.*, **14,** 424 (1975).

1695. L. Henry, *C. R. Hebd. Seances Acad. Sci.*, **120,** 1265 (1895).

1696. G. Rosini, R. Ballini, and P. Sorrenti, *Synthesis,* 1014 (1983).

1697. K. Matsumoto, *Angew. Chem.*, **96,** 599 (1984); *Int. Ed. Engl.*, **23,** 617 (1984).

1698. E. Colvin, A. K. Beck, and D. Seebach, *Helv. Chim. Acta,* **64,** 2264 (1981).

1699. D. Seebach, A. K. Beck, T. Mukhopadhyay, and E. Thomas, *Helv. Chim. Acta,* **65,** 1101 (1982).

1700. F. M. Schell, J. P. Carter, and Wiaux-Zamar, *J. Am. Chem. Soc.*, **100,** 2894 (1978).

1701. R. E. Lyle, H. M. Fribush, S. Singh, J. E. Saavedra, G. G. Lyle, R. Barton, S. Yoder, and M. K. Jacobson, "N-Nitrosamines", J. P. Anselme, Ed., *ACS. Symp. Ser.*, **101,** 39 (1979).

1702. A. L. Fridman, F. M. Mukhametshin, and S. S. Novikov, *Russ. Chem. Rev.*, **40,** 34 (1971).

1703. L. K. Keefer and C. H. Fodor, *J. Am. Chem. Soc.*, **92,** 5747 (1970).

1704. D. Seebach and D. Enders, *Angew. Chem.*, **84,** 350 (1972); *Int. Ed. Engl.*, **11,** 301 (1972).

1705. D. Seebach and D. Enders, *J. Med. Chem.*, **17,** 1225 (1974).

1706. D. Seebach, D. Enders, and B. Renger, *Chem. Ber.*, **110,** 1852 (1977).

1707. D. Seebach, R. Dach, D. Enders, B. Renger, M. Jansen, and G. Brachtel, *Helv. Chim. Acta,* **61,** 1622 (1978).

1708. R. E. Lyle, J. E. Saavedra, G. G. Lyle, H. M. Fribush, J. L. Marshall, W. Lijinsky, and G. M. Singer, *Tetrahedron Lett.*, 4431 (1976).

1709. R. R. Fraser and L. K. Ng, *J. Am. Chem. Soc.*, **98,** 5895 (1976).

1710. B. Renger, H.-O. Kalinowski, and D. Seebach, *Chem. Ber.*, **110,** 1866 (1977).

1711. R. R. Fraser, T. B. Grindley, and S. Passannanti, *Can J. Chem.*, **53,** 2473 (1975).

1712. G. Lunn, E. B. Sansone, and L. K. Keefer, *Carcinogenesis,* **4,** 315 (1983).

1713. J. E. Saavedra, *J. Org. Chem.*, **44,** 4516 (1979).

1714. J. E. Baldwin, A. Scott, S. E. Branz, S. R. Tannenbaum, and L. Green, *J. Org. Chem.*, **43,** 2427 (1978).

1715. M. Wiessler, "N-Nitrosamines", J.-P. Anselme, Ed., *ACS Symp. Ser.*, **101,** 57, (1979).

1716. A. I. Meyers and W. Ten Hoeve, *J. Am. Chem. Soc.*, **102,** 7125 (1980).

1717. A. I. Meyers, S. Hellring, and W. Ten Hoeve, *Tetrahedron Lett.*, **22,** 5115 (1981).

1718. A. I. Meyers, M. Boes, and D. A. Dickman, *Angew. Chem.*, **96,** 448 (1984); *Int. Ed. Engl.*, **23,** 458 (1984).

1719. A. I. Meyers and M. F. Loewe, *Tetrahedron Lett.*, **25,** 2641 (1984).

1720. A. I. Meyers and S. Hellring, *Tetrahedron Lett.*, **22,** 5119 (1981).

1721. A. I. Meyers and S. Hellring, *J. Org. Chem.*, **47,** 2229 (1982).
1722. R. F. Fraser, G. Boussard, I. D. Postescu, J. J. Whiting, and Y. Y. Wigfield, *Can. J. Chem.*, **51,** 1109 (1973).
1723. P. Beak, W. J. Zajdel, and D. B. Reitz, *Chem. Rev.*, **84,** 471 (1984).
1724. R. Schlecker, D. Seebach, and W. Lubosch, *Helv. Chim. Acta,* **61,** 512 (1978).
1725. P. Beak, B. G. McKinnie, and D. B. Reitz, *Tetrahedron Lett.*, 1839 (1977).
1726. W. Wykypiel, J.-J. Lohmann, and D. Seebach, *Helv. Chim. Acta,* **64,** 1337 (1981).
1727. D. Seebach and T. Hassel, *Angew. Chem.*, **90,** 296 (1978); *Int. Ed. Engl.*, **17,** 274 (1978).
1728. T. Hassel and D. Seebach, *Helv. Chim. Acta,* **61,** 2237 (1978).
1729. D. Seebach, J.-J. Lohmann, M. A. Syfrig, and M. Yoshifuji, *Tetrahedron,* **39,** 1963 (1983).
1730. D. Seebach and M. A. Syfrig, *Angew. Chem.*, **96,** 235 (1984); *Int. Ed. Engl.*, **23,** 248 (1984).
1731. D. Seebach and W. Lubosch, *Angew. Chem.*, **88,** 339 (1976); *Int. Ed. Engl.*, **15,** 313 (1976).
1732. P. Savignac and M. Dreux, *Tetrahedron Lett.*, 2025 (1976).
1733. P. Magnus and G. Roy, *Synthesis,* 575 (1980).
1734. P. Savignac, Y. Leroux, and H. Normant, *Tetrahedron,* **31,** 877 (1975).
1735. G. Wittig and R. Polster, *Justus Liebigs Ann. Chem.*, **599,** 1 (1956).
1736. D. H. R. Barton, R. Beugelmans, and R. N. Young, *Nouv. J. Chim.*, **2,** 363 (1978).
1737. D. J. Peterson, *J. Am. Chem. Soc.*, **93,** 4027 (1971).
1738a. G. Wittig and M. H. Wetterling, *Justus Liebigs Ann. Chem.*, **557,** 193 (1947).
1738b. J.-P. Quintard, B. Elissondo, and B. Jousseaume, *Synthesis,* 495 (1984).
1739. E. C. Ashby, J. Laemmle, and H. M. Neumann, *Acc. Chem. Res.*, **7,** 272 (1974).
1740. M. W. Rathke, *Org. React.*, **22,** 423 (1975).
1741. T. Cohen and J. R. Matz, *J. Am. Chem. Soc.*, **102,** 6900 (1980).
1742. K. Tamao, N. Ishida, and M. Kumada, *J. Org. Chem.*, **48,** 2120 (1983).
1743. D. Hoppe and A. Brönneke, *Synthesis,* 1045 (1982).
1744. M. W. Rathke and R. Kow, *J. Am. Chem. Soc.*, **94,** 6854 (1972).
1745. J. W. Wilson, *J. Organomet. Chem.*, **186,** 297 (1980).
1746. W. C. Still and C. Sreekumar, *J. Am. Chem. Soc.*, **102,** 1201 (1980).
1747. M. K. Yeh, *J. Chem. Soc., Perkin Trans. 1,* 1652 (1981).
1748. S. Wolfe, L. A. LaJohn, and D. F. Weaver, *Tetrahedron Lett.*, **25,** 2863 (1984).
1749. P. Beak and J. W. Worley, *J. Am. Chem. Soc.*, **94,** 597 (1972).
1750. P. Beak and P. D. Becker, *J. Org. Chem.*, **47,** 3855 (1982).
1751. D. B. Reitz, P. Beak, R. F. Farney, and L. S. Helmick, *J. Am. Chem. Soc.*, **100,** 5428 (1978).
1752. L. Wartski, M. El Bouz, J. Seyden-Penne, W. Dumont, and A. Krief, *Tetrahedron Lett.*, 1543 (1979).
1753. K. Ogura, M. Fujita, K. Takahashi, and H. Iida, *Chem. Lett.*, 1697 (1982).
1754. K. Hirai, H. Matsuda, and Y. Kishida, *Tetrahedron Lett.*, 4359 (1971).
1755. D. J. Peterson, *J. Org. Chem.*, **32,** 1717 (1967).
1756. A. I. Meyers and M. E. Ford, *J. Org. Chem.*, **41,** 1735 (1976).
1757. C. R. Johnson, A. Nakanishi, N. Nakanishi, and K. Tanaka, *Tetrahedron Lett.*, 2865 (1975).
1758. T. Hayashi, *Tetrahedron Lett.*, 339 (1974).

1759. K. Hirai and Y. Kishida, *Heterocycles,* **2,** 185 (1974).
1760. G. Köbrich, A. Akhtar, F. Ansari, W. E. Breckoff, H. Büttner, W. Drischel, R. H. Fischer, K. Flory, H. Fröhlich, W. Goyert, H. Heinemann, I. Hornke, H. R. Merkle, H. Trapp, and W. Zündorf, *Angew. Chem.,* **79,** 15 (1967); *Int. Ed. Engl.,* **6,** 41 (1967).
1761. G. Köbrich and H. Trapp, *Z. Naturforsch.* **18b,** 1125 (1963).
1762. J. Villieras, M. Rambaud, R. Tarhouni, and B. Kirschleger, *Synthesis,* 68 (1981).
1763. G. Köbrich and R. H. Fischer, *Tetrahedron,* **24,** 4343 (1968).
1764. R. Tarhouni, B. Kirschleger, M. Rambaud, and J. Villieras, *Tetrahedron Lett.,* **25,** 835 (1984).
1765. T. Kauffmann, K.-J. Echsler, A. Hamsen, R. Kriegesmann, F. Steinseifer, and A. Vahrenhorst, *Tetrahedron Lett.,* 4391 (1978).
1766. T. Kauffmann, A. Hamsen, R. Kriegesmann, and A. Vahrenhorst, *Tetrahedron Lett.,* 4395 (1978).
1767. V. Reutrakul and W. Kanghae, *Tetrahedron Lett.,* 1225 (1977).
1768. V. Reutrakul and P. Thamnusan, *Tetrahedron Lett.,* 617 (1979).
1769. V. Reutrakul and W. Kanghae, *Tetrahedron Lett.,* 1377 (1977).
1770. A. Krief, "α-Selenoorganometallics. Synthesis and Synthetic Usefulness," in *The Chemistry of Functional Groups,* S. Patai, Ed., Interscience, London, 1985.
1771. D. Liotta, *Acc. Chem. Res.,* **17,** 28 (1984).
1772. H. J. Reich, *Acc. Chem. Res.,* **12,** 22 (1979).
1773. K. C. Nicolaou and N. A. Petasis, *Selenium in Natural Products Synthesis,* CIS, Inc., Philadelphia, 1984.
1774. K. B. Sharpless, M. W. Young, and R. F. Lauer, *Tetrahedron Lett.,* 1979 (1973).
1775. D. N. Jones, D. Mundy, and R. D. Whitehouse, *J. Chem. Soc., Chem. Commun.,* 86 (1970).
1776. R. Walter and J. Roy, *J. Org. Chem.,* **36,** 2561 (1971).
1777. K. B. Sharpless and R. F. Lauer, *J. Am. Chem. Soc.,* **95,** 2697 (1973).
1778. H. J. Reich and S. K. Shah, *J. Am. Chem. Soc.,* **97,** 3250 (1975).
1779. R. H. Mitchell, *J. Chem. Soc., Chem. Commun.,* 990 (1974).
1780. H. J. Reich, *J. Org. Chem.,* **40,** 2570 (1975).
1781. S. Raucher and G. A. Koolpe, *J. Org. Chem.,* **43,** 4252 (1978).
1782. W. Dumont, P. Bayet, and A. Krief, *Angew. Chem.,* **86,** 857 (1974); *Int. Ed. Engl.,* **13,** 804 (1974).
1783. O. Van Ende, W. Dumont, and A. Krief, *Angew. Chem.,* **87,** 709 (1975); *Int. Ed. Engl.,* **14,** 700 (1975).
1784. S. Halazy and A. Krief, *J. Chem. Soc., Chem. Commun.,* 1136 (1979).
1785. S. Halazy and A. Krief, *J. Chem. Soc., Chem. Commun.,* 1200 (1982).
1786. S. Halazy and A. Krief, *Tetrahedron Lett.,* 4233 (1979).
1787. J. L. Laboureur and A. Krief, *Tetrahedron Lett.,* **25,** 2713 (1984).
1788. D. Labar and A. Krief, *J. Chem. Soc., Chem. Commun.,* 564 (1982).
1789. J. N. Denis, W. Dumont, and A. Krief, *Tetrahedron Lett.,* 453 (1976).
1790. W. Dumont, M. Sevrin, and A. Krief, *Angew. Chem.,* **89,** 561 (1977); *Int. Ed. Engl.,* **16,** 541 (1977).
1791. D. Seebach and A. K. Beck, *Chem. Ber.,* **108,** 314 (1975).
1792. T.-H. Chan, *Acc. Chem. Res.,* **10,** 442 (1977).
1793. T.-H. Chan and E. Chang, *J. Org. Chem.,* **39,** 3264 (1974).
1794. A. G. Brook, J. M. Duff, and D. G. Anderson, *Can. J. Chem.,* **48,** 561 (1970).

1795. C. R. Hauser and C. R. Hance, *J. Am. Chem. Soc.*, **74,** 5091 (1952).
1796. A. R. Bassindale, R. J. Ellis, and P. G. Taylor, *Tetrahedron Lett.*, **25,** 2705 (1984).
1797. L. F. Cason and H. G. Brooks, *J. Org. Chem.*, **19,** 1278 (1954).
1798. T. H. Chan, E. Chang, and E. Vinokur, *Tetrahedron Lett.*, 1137 (1970).
1799. T. Kauffmann, R. Kriegesmann, and A. Wolterman, *Angew. Chem.*, **89,** 900 (1977); *Int. Ed. Engl.*, **16,** 862 (1977).
1800. D. J. Peterson and H. R. Hays, *J. Org. Chem.*, **30,** 1939 (1965).
1801. A. M. Aguiar, J. Giacin, and A. Mills, *J. Org. Chem.*, **27,** 674 (1962).
1802. D. J. Peterson, *J. Organomet. Chem.*, **8,** 199 (1967).
1803. D. J. Peterson, and J. H. Collins, *J. Org. Chem.*, **31,** 2373 (1966).
1804. E. P. Kyba and C. W. Hudson, *Tetrahedron Lett.*, 1869 (1975).
1805. M. J. O'Donnell, J. M. Boniece, and S. E. Earp *Tetrahedron Lett.*, 2641 (1978).
1806. A. P. Krapcho and E. A. Dundulis, *Tetrahedron Lett.*, 2205 (1976).
1807. U. Schöllkopf and P.-H. Porsch, *Angew. Chem.*, **84,** 478 (1972); *Int. Ed. Engl.*, **11,** 429 (1972).
1808. D. Seebach and J. D. Aebi, *Tetrahedron Lett.*, **25,** 2545 (1984).
1809. D. Hoppe, *Angew. Chem.*, **87,** 450 (1975); *Int. Ed. Engl.*, **14,** 426 (1975).
1810. J.-F. F. Biellmann and J.-B. Ducep, *Org. React.*, **27,** 1 (1982).
1811. B. Renger and D. Seebach, *Chem. Ber.*, **110,** 2334 (1977).
1812. R. Kupper and C. J. Michejda, *J. Org. Chem.*, **44,** 2326 (1979).
1813. T. L. Macdonald, *J. Org. Chem.*, **45,** 193 (1980).
1814. T. Shono, Y. Matsumura, and T. Kanazawa, *Tetrahedron Lett.*, **24,** 4577 (1983).
1815. B. W. Metcalf and P. Casara, *Tetrahedron Lett.*, 3337 (1975).
1816. U. Schöllkopf, *Angew. Chem.*, **82,** 795 (1970); *Int. Ed. Engl.*, **9,** 763 (1970).
1817. T. R. Kelly and A. Arvanitis, *Tetrahedron Lett.*, **25,** 39 (1984).
1818. D. A. Evans, G. C. Andrews, and B. Buckwalter, *J. Am. Chem. Soc.*, **96,** 5560 (1974).
1819. W. C. Still and T. L. Macdonald, *J. Am. Chem. Soc.*, **96,** 5561 (1974).
1820. M. R. Binns and R. K. Haynes, *J. Org. Chem.*, **46,** 3790 (1981).
1821. K. Tanaka, M. Terauchi, and A. Kaji, *Chem. Lett.*, 315 (1981).
1822. S. Torii, H. Tanaka, and Y. Tomotaki, *Chem. Lett.*, 1541 (1974).
1823. Y. Yamamoto, H. Yatagai, and K. Maruyama, *Chem. Lett.*, 385 (1979).
1824. J. Lucchetti and A. Krief, *J. Chem. Soc., Chem. Commun.*, 127 (1982).
1825. T. Kauffmann, R. König, and M. Wensing, *Tetrahedron Lett.*, **25,** 637 (1984).
1826. T. Kauffmann, R. König, R. Kriegesmann, and M. Wensing, *Tetrahedron Lett.*, **25,** 641 (1984).
1827. T. Kauffmann, R. Kriegesmann, and A. Hamsen, *Chem. Ber.*, **115,** 1818 (1982).
1828. D. B. Reitz, P. Beak, and A. Tse, *J. Org. Chem.*, **46,** 4316 (1981).
1829. P. M. Warner and D. Le, *Synth Commun.*,**14,** 1341 (1984).
1830. H. Mattes and C. Benezra, *Tetrahedron Lett.*, **26,** 5697 (1985).
1831. A. C. Hopkinson, L. H. Dao, P. Duperrouzel, M. Maleki, and E. Lee-Ruff, *J. Chem. Soc., Chem. Commun.*, 727 (1983).
1832. R. Verhé and N. De Kimpe, in *The Chemistry Functional Groups, Supplement D,* S. Patai and Z. Rappoport, Eds., Wiley, 1983. p. 813.
1833. T. Tidwell, *Angew. Chem.*, **96,** 16 (1984); *Int. Ed. Engl.*, **23,** 20 (1984).
1834. D. J. McLennan and A. Pross, *J. Chem. Soc., Perkin Trans. 2,* 981 (1984).
1835. P. Gassman and T. T. Tidwell, *Acc. Chem. Res.*, **16,** 279 (1983).

1836. A. Pross, K. Aviram, R. C. Klix, D. Kost, and R. D. Bach, *Nouv. J. Chim.*, **8,** 711 (1984).
1837. X. Creary, *Acc. Chem. Res.*, **18,** 3 (1985).
1838. M. Maleki, A. C. Hopkinson, and E. Lee-Ruff, *Tetrahedron Lett.*, **24,** 4911 (1983).
1839. R. H. Nobes, W. J. Bouma, and L. Radom, *J. Am. Chem. Soc.*, **105,** 309 (1983).
1840. X. Creary, S. R. McDonald, and M. D. Eggers, *Tetrahedron Lett.*, **26,** 811 (1985).
1841. G. A. Russell and F. Ros, *J. Am. Chem. Soc.*, **104,** 7349 (1982).
1842. R. Verhé, N. De Kimpe, L. De Buyck, and N. Schamp, *Synthesis,* 46 (1984).
1843. C. Mioskowski, S. Manna, and J. R. Falck, *Tetrahedron Lett.*, **25,** 519 (1984).
1844. K. Tamao, M. Zembayashi, and M. Kumada, *Chem. Lett.*, 1239 (1976).
1845. G. R. Kieczykowski, C. S. Pogonowski, J. E. Richman, and R. H. Schlessinger, *J. Org. Chem.*, **42,** 175 (1977).
1846. O. De Lucchi, L. Pasquato, and G. Modena, *Tetrahedron Lett.*, **25,** 3643 (1984).
1847. O. De Lucchi, L. Pasquato, and G. Modena, *Tetrahedron Lett.*, **25,** 3647 (1984).
1848. B. M. Trost and M. Shimizu, *J. Am. Chem. Soc.*, **105,** 6757 (1983).
1849. J. N. Gardner, S. Kaiser, A. Krubiner, and H. Lucas, *Can. J. Chem.*, **51,** 1419 (1973).
1850. K. Seki, T. Ohnuma, T. Oishi, and Y. Ban, *Tetrahedron Lett.*, 723 (1975).
1851. G. A. Koppel and M. D. Kinnick, *J. Chem. Soc., Chem. Commun.*, 473 (1975).
1852. R. Tanikaga, H. Sugihara, K. Tanaka, and A. Kaji, *Synthesis,* 299 (1977).
1853. K. Praefcke, *Methoden der organischen Chemie* (*Houben-Weyl*), Vol. E3, G. Thieme, Stuttgart, 1983, p. 312.
1854. G. Kresze, *Methoden der organischen Chemie* (*Houben-Weyl*), Vol. E11/1, G. Thieme, Stuttgart, 1985, p. 872.
1855. P. J. Brown, D. N. Jones, M. A. Khan, N. A. Meanwell, and P. J. Richards, *J. Chem. Soc., Perkin Trans. 1,* 2049 (1984).
1856. G. Tsuchihashi, S. Mitamura, S. Inoue, and K. Ogura, *Tetrahedron Lett.*, 323 (1973).
1857. M. E. Kuehne and L. Foley, *J. Org. Chem.*, **30,** 4280 (1965).
1858. A. Risaliti, M. Forchiassin, and E. Valentin, *Tetrahedron Lett.*, 6331 (1966).
1859. E. Valentin, G. Pitacco, F. P. Colonna, and A. Risaliti, *Tetrahedron,* **30,** 2741 (1974).
1860. M. Züger, T. Weller, and D. Seebach, *Helv. Chim. Acta,* **63,** 2005 (1980).
1861. R. Häner, T. Laube, and D. Seebach, *Chimia,* **38,** 255 (1984).
1862. R. Cloux and M. Schlosser, *Helv. Chim. Acta,* **67,** 1470 (1984).
1863. J. Colonge and J. C. Dubin, *Bull. Soc. Chim. Fr.*, 1180 (1960).
1864. D. Bartlett and G. F. Woods, *J. Am. Chem. Soc.*, **62,** 2933 (1940).
1865. C. L. Stevens and E. Farkas, *J. Am. Chem. Soc.*, **74,** 5352 (1952).
1866. R. N. McDonald and P. A. Schwab, *J. Am. Chem. Soc.*, **85,** 4004 (1963).
1867. C. A. Buehler, H. A. Smith, K. V. Nayak, and T. A. Magee, *J. Org. Chem.*, **26,** 1573 (1961).
1868. C. L. Stevens, P. Blumbergs, and M. Munk, *J. Org. Chem.*, **28,** 331 (1963).
1869. A. Hassner and P. Catsoulacos, *J. Org. Chem.*, **31,** 3149 (1966).
1870. W. Ziegenbein, *Chem. Ber.*, **94,** 2989 (1961).
1871. F. Asinger, W. Schäfer, M. Baumann, and H. Römgens, *Justus Liebigs Ann. Chem.*, **672,** 103 (1964).
1872. F. G. Bordwell and R. G. Scamehorn, *J. Am. Chem. Soc.*, **90,** 6751 (1968).
1873. D. G. Holland and E. D. Amstutz, *Recl. Trav. Chim. Pays-Bas,* **83,** 1047 (1964).
1874. K. von Auwers, H. Ludewig, and A. Müller, *Justus Liebigs Ann. Chem.*, **526,** 143 (1936).
1875. V. Caplar, A. Lisini, F. Kajfez, D. Kolbah, and V. Sunjic, *J. Org. Chem.*, **43,** 1355 (1978).

1876. E. B. Reid, R. B. Portenbaugh, and H. R. Patterson, *J. Org. Chem.*, **15,** 572 (1950).
1877. S. T. Vijayaraghavan and T. R. Balasubramanian, *J. Organomet. Chem.*, **282,** 17 (1985).
1878. P. A. Levene and A. Walti, *Org. Synth., Collect. Vol. 2,* 5 (1943).
1879. P. J. Boyle and J. F. W. Keana, *Org. Prep. Proced. Int.*, **10,** 101 (1978).
1880. A. Pusino, V. Rosnati, and A. Saba, *Tetrahedron,* **40,** 1893 (1984).
1881. C. L. Stevens, M. L. Weiner, and R. C. Freeman, *J. Am. Chem. Soc.*, **75,** 3977 (1953).
1882. F. G. Bordwell and M. W. Carlson, *J. Am. Chem. Soc.*, **92,** 3370 (1970).
1883. M. Mousseron and R. Jacquir, *Bull. Soc. Chim. Fr.*, 689 (1949).
1884. M. Kopp, *Bull. Soc. Chim. Fr.*, 628 (1954).
1885. F. G. Bordwell and J. Almy, *J. Org. Chem.*, **38,** 571 (1973).
1886. C. L. Stevens, J. J. Beereboom, Jr., and K. G. Rutherford, *J. Am. Chem. Soc.*, **77,** 4590 (1955).
1887. N. J. Turro, R. B. Gagosian, C. Rappe, and L. Knutsson, *J. Chem. Soc., Chem. Commun.*, 270 (1969).
1888. D. Baudry and M. Charpentier-Morize, *Tetrahedron Lett.*, 3013 (1973).
1889. D. Mayer, *Methoden der organischen Chemie (Houben Weyl),* Vol. 7/2c, 4th ed., G. Thieme, Stuttgart, 1977, p. 2253.
1890. G. Kempter, J. Spindler, H. J. Fiebig, and G. Sarodnick, *J. Prakt. Chem.*, **313,** 977 (1971).
1891. E. W. Warnhoff, *J. Org. Chem.*, **27,** 4587 (1962).
1892. A. Halvorsen and J. Songstad, *J. Chem. Soc., Chem. Commun.*, 327 (1978).
1893. O. E. Edwards and C. Grieco, *Can. J. Chem.*, **52,** 3561 (1974).
1894. G. Pasquet, D. Boucherot, W. R. Pilgrim, and B. Wright, *Tetrahedron Lett.*, **21,** 931 (1980).
1895. Belg. Pat. 621,456; *Chem. Abstr.*, **59,** 9835d (1963).
1896. R. F. Abdulla and J. C. Williams, Jr., *Tetrahedron Lett.*, 997 (1980).
1897. A. N. Mirskova, G. G. Levkovskaya, I. D. Kalikhman, and M. G. Voronkov, *Ah. Org. Khim.*, **15,** 2301 (1979); *Chem. Abstr.*, **92,** 128792b (1980).
1898. W. Ried and L. Kaiser, *Justus Liebigs Ann. Chem.*, 958 (1975).
1899. J. C. Meslin, Y. T. N'Guessan, H. Quinion, and F. Tonnard, *Tetrahedron,* **31,** 2679 (1975).
1900. A. Babadjamian and J. Metzger, *J. Heterocycl. Chem.*, **12,** 643 (1975).
1901. A. Babadjamian, R. Gallo, and J. Metzger, *J. Heterocycl. Chem.*, **13,** 1205 (1976).
1902. D. Martinetz and A. Hiller, *Z. Chem.*, **16,** 320 (1976).
1903. G. Ege, P. Arnold, and R. Noronha, *Justus Liebigs Ann. Chem.*, 656 (1979).
1904. J. Gierer and B. Alfredsson, *Chem. Ber.*, **90,** 1240 (1957).
1905. T. Terasawa and T. Okada, *J. Org. Chem.*, **42,** 1163 (1977).
1906. B. M. Trost, W. C. Vladuchick, and A. J. Bridges, *J. Am. Chem. Soc.*, **102,** 3548 (1980).
1907. M. Lissel, *J. Chem. Res. (S),* 286 (1982).
1908. H. Ishihara and Y. Hirabayashi, *Chem. Lett.*, 1007 (1978).
1909. M. H. Elnagdi, M. R. H. Elmoghayar, and G. E. H. Elgemeie, *Synthesis,* 1 (1984).
1910. O. Widman and E. Wahlberg, *Ber. Dtsch. Chem. Ges.*, **44,** 2065 (1911).
1911. H. Behringer, M. Ruff, and R. Wiedenmann, *Chem. Ber.*, **97,** 1732 (1964).
1912. H. Kobler, K.-H. Schuster, and G. Simchen, *Justus Liebigs Ann. Chem.*, 1946 (1978).
1913. German Pat. 2,819,264; *Chem. Abstr,* **90** 103450j (1979).
1914. R. Justoni, *Gazz. Chim. Ital.*, **69,** 378 (1939).
1915. H. Behringer, M. Ruff, and R. Wiedenmann, *Chem. Ber.*, **97,** 1732 (1964).
1916. L. Nilsson and C. Rappe, *Acta Chem. Scand., Ser. B,* **30,** 1000 (1976).

1917. M. M. Shemyakin, M. N. Kolosov, Yu. A. Arbuzov, V. V. Onoprienko, and H. Yü-Yüan, *Zh. Obshch. Khim.,* **30,** 545 (1960).

1918. F. Ebel, F. Huber, and A. Brunner, *Helv. Chim. Acta,* **12,** 16 (1929).

1919. S. R. Ramadas and S. Padmanabhan, *Curr. Sci.,* **48,** 52 (1979).

1920. T. Sakai, E. Amano, A. Kawabata, and A. Takeda, *J. Org. Chem.,* **45,** 43 (1980).

1921. R. T. LaLonde, N. Muhammad, C. F. Wong, and E. R. Sturiale, *J. Org. Chem.,* **45,** 3664 (1980).

1922. M. Kosugi, I. Takano, M. Sakurai, H. Sano, and T. Migita, *Chem. Lett.,* 1221 (1984).

1923. R. Verhé, N. De Kimpe, L. De Buyck, R. Thierie, and N. Schamp, *Bull. Soc. Chim. Belg.,* **89,** 563 (1980).

1924. P.-E. Sum and L. Weiler, *Can. J. Chem.,* **56,** 2301 (1978).

1925. A. S. Hussey and R. R. Herr, *J. Org. Chem.,* **24,** 843 (1959).

1926. H. H. Ong, V. B. Anderson, J. C. Wilker, T. C. Spaulding, and L. R. Meyerson, *J. Med. Chem.,* **23,** 726 (1980).

1927. A. J. Sisti and A. C. Vitale, *J. Org. Chem.,* **37,** 4090 (1972).

1928. J. E. Dubois, C. Lion, and C. Moulineau, *Tetrahedron Lett.,* 177 (1971).

1929. J. E. Dubois and C. Lion, *Tetrahedron,* **31,** 1227 (1975).

1930. L. Hamon and J. Levisalles, *J. Organomet. Chem.,* **253,** 259 (1983).

1931. E. M. Schultz and S. Mickey, *Org. Synth., Collect. Vol. 3,* 343 (1955).

1932. D. Seebach and P. Knochel, *Helv. Chim. Acta,* **67,** 261 (1984).

1933. D. Monti, P. Gramatica, G. Speranza, and P. Manitto, *Tetrahedron Lett.,* **24,** 417 (1983).

1934. T. Shono, H. Hamaguchi, H. Mikami, H. Nogusa, and S. Kashimura, *J. Org. Chem.,* **48,** 2103 (1983).

1935. R. S. Varma, M. Varma, and G. W. Kabalka, *Synth. Commun.,* **15,** 1325 (1985).

1936. S. Cacchi, M. Felici, and G. Rosini, *J. Chem. Soc., Perkin Trans. 1,* 1260 (1977).

1937. P. Laszlo and E. Polla, *Synthesis,* 439 (1985).

1938. O. Attanasi, M. Grossi, and F. Serra-Zanetti, *J. Chem. Res. (S),* 322 (1983).

1939. C. E. Sacks and P. L. Fuchs, *J. Am. Chem. Soc.,* **97,** 7372 (1975).

1940. S. Cacchi, F. La Torre, and D. Misiti, *Chim. Ind. (Milan),* **60,** 715 (1978).

1941. R. A. Amos and J. A. Katzenellenbogen, *J. Org. Chem.,* **42,** 2537 (1977).

1942. P. A. Wender, J. M. Erhardt, and L. J. Letendre, *J. Am. Chem. Soc.,* **103,** 2114 (1981).

1943. R. M. Moriarty, O. Prakash, and M. P. Duncan, *J. Chem. Soc., Chem. Commun.,* 420 (1985).

1944. M. Miyashita, T. Yanami, T. Kumazawa, and A. Yoshikoshi, *J. Am. Chem. Soc.,* **106,** 2149 (1984).

1945. A. Yoshikoshi and M. Miyashita, *Acc. Chem. Res.,* **18,** 284 (1985).

1946. E. Negishi, F. T. Luo, A. J. Pecora, and A. Silveira, Jr., *J. Org. Chem.,* **48,** 2427 (1983).

1947. M. Horton and G. Pattenden,*Tetrahedron Lett.,* **24,** 2125 (1983).

1948. M. Miyashita, R. Yamaguchi, and A. Yoshikoshi, *J. Org. Chem.,* **49,** 2857 (1984).

1949. T. Yanami, A. Ballatore, M. Miyashita, M. Kato, and A. Yoshikoshi, *Synthesis,* 407 (1980).

1950. A. Hosomi, A. Shirahata, Y. Araki, and H. Sakurai, *J. Org. Chem.,* **46,** 4631 (1981).

1951. P. F. Schuda and B. Bernstein, *Synth. Commun.,* **14,** 293 (1984).

1952. B. M. Trost and F. W. Gowland, *J. Org. Chem.,* **44,** 3448 (1979).

1953. W. R. Jackson and J. U. G. Strauss, *Tetrahedron Lett.,* 2591 (1975).

1954. A. J. Pearson, *Chem. Ind. (London),* 741 (1982).

1955. A. J. Birch and L. F. Kelly, *J. Organomet. Chem.*, **285,** 267 (1985).

1956. A. J. Birch, L. F. Kelly, and A. S. Narula, *Tetrahedron,* **38,** 1813 (1982).

1957. S. Cacchi, D. Misiti, and M. Felici, *Synthesis,* 147 (1980).

1958. L. Bernardi, P. Masi, and G. Rosini, *Ann. Chim. (Rome),* **63,** 601 (1973).

1959. A. Bell, A. H. Davidson, C. Earnshaw, H. K. Norrish, R. S. Torr, D. B. Trowbridge, and S. Warren, *J. Chem. Soc., Perkin Trans. 1,* 2879 (1983).

1960. P. Savignac, A. Breque, F. Mathey, J.-M. Varlet, and N. Collignon, *Synth. Commun.*, **9,** 287 (1979).

1961. Y. Tamura, H. Shindo, J. Uenishi, and H. Ishibashi, *Tetrahedron Lett.,* **21,** 2547 (1980).

1962. S. Shatzmiller, R. Lidor, E. Shalom, and E. Bahar, *J. Chem. Soc., Chem. Commun.,* 795 (1984).

1963. S. Halazy and L. Hevesi, *J. Org. Chem.,* **48,** 5242 (1983).

1964. Y. Watanabe, M. Idei, and Y. Takegami, *Tetrahedron Lett.,* 3523 (1979).

1965. J. Gasteiger and C. Herzig, *Angew. Chem.,* **93,** 933 (1981); *Int. Ed. Engl.,* **20,** 868 (1981).

1966. L. Caglioti, A. Dondoni, and G. Rosini, *Chim. Ind. (Milan),* **50,** 122 (1968).

1967. J. J. Partridge, N. K. Chadha, and M. R. Uskokovic, *J. Am. Chem. Soc.,* **95,** 7171 (1973).

1968. S. Inabe and R. D. Rieke, *Synthesis,* 842 (1984).

1969. P. Canonne and M. Akssira, *Tetrahedron Lett.,* **25,** 3453 (1984).

1970. I. Fleming and J. Iqbal, *Tetrahedron Lett.,* **24,** 327 (1983).

1971. T. V. Lee and J. O. Okonkwo, *Tetrahedron Lett.,* **24,** 323 (1983).

1972. F. Orsini and F. Pelizzoni, *Synth. Commun.,* **13,** 523 (1983).

1973. P. Prempree, S. Radviroongit, and Y. Thebtaranonth, *J. Org. Chem.,* **48,** 3553 (1983).

1974. N. Kornblum, H. K. Singh, and S. D. Boyd, *J. Org. Chem.,* **49,** 358 (1984).

1975. Y. Tamura, Hong Dae Choi, H. Shindo, and H. Ishibashi, *Chem. Pharm. Bull.,* **30,** 915 (1982).

1976. F. Effenberger and K. Drauz, *Angew. Chem.,* **91,,** 504 (1979); *Int. Ed. Engl.,* **18,** 474 (1979).

1977. J. Kihlberg, R. Bergman, and R. Wickberg, *Acta Chem. Scand. Ser. B,* **37,** 911 (1983).

1978. T. Mukaiyama, *Tetrahedron,* **37,** 4111 (1981).

1979. S. Inaba and R. D. Rieke, *Synthesis,* 844 (1984).

1980. T. Fujisawa, M. Watanabe, and T. Sato, *Chem. Lett.,* 2055 (1984).

1981. M. Shimizu, T. Akiyama, and T. Mukaiyama, *Chem. Lett.,* 1531 (1984).

1982. A. Solladie-Cavallo and J. Suffert, *Tetrahedron Lett.,* **25,** 1897 (1984).

1983. J. K. Whitesell, A. Bhattacharya, and K. Henke, *J. Chem. Soc., Chem. Commun.,* 988 (1982).

1984. J. K. Whitesell, D. Deyo, and A. Bhattacharya, *J. Chem. Soc., Chem. Commun.,* 802 (1983).

1985. Y. Yamamoto, T. Komatsu, and K. Maruyama, *J. Chem. Soc., Chem. Commun.,* 191 (1983).

1986. A. Citterio, M. Gandolfi, O. Piccolo, L. Filippini, L. Tinucci, and E. Valotti, *Synthesis,* 760 (1984).

1987. M. J. O'Donnell, W. D. Bennett, and R. L. Polt, *Tetrahedron Lett.,* **26,** 695 (1985).

1988. M. J. O'Donnell, and J.-B. Falmagne, *Tetrahedron Lett.,* **26,** 699 (1985).

1989. J. L. Colin and B. Loubinoux, *Synthesis,* 568 (1983).

1990. K. Soai and M. Ishizaki, *J. Chem. Soc., Chem. Commun.,* 1016 (1984).

1991. E. J. Corey, L. S. Melvin, Jr., and M. F. Haslanger, *Tetrahedron Lett.,* 3117 (1975).

1992. H. J. Sattler, H. G. Lennartz, and W. Schunack, *Arch. Pharm.,* **312,** 107 (1979).
1993. M. Kawashima and T. Fujisawa, *Chem. Lett.,* 1851 (1984).
1994. S. Ghosh, S. N. Pardo, and R. G. Salomon, *J. Org. Chem.,* **47,** 4692 (1982).
1995. E. V. Dehmlow and E. Kunesch, *Synthesis,* 320 (1985).
1996. K. Sato, S. Inoue, S. Kuranami, and M. Ohashi, *J. Chem. Soc., Perkin Trans. 1,* 1666 (1977).
1997. B. Föhlisch and W. Gottstein, *Justus Liebigs Ann. Chem.,* 1768 (1979).
1998. L. Schotte, *Arkiv för Kemi,* **5,** 533 (1952).
1999. J. P. Marino and J. C Jaén, *J. Am. Chem. Soc.,* **104,** 3165 (1982).
2000. J. P. Marino and N. Hatanaka, *J. Org. Chem.,* **44,** 4467 (1979).
2001. P. B. Anzeveno, D. P. Matthews, C. L. Barney, and R. J. Barbuch, *J. Org. Chem.,* **49,** 3134 (1984).
2002. G. H. Posner and J. J. Sterling, *J. Am. Chem. Soc.,* **95,** 3076 (1973).
2003. E. Buncel, *Reaction Mechanisms in Organic Chemistry,* Monograph 9, Elsevier, Amsterdam, 1975, pp. 116–122, 161–167.
2004. D. H. Hunter, in *Isotopes in Organic Chemistry,* Vol. 1, E. Buncel and C. C. Lee, Eds., Elsevier, Amsterdam, 1975, pp. 197–203.
2005. D. H. Hunter, J. B. Strothers and E. W. Warnhoff, in *Rearrangements in Ground and Excited States,* Vol. 1, P. de Mayo, Ed., Academic, New York, 1980, pp. 393–461.
2006. N. H. Werstiuk, *Tetrahedron,* **39,** 205 (1983).
2007. G. Büchi and H. Wüest, *J. Org. Chem.,* **34,** 1122 (1969).
2008. H. J. J. Loozen and E. F. Godefroi, *J. Org. Chem.,* **38,** 1056 (1973).
2009. H. J. J. Loozen and E. F. Godefroi, *J. Org. Chem.,* **38,** 3495 (1973).
2010. H. J. J. Loozen, *J. Org. Chem.,* **40,** 520 (1975).
2011. H. J. J. Loozen, E. F. Godefroi, and J. S. M. M. Besters, *J. Org. Chem.,* **40,** 892 (1975).
2012. P. E. Eaton, R. H. Mueller, G. R. Carlson, D. A. Cullison, G. F. Cooper, T.-C. Chou, and E.-P. Krebs, *J. Am. Chem. Soc.,* **99,** 2751 (1977).
2013. R. A. Volkmann, J. T. Davis, and C. N. Meltz, *J. Org. Chem.,* **48,** 1767 (1983).
2014. J. C. Stowell, *J. Org. Chem.,* **41,** 560 (1976).
2015. T. Sato, K. Naruse, and T. Fujisawa, *Tetrahedron Lett.,* **23,** 3587 (1982).
2016. J. C. Stowell and B. T. King, *Synthesis,* 278 (1984).
2017. K. Kondo and D. Tunemoto, *Tetrahedron Lett.,* 1397 (1975).
2018. K. Kondo, E. Saito, and D. Tunemoto, *Tetrahedron Lett.,* 2275 (1975).
2019. R. Quelet and J. D'Angelo, *Bull. Soc. Chim. Fr.,* 1503 (1967).
2020. J. F. Normant, A. Commercon, M. Bourgain, and J. Villieras, *Tetrahedron Lett.,* 3833 (1975).
2021. S. F. Martin, *Synthesis,* 656 (1979).
2022. R. Gompper and H.-U. Wagner, *Angew. Chem.,* **15,** 389 (1976); *Int. Ed. Engl.,* **15,** 321 (1976).
2023. M. Schlosser, *Angew. Chem.,* **13,** 751 (1974); *Int. Ed. Engl.,* **13,** 701 (1974).
2024. A. Pelter, B. Singaram, and J. W. Wilson, *Tetrahedron Lett.,* **24,** 631 (1983).
2025. H. Ahlbrecht, *Chimia,* **31,** 391 (1977).
2026. S. F. Martin and M. T. Du Priest, Tetrahedron Lett., 3925 (1977).
2027. H. Ahlbrecht and C. S. Sudheendranath, *Synthesis,* 717 (1982).
2028. I. Fleming and J. Goldhill, *J. Chem. Soc., Perkin Trans. 1,* 1493 (1980).
2029. D. J. Ager, I. Fleming, and S. K. Patel, *J. Chem. Soc., Perkin Trans. 1,* 2520 (1981).

2030. P. Coutrot and P. Savignac, *J. Chem. Res. (S),* 308 (1977).

2031. P. Coutrot. J. R. Dormoy, and A. Moukimou, *J. Organomet. Chem.,* **258,** C25 (1983).

2032. W. C. Still and T. L. Macdonald, *J. Org. Chem.,* **22,** 3620 (1976).

2033. J. Hartmann, R. Muthukrishnan, and M. Schlosser, *Helv. Chim. Acta,* **57,** 2261 (1974).

2034. A. Hosomi, H. Hashimoto, and H. Sakurai, *J. Org. Chem.,* **43,** 2551 (1978).

2035. Y. Yamamoto, H. Yatagai, and K. Maruyama, *J. Org. Chem.,* **45,** 195 (1980).

2036. T. Mukaiyama and M. Yamaguchi, *Chem. Lett.,* 657 (1979).

2037. M. Yamaguchi and T. Mukaiyama, *Chem. Lett.,* 1279 (1979).

2038. D. Hoppe, R. Hanko, and A. Brönneke, *Angew. Chem.,* **92,** 637 (1980); *Int. Ed. Engl.,* **19,** 625 (1980).

2039. R. Hanko and D. Hoppe, *Angew. Chem.,* **93,** 115 (1981); *Int. Ed. Engl.,* **20,** 127 (1981).

2040. D. Hoppe, R. Hanko, A. Brönneke, and F. Lichtenberg, *Angew. Chem.,* **93,** 1106 (1981); *Int. Ed. Engl.,* **20,** 1024 (1981).

2041. D. Hoppe and A. Brönneke, *Tetrahedron Lett.,* **24,** 1687 (1983).

2042. P. M. Atlani, J. F. Biellmann, S. Dube, and J. J. Vicens, *Tetrahedron Lett.,* 2665 (1974).

2043. D. D. Ridley and M. A. Smal, *Aust. J. Chem.,* **33,** 1345 (1980).

2044. M. R. Binns, R. K. Haynes, T. L. Houston, and W. R. Jackson, *Tetrahedron Lett.,* **21,** 573 (1980).

2045. K. Oshima, H. Yamamoto, and H. Nozaki, *J. Am. Chem. Soc.,* **95,** 7926 (1973).

2046. Y. Yamamoto, H. Yatagai, and K. Maruyama, *J. Chem. Soc., Chem. Commun.,* 157 (1979).

2047. J. Geiss, B. Seuring, R. Pieter, and D. Seebach, *Angew. Chem.,* **13,** 484 (1974); *Int. Ed. Engl.,* **13,** 479 (1974).

2048. R. J. P. Corriu and J. Masse, *J. Organomet. Chem.,* **57,** C5 (1973).

2049. R. J. P. Corriu, J. Masse, and D. Samate, *J. Organomet. Chem.,* **93,** 71 (1975).

2050. R. J. P. Corriu, G. F. Lanneau, D. Leclercq, and D. Samate, *J. Organomet. Chem.,* **144,** 155 (1978).

2051. P. W. K. Lau and T. H. Chan, *Tetrahedron Lett.,* 2383 (1978).

2052. E. J. Corey and P. Ulrich, *Tetrahedron Lett.,* 3685 (1975).

2053. S. F. Martin and P. J. Garrison, *Tetrahedron Lett.,* 3875 (1977).

2054. A. M. Tischler and M. H. Tischler, *Tetrahedron Lett.,* 3407 (1978).

2055. A. P. Kozikowski and K. Isobe, *Tetrahedron Lett.,* 833 (1979).

2056. T. Hayashi, N. Fujitaka, T. Oishi, and T. Takeshima, *Tetrahedron Lett.,* **21,** 303 (1980).

2057. D. Hoppe and F. Lichtenberg, *Angew. Chem.,* **94,** 378 (1982); *Int. Ed. Engl.,* **21,** 372 (1982).

2058. F. Sato, H. Uchiyama, K. Iida, Y. Kobayashi, and M. Sato, *J. Chem. Soc., Chem. Commun.,* 921 (1983).

2059. A. J. Pratt and E. J. Thomas, *J. Chem. Soc., Chem. Commun.,* 1115 (1982).

2060. H. Ahlbrecht, G. Bonnet, D. Enders, and G. Zimmermann, *Tetrahedron Lett.,* **21,** 3175 (1980).

2061. T. Mukaiyama, H. Hayashi, T. Miwa, and K. Narasaka, *Chem. Lett.,* 1637 (1982).

2062. A. S. Kende, D. Constantinides, S. J. Lee, and L. Liebeskind, *Tetrahedron Lett.,* 405 (1975).

2063. P. Brownbridge, P. G. Hunt, and S. Warren, *Tetrahedron Lett.,* **24,** 3391 (1983).

2064. E. J. Corey, I. Vlattas, N. H. Andersen, and K. Harding, *J. Am. Chem. Soc.,* **90,** 3247 (1968).

2065. K. Kondo, K. Matsui, and A. Negishi, *Chem. Lett.,* 1371 (1974).

2066. G. K. Cooper and L. J. Dolby, *Tetrahedron Lett.*, 4675 (1976).
2067. I. Böhm, E. Hirsch, and H.-U. Reissig, *Angew. Chem.*, **93,** 593 (1981); *Int. Ed. Engl.*, **20,** 574 (1981).
2068. D. A. Evans, D. J. Baillargeon, and J. V. Nelson, *J. Am. Chem. Soc.,* **100,** 2242 (1978).
2069. A. A. Ponaras, *Tetrahedron Lett.*, 3105 (1976).
2070. T. Fujisawa, T. Sato, T. Kawara, and A. Noda, *Tetrahedron Lett.,* **23,** 3193 (1982).
2071. A. Bell, A. H. Davidson, C. Earnshaw, H. K. Norrish, R. S. Torr, and S. Warren, *J. Chem. Soc., Chem. Commun.*, 988 (1978).
2072. C. Earnshaw, R. S. Torr, and S. Warren, *J. Chem. Soc., Perkin Trans. 1,* 2893 (1983).
2073. E. Wada, M. Okawara, and T. Nakai, *J. Org. Chem.,* **44,** 2952 (1979).
2074. Y. Masaki, K. Nagata, Y. Serizawa, and K. Kaji, *Tetrahedron Lett.,* **23,** 5553 (1982).
2075. J. L. Moreau and R. Couffignal, *Tetrahedron Lett.,* **23,** 5271 (1982).
2076. G. Rosini, R. Ballini, and P. Sorrenti, *Tetrahedron,* **39,** 4127 (1983).
2077. I. Kuwajima and M. Kato, *J. Chem. Soc., Chem. Commun.*, 708 (1979).
2078. S. Murai, Y. Seki, and N. Sonoda, *J. Chem. Soc., Chem. Commun.,* 1032 (1974).
2079. B. M. Trost and L. H. Latimer, *J. Org. Chem.,* **42,** 3212 (1977).
2080. W. Oppolzer and R. L. Snowden, *Helv. Chim. Acta,* **63,** 967 (1980).
2081. W. Oppolzer, R. C. Snowden, and D. P. Simmons, *Helv. Chim. Acta,* **64,** 2002 (1981).
2082. W. Oppolzer, R. C. Snowden, and P. H. Briner, *Helv. Chim. Acta,* **64,** 2022 (1981).
2083. R. Goswami, *J. Am. Chem. Soc.,* **102,** 5973 (1980).
2084. I. Ryu, M. Ando, A. Ogawa, S. Murai, and N. Sonoda, *J. Am. Chem. Soc.,* **105,** 7192 (1983).
2085. H.-U. Reissig, *Tetrahedron Lett.,* **22,** 2981 (1981).
2086. H. W. Thompson and B. S. Huegi, *J. Chem. Soc., Perkin Trans. 1,* 1603 (1976).
2087. R. Oda, T. Kawabata, and S. Tanimoto, *Tetrahedron Lett.*, 1653 (1964).
2088. G. Stork, R. A. Kretchmer, and R. H. Schlessinger, *J. Am. Chem. Soc.,* **90,** 1647 (1968).
2089. H. S. Corey, Jr., J. R. D. McCormick, and W. E. Swenson, *J. Am. Chem. Soc.,* **86,** 1884 (1964).
2090. M. J. Bogdanowicz, T. Ambelang, and B. M. Trost, *Tetrahedron Lett.,* 923 (1973).
2091. B. M. Trost, *Acc. Chem. Res.,* **7,** 85 (1974).
2092. G. Sturz, B. Corbel, and J.-P. Paugam, *Tetrahedron Lett.,* 47 (1976).
2093. D. Caine and A. S. Frobese, *Tetrahedron Lett.,* 883 (1978).
2094. B. Costisella and H. Gross, *Tetrahedron,* **38,** 139 (1982).
2095. B. Costisella, H. Gross, and H. Schick, *Tetrahedron,* **40,** 733 (1984).
2096. H. Ahlbrecht and W. Farnung, *Chem. Ber.,* **117,** 1 (1984).
2097. E. Nakamura and I. Kuwajima, *J. Am. Chem. Soc.,* **99,** 7360 (1977).
2098. E. Nakamina and I. Kuwajima, *J. Am. Chem. Soc.,* **105,** 651 (1983).
2099. T. Hata, M. Nakajima, and M. Sekine, *Tetrahedron Lett.,* 2047 (1979).
2100. D. A. Evans, J. M. Takacs, and K. M. Hurst, *J. Am. Chem. Soc.,* **101,** 371 (1979).
2101. M. Maleki, J. A. Miller, and O. W. Lever, Jr., *Tetrahedron Lett.,* **22,** 3789 (1981).
2102. R. M. Jacobson, G. M. Lahm, and J. W. Clader, *J. Org. Chem.,* **45,** 345 (1980).
2103. D. Seyferth, R. E. Mammarella, and H. A. Klein, *J. Organomet. Chem.,* **194,** 1 (1980).
2104. E. J. Corey and D. E. Cane, *J. Org. Chem.,* **35,** 3405 (1970).
2105. R. Goswami and D. E. Corcoran, *Tetrahedron Lett.,* **23,** 1463 (1982).
2106. R. Goswami and D. E. Corcoran, *J. Am. Chem. Soc.,* **105,** 7182 (1983).
2107. R. Hanko and D. Hoppe, *Angew. Chem.,* **94,** 378 (1982); *Int. Ed. Engl.,* **21,** 372 (1982).

2108. F. E. Ziegler and J. J. Mencel, *Tetrahedron Lett.*, **24,** 1859 (1983).
2109. T. Yamamoto, M. Kakimoto, and M. Okawara, *Tetrahedron Lett.*, 1659 (1977).
2110. G. Tsuchihashi, S. Mitamura, and K. Ogura, *Tetrahedron Lett.*, 855 (1976).
2111. M. Miyashita, T. Yanami, and A. Yoshikoshi, *Org. Synth.*, **60,** 117 (1981).
2112. L. P. Ellinger and A. A. Goldberg, *J. Chem. Soc.*, 263 (1949).
2113. O. Dahl, *J. Chem. Soc., Perkin Trans. 1,* 947 (1978).
2114. A. Pelter, K. J. Gould, and C. R. Harrison, *J. Chem. Soc., Perkin Trans. 1,* 2428 (1976).
2115. P. A. Chopard, V. M. Clark, R. F. Hudson, and A. J. Kirby, *Tetrahedron,* **21,** 1961 (1965).
2116. H. C. Brown, M. M. Rogic, H. Nambu, and M. W. Rathke, *J. Am. Chem. Soc.*, **91,** 2147 (1969).
2117. I. J. Borowitz and R. Virkhaus, *J. Am. Chem. Soc.*, **85,** 2183 (1963).
2118. S. H. McAllister, W. A. Bailey, Jr., and C. M. Bouton, *J. Am. Chem. Soc.*, **62,** 3210 (1940).
2119. R. Weidenhagen and R. Herrmann, *Ber. Dtsch. Chem. Ges.*, **68,** 1953 (1935).
2120. J. J. Katz, J. E. Dubois, and C. Lion, *Bull. Soc. Chim. Fr.*, 683 (1977).
2121. C. L. Stevens and J. J. De Young, *J. Am. Chem. Soc.*, **76,** 718 (1954).
2122. M. Miyashita, T. Yanami, and A. Yoshikoshi, *J. Am. Chem. Soc.*, **98,** 4679 (1976).
2123. H.-J. Altenbach, *Angew. Chem.*, **91,** 1005 (1979); *Int. Ed. Engl.*, **18,** 940 (1979).
2124. E. Piers and B. Abeysekera, *Can. J. Chem.*, **60,** 1114 (1982).
2125. K. Klager, *Monatsh. Chem.*, **96,** 1 (1965).
2126. B. Bravo, G. Gaudiano, C. Ticozzi, and A. Umani-Ronchi, *Tetrahedron Lett.*, 4481 (1968).
2127. K. Kondo and D. Tunemoto, *Tetrahedron Lett.*, 1007, (1975).
2128. H. Roder, G. Helmchen, E.-M. Peters, K. Peters, and H.-G. von Schnering, *Angew. Chem.*, **96,** 895 (1984); *Int. Ed. Engl.*, **23,** 898 (1984).
2129. G. Rosini, R. Ballini, M. Petrini, and P. Sorrenti, *Tetrahedron,* **40,** 3809 (1984).
2130. N. A. R. Hatam and D. A. Whiting, *J. Chem. Soc., Perkin Trans. 1,* 461 (1982).
2131. E. Götschi, F. Schneider, H. Wagner, and K. Bernauer, *Helv. Chim. Acta,* **60,** 1416 (1977).
2132. W. R. Roush, H. R. Gillis, and A. I. Ko, *J. Am. Chem. Soc.* **104,** 2269 (1982).
2133. G. Pattenden and D. Whybrow, *Tetrahedron Lett.*, 1885 (1979).
2134. J. L. Gras and M. Bertrand, *Tetrahedron Lett.*, 4549 (1979).
2135. J. I. Levin and S. M. Weinreb, *J. Am. Chem. Soc.*, **105,** 1397 (1983).
2136. G. Feldstein and P. J. Kocienski, *Synth. Commun.*, **7,** 27 (1977).
2137. R. G. Almquist, Wan-Ru Chao, M. E. Ellis, and H. L. Johnson, *J. Med. Chem.*, **23,** 1392 (1980).
2138. J. H. Rigby, *Tetrahedron Lett.*, **23,** 1863 (1982).
2139. L. A. Paquette and Yeun-Kwei Han, *J. Am. Chem. Soc.*, **103,** 1835 (1981).
2140. W. R. Roush, A. I. Ko, and H. R. Gillis, *J. Org. Chem.*, **45,** 4264 (1980).
2141. T. Tsunoda, M. Kodama, and S. Ito, *Tetrahedron Lett.*, **24,** 83 (1983).
2142. J. Dodge, W. Hedges, J. W. Timberlake, and L. M. Trefonas, *J. Org. Chem.*, **43,** 3615 (1978).
2143. D. J. Hart and T.-K. Yang, *Tetrahedron Lett.*, **23,** 2761 (1982).
2144. T. Sato, T. Kawara, K. Sakata, and T. Fujisawa, *Bull. Chem. Soc. Jpn.*, **54,** 505 (1981).
2145. R. Menicagli, M. L. Wis, L. Lardicci, C. Botteghi, and G. Caccia, *J. Chem. Soc., Perkin Trans. 1,* 847 (1979).
2146. L. A. Paquette and A. Leone-Bay, *J. Am. Chem. Soc.*, **105,** 7352 (1983).

2147. A. Leone-Bay and L. A. Paquette, *J. Org. Chem.*, **47,** 4173 (1982).
2148. D. N. Brattesani and C. H. Heathcock, *J. Org. Chem.*, **40,** 2165 (1975).
2149. M. Julia and B. Badet, *Bull. Soc. Chim. Fr.*, 1363 (1975).
2150. R. L. Crumbie, J. S. Nimitz, and H. S. Mosher, *J. Org. Chem.*, **47,** 4040 (1982).
2151. H. Ahlbrecht and J. Eichler, *Synthesis,* 672 (1974).
2152. M. Julia, A. Schouteeten, and M. Baillarge, *Tetrahedron Lett.*, 3433 (1974).
2153. D. Hoppe, R. Hanko, A. Brönneke, F. Lichtenberg, and E. van Hülsen, *Chem. Ber.*, **118,** 2822 (1985).
2154. B. Weidmann and D. Seebach, *Angew. Chem.*, **95,** 12 (1983); *Int. Ed. Engl.*, **22,** 31 (1983).
2155. M. R. Binns, R. K. Haynes, T. L. Houston, and W. R. Jackson, *Aust. J. Chem.*, **34,** 2465 (1981).
2156. T. Jeffery, *J. Chem. Soc., Chem. Commun.*, 1287 (1984).
2157. K. Inomata, Y. Nakayama, M. Tsutsumi, and H. Katake, *Heterocycles,* **12,** 1467 (1974).
2158. G. Rauchschwalbe and H. Ahlbrecht, *Synthesis,* 663 (1974).
2159. S. Cacchi, F. La Torre, and G. Palmieri, *J. Organomet. Chem.*, **268,** C48 (1984).
2160. E. J. Corey, *Ann. N. Y. Acad. Sci.*, **180,** 24 (1971).
2161. J. Fayos, T. Clardy, L. T. Dolby, and T. Farnham, *J. Org. Chem.*, **42,** 1349 (1977).
2162. G. Trimitsis, S. Beers, J. Ridella, M. Carlon, D. Cullin, J. High, and D. Brutts, *J. Chem. Soc., Chem. Commun.*, 1088 (1984).
2163. J. Enda, T. Matsutani, and I. Kuwajima, *Tetrahedron Lett.*, **25,** 5307 (1984).
2164. B. Giese, H. Horler, and W. Zwick, *Tetrahedron Lett.*, **23,** 931 (1982).
2165. H. Ahlbrecht and G. Rauchschwalbe, *Synthesis,* 417 (1973).
2166. H.-U. Reissig and I. Reichelt, *Tetrahedron Lett.*, **25,** 5879 (1984).
2167. G. Sturtz and J. J. Yaouanc, *Synthesis,* 289 (1980).
2168. K. Iwai, H. Kosugi, A. Miyazaki, and H. Uda, *Synth. Commun.*, **6,** 357 (1976).
2169. P. Bravo, P. Carrera, G. Resnati, and C. Ticozzi, *J. Chem. Soc., Chem. Commun.*, 19, (1984).
2170. S. De Lombaert and L. Ghosez, *Tetrahedron Lett.*, **25,** 3475 (1984).
2171. E. Nakamura and I. Kuwajima, *J. Am. Chem. Soc.*, **106,** 3368 (1984).
2172. T. Shono, Y. Matsumura, and S. Kashimura, *J. Chem. Res. (S),* 216 (1984).
2173. G. Sturtz and B. Gorbel, *C. R. Hebd. Seances Acad. Sci.*, **277,** 395 (1977).
2174. F. Z. Basha, J. F. DeBernardis, and S. Spanton, *J. Org. Chem.*, **50,** 4160 (1985).
2175. H. Redlich and S. Thormählen, *Tetrahedron Lett.*, **26,** 3685 (1985).
2176. H. Paulsen, M. Schüller, M. A. Nashed, A. Heitmann, and H. Redlich, *Tetrahedron Lett.*, **26,** 3689 (1985).
2177. K. Ogura, M. Yamashita, M. Suzuki, S. Furukawa, and G. Tsuchihashi, *Bull. Chem. Soc. Jpn.*, **57,** 1637 (1984).
2178. A. Ricci, M. Fiorenza, A. Degl'Innocenti, G. Seconi, P. Dembech, K. Witzgall, and H. J. Bestmann, *Angew. Chem.*, **97,** 1068 (1985); *Int. Ed. Engl.*, **24,** 1068 (1985).
2179. N. Maigrot, J.-P. Mazaleyrat, and Z. Welvart, *J. Org. Chem.*, **50,** 3916 (1985).
2180. A. I. Meyers, P. D. Edwards, T. R. Bailey, and G. E. Jagdmann, Jr., *J. Org. Chem.*, **50,** 1019 (1985).
2181. H. Stetter and B. Jansen, *Chem. Ber.*, **118,** 4877 (1985).
2182. H. Stetter and L. Simons, *Chem. Ber.*, **118,** 3172 (1985).
2183. D. Seyferth and R. C. Hui, *J. Am. Chem. Soc.*, **107,** 4551 (1985).
2184. T. Takeda, H. Furukawa, M. Fujimori, K. Suzuki, and T. Fujiwara, *Bull. Chem. Soc. Jpn.*, **57,** 1863 (1984).

2185. T. Satoh, T. Kumagawa, and K. Yamakawa, *Bull. Chem. Soc. Jpn.*, **58,** 2849 (1985).

2186. M. D. Rozwadowska, *Tetrahedron,* **41,** 3135 (1985).

2187. O. Ruel, C. Bibang Bi Ekogha, R. Lorne, and S. A. Julia, *Bull. Soc. Chim. Fr.*, 1250 (1985).

2188. M. Fetizon, P. Goulaouic, and I. Hanna, *Tetrahedron Lett.*, **26,** 4925 (1985).

2189. V. Hedtmann and P. Welzel, *Tetrahedron Lett.*, **26,** 2773 (1985).

2190. Y. Tamura, T. Yakura, and J. Haruta, *Tetrahedron Lett.*, **26,** 3837 (1985).

2191. T. Cohen and L.-C. Yu, *J. Org. Chem.*, **50,** 3266 (1985).

2192. M. Tanaka, T. Kobayashi, and T. Sakakura, *J. Chem. Soc., Chem. Commun.*, 837 (1985).

2193. J.-C. Combret, J. Tekin, and D. Postaire, *Bull. Soc. Chim. Fr.*, **II,** 371 (1984).

2194. M. T. Reetz and S.-H. Kyung, *Tetrahedron Lett.*, **26,** 6333 (1985).

2195. F. Ozawa, H. Soyama, H. Yanagihara, I. Aoyama, H. Takino, K. Izawa, T. Yamamoto, and A. Yamamoto, *J. Am. Chem. Soc.*, **107,** 3235 (1985).

2196. H. Ahlbrecht and M. Dietz, *Synthesis,* 417 (1985).

2197. H. Ahlbrecht and M. Ibe, *Synthesis,* 421 (1985).

2198. Y. Ikeda and E. Manda, *Chem. Lett.*, 453 (1984).

2199. G. Silvestri, S. Gambino, G. Filardo, and A. Gulotta, *Angew. Chem.*, **96,** 978 (1984); *Int. Ed. Engl.*, **23,** 979 (1984).

2200. J.-C. Folest, J.-M. Duprilot, J. Perichon, Y. Robin, and J. Devynck, *Tetrahedron Lett.*, **26,** 2633 (1985).

2201. F. Francalanci, A. Gardano, and M. Foà, *J. Organomet. Chem.*, **282,** 277 (1985).

2202. O. Sock, M. Troupel, and J. Perichon, *Tetrahedron Lett.*, **26,** 1509 (1985).

2203. H. Alper, N. Hamel, D. J. Smith, and J. B. Woell, *Tetrahedron Lett.*, **26,** 2273 (1985).

2204. J. B. Woell and H. Alper, *Tetrahedron Lett.*, **25,** 3791 (1984).

2205. J. E. Hengeveld, V. Grief, J. Tadanier, C.-M. Lee, D. Riley, and P. A. Larkey, *Tetrahedron Lett.*, **25,** 4075 (1984).

2206. H. Alper, S. Antebi, and J. B. Woell, *Angew. Chem.*, **96,** 710 (1984); *Int. Ed. Engl.*, **23,** 732 (1984).

2207. M. Foà, F. Francalanci, E. Bencini, and A. Gardano, *J. Organomet. Chem.*, **285,** 293 (1985).

2208. N. A. Bumagin, Y. V. Gulevich, and I. P. Beletskaya, *J. Organomet. Chem.*, **285,** 415 (1985).

2209. S. Cacchi, E. Morera, and G. Ortar, *Tetrahedron Lett.*, **26,** 1109 (1985).

2210. H. Oshino, E. Nakamura, and I. Kuwajima, *J. Org. Chem.*, **50,** 2802 (1985).

2211. B. J. Banks, A. G. M. Barrett, and M. A. Russell, *J. Chem. Soc., Chem. Commun.*, 670 (1984).

2212. O. Tsuge, J. Tanaka, and S. Kanemasa, *Bull. Chem. Soc. Jpn.*, **58,** 1991 (1985).

2213. K. Utimoto, Y. Wakabayashi, Y. Shishiyama, M. Inoue, and H. Nozaki, *Tetrahedron Lett.*, **22,** 4279 (1981).

2214. A. I. Meyers, *Heterocycles,* **21,** 360 (1984).

2215. S. Ingemann and N. M. M. Nibbering, *J. Org. Chem.*, **50,** 682 (1985).

2216. A. R. Katritzky, F. Saczewski, and C. M. Marson, *J. Org. Chem.*, **50,** 1351 (1985).

2217. K. Achiwa, N. Imai, T. Motoyama, and M. Seikya, *Chem. Lett.*, 2041 (1984).

2218. M. F. Loewe and A. I. Meyers,, *Tetrahedron Lett.*, **26,** 3291 (1985).

2219. M. F. Loewe, M. Boes, and A. I. Meyers, *Tetrahedron Lett.*, **26,** 3295 (1985).

2220. A. I. Meyers and J. M. Marra, *Tetrahedron Lett.*, **26,** 5863 (1985).

2221. A. I. Meyers, L. M. Fuentes, M. Bös, and D. A. Dickman, *Chem. Scr.*, **25,** 25 (1985).

2222. B. H. Bakker, D. S. T. Alim, and A. Vandergen, *Tetrahedron Lett.,* **25,** 4259 (1984).
2223. J. H. Boyer and G. Kumar, *Heterocycles,* **22,** 2351 (1984).
2224. T. L. Weber and D. Seebach, *Helv. Chim. Acta,* **68,** 155 (1985).
2225. T. Tsushima and K. Kawado, *Tetrahedron Lett.,* **26,** 2445 (1985).
2226. M. J. O'Donnell, C. L. Barney, and J. R. McCarthy, *Tetrahedron Lett.,* **26,** 3067 (1985).
2227. T. Mukaiyama, R. Tsuzuki, and J. Kato, *Chem. Lett.,* 1825 (1983).
2228. R. C. Gadwood, M. R. Rubino, S. C. Nagarajan, and S. T. Michel, *J. Org. Chem.,* **50,** 3255 (1985).
2229. G. J. McGarvey and M. Kimura, *J. Org. Chem.,* **50,** 4655 (1985).
2230. A. Pelter, D. Buss, and A. Pitchford, *Tetrahedron Lett.,* **26,** 5093 (1985).
2231. A. Pelter, G. Bugden, and R. Rosser, *Tetrahedron Lett.,* **26,** 5097 (1985).
2232. M. Kosugi, T. Sumiya, T. Ogata, H. Sano, and T. Migita, *Chem. Lett.,* 1225 (1984).
2233. A. Duchene, D. Mouko-Mpegna, and J.-P. Quintard, *Bull. Soc. Chim. Fr.,* 787 (1985).
2234. J.-P. Quintard, B. Elissondo, T. Hattich, and M. Pereyre, *J. Organomet. Chem.,* **285,** 149 (1985).
2235. K. Tamao and N. Ishida, *Tetrahedron Lett.,* **25,** 4245 (1984).
2236. K. Tamao and N. Ishida, *Tetrahedron Lett.,* **25,** 4249 (1984).
2237. T. Murai and S. Kato, *J. Am. Chem. Soc.,* **106,** 6093 (1984).
2238. J. S. Sawyer, T. L. Macdonald, and G. J. McGarvey, *J. Am. Chem. Soc.,* **106,** 3376 (1984).
2239. E. van Hüslen and D. Hoppe, *Tetrahedron Lett.,* **26,** 411 (1985).
2240. K. Tanaka, H. Yoda, and A. Kaji, *Tetrahedron Lett.,* **26,** 4751 (1985).
2241. R. Metternich and R. W. Hoffman, *Tetrahedron Lett.,* **25,** 4095 (1984).
2242. R. W. Hoffman, B. Kemper, R. Metternich, and T. Lehmeier, *Justus Liebigs Ann. Chem.,* 2246 (1985).
2243. B. D. Gray and J. D. White, *J. Chem. Soc., Chem. Commun.,* 20 (1985).
2244. M. Reglier and S. A. Julia, *Tetrahedron Lett.,* **26,** 2319 (1985).
2245. E. Block and M. Aslam, *J. Am. Chem. Soc.,* **107,** 6729 (1985).
2246. K. Furuta, Y. Ikeda, N. Meguriya, N. Ikeda, and H. Yamamoto, *Bull. Chem. Soc. Jpn.,* **57,** 2781 (1984).
2247. K. M. Sadhu and D. S. Matteson, *Organometallics,* **4,** 1687 (1985).
2248. P. Ongoka, B. Mauzé, and L. Miginiac, *J. Organomet. Chem.,* **284,** 139 (1985).
2249. R. Tarhouni, B. Kirschleger, and J. Villieras, *J. Organomet. Chem.,* **272,** C1 (1984).
2250. R. Tarhouni, B. Kirschleger, and J. Villieras, *Bull. Soc. Chim. Fr.,* 825 (1985).
2251. E. Murayama, T. Kikuchi, K. Sasaki, N. Sootome, and T. Sato, *Chem. Lett.,* 1897 (1984).
2252. T. Sato, T. Kikuchi, N. Sootome, and E. Murayama, *Tetrahedron Lett.,* **26,** 2205 (1985).
2253. T. Kauffmann, B. Altepeter, N. Klas, and R. Kriegesmann, *Chem. Ber.,* **118,** 2356 (1985).
2254. O. G. Kulinkovich, I. G. Tischenko, and V. L. Sorokin, *Synthesis,* 1058 (1985).
2255. D. Seebach, G. Calderari, and P. Knochel, *Tetrahedron,* **41,** 4861 (1985).
2256. T. K. Jones and S. E. Denmark, *J. Org. Chem.,* **50,** 4037 (1985).
2257. U. Schöllkopf, H.-J. Neubauer, and M. Hauptreif, *Angew. Chem.,* **97,** 1065 (1985); *Int. Ed. Engl.,* **24,** 1066 (1985).
2258. D. Hoppe, R. Hanko, A. Brönneke, F. Lichtenberg, and E. van Hülsen, *Chem. Ber.,* **118,** 2822 (1985).
2259. L. Hevesi, K. M. Nsunda, and M. Renard, *Bull. Soc. Chim. Belg.,* **94,** 1039 (1985).
2260. J. M. Muchowski, R. Naef, and M. L. Maddox, *Tetrahedron Lett.,* **26,** 5375 (1985).

2261. J. Enda and I. Kuwajima, *J. Am. Chem. Soc.,* **107,** 5495 (1985).
2262. I. Kuwajima, *J. Organomet. Chem.,* **285,** 137 (1985).
2263. B. Giese, J. A. Gonzalez-Gomez, and T. Witzel, *Angew. Chem.,* **96,** 51 (1984); *Int. Ed. Engl.,* **23,** 69 (1984).
2264. J.-L. Moreau and R. Couffignal, *J. Organomet. Chem.,* **294,** 139 (1985).
2265. J.-L. Moreau and R. Couffignal, *J. Organomet. Chem.,* **297,** 1 (1985).
2266. J. B Hendrickson, and P. S. Palumbo, *J. Org. Chem.,* **50,** 2110 (1985).
2267. E. Nakamura, J. Shimada, and I. Kuwajima, *Organometallics,* **4,** 641 (1985).
2268. C. Buchan, N. Hamel, J. B. Woell, and H. Alper, *Tetrahedron Lett.,* **26,** 5743 (1985).
2269. Y. Tamaru, H. Ochiai, T. Nakamura, K. Tsubaki, and Z. Yoshida, *Tetrahedron Lett.,* **26,** 5559 (1985).
2270. T. Kauffmann, E. Köppelmann, and H. Berg, *Angew. Chem.,* **82,** 138 (1970); *Int. Ed. Engl.,* **9,** 163 (1970).
2271. A. R. Katritzky, M. A. Kashmiri, and D. K. Wittman, **40,** 1501 (1984).
2272. D. H. R. Barton, R. D. Bracho, A. A. L. Gunatilaka, and D. A. Widdowson, *J. Chem. Soc., Perkin Trans. 1,* 579 (1975).
2273. D. Enders, R. Pieter, B. Renger, and D. Seebach, *Org. Synth.,* **58,** 113 (1978).
2274. D. Seebach and D. Enders, *Chem. Ber.,* **108,** 1293 (1975).
2275. K. Piotrowska, *Synth. Commun.,* **9,** 765 (1979).
2276. R. R. Fraser and S. Passannanti, *Synthesis,* 540 (1976).
2277. W. Wykypiel and D. Seebach, *Tetrahedron Lett.,* **21,** 1927 (1980).
2278. D. Seebach and W. Wykypiel, *Synthesis,* 423 (1979).
2279. J.-J. Lohmann, D. Seebach, M. A. Syfrig, and M. Yoshifuji, *Angew. Chem.,* **93,** 125 (1981); *Int. Ed. Engl.,* **20,** 128 (1981).
2280. A. I. Meyers and L. N. Fuentes, *J. Am. Chem. Soc.,* **105,** 117 (1983).
2281. G. Wittig and W. Tochtermann, *Chem. Ber.,* **94,** 1692 (1961).
2282. D. H. R. Barton, R. Beugelmans, and R. N. Young, *Nouv. J. Chim.,* **2,** 363 (1978).
2283. P. Beak and B. G. McKinnie, *J. Am. Chem. Soc.,* **99,** 5213 (1977).
2284. Z. Welvart, *Bull. Soc. Chim. Fr.,* 1653 (1961).
2285. P. Beak, H. Baillargeon, and L. G. Carter, *J. Org. Chem.,* **43,** 4255 (1978).
2286. T. Nakai, T. Mimura, and T. Kurokawa, *Tetrahedron Lett.,* 2895 (1978).
2287. J. Rémion, W. Dumont, and A. Krief, *Tetrahedron Lett.,* 1385 (1976).
2288. W. Dumont and A. Krief, *Angew. Chem.,* **87,** 347 (1975); *Int. Ed. Engl.,* **14,** 350 (1975).
2289. W. Dumont and A. Krief, *Angew. Chem.,* **88,** 184 (1976); *Int. Ed. Engl.,* **15,** 161 (1976).
2290. M. Sevrin, D. Van Ende and A. Krief, *Tetrahedron Lett.,* 2643 (1976).
2291. D. Van Ende and A. Krief, *Tetrahedron Lett.,* 457 (1976).
2292. P. L. Stotter and R. E. Hornish, *J. Am. Chem. Soc.,* **95,** 4444 (1973).
2293. J. Nokami, T. Ono, A. Iwao, and S. Wakabayashi, *Bull. Chem. Soc. Jpn.,* **55,** 3043 (1982).
2294. T. Mukaiyma, K. Narasaka, K. Maekawa, and M. Furusato, *Bull. Chem. Soc. Jpn.,* **44,** 2285 (1971).
2295. H. Yatagai, Y. Yamamoto, and K. Maruyama, *J. Chem. Soc., Chem. Commun.,* 702 (1978).
2296. A. Anciaux, A. Eman, W. Dumont, D. Van Ende, and A. Krief, *Tetrahedron Lett.,* 1613 (1975).
2297. S. Halazy, J. Lucchetti, and A. Krief, *Tetrahedron Lett.,* 3971 (1978).
2298. D. Seebach and A. K. Beck, *Angew. Chem.,* **86,** 859 (1974); *Int. Ed. Engl.,* **13,** 806 (1974).

2299. D. Labar, W. Dumont, L. Havesi, and A. Krief, *Tetrahedron Lett.*, 1145 (1978).
2300. A. G. Brook, J. M. Duff, and D. G. Anderson, *Can. J. Chem.*, **48,** 561 (1970).
2301. L. H. Sommer, J. R. Gold, G. M. Goldberg, and N. S. Marans, *J. Am. Chem. Soc.*, **71,** 1509 (1949).
2302. F. C. Whitmore, L. H. Sommer, J. Gold, and R. E. Van Strien, *J. Am. Chem. Soc.*, **69,** 1551 (1947).
2303. J. E. Mulvaney and Z. G. Gardlund, *J. Org. Chem.*, **30,** 917 (1965).
2304. D. A. Evans and P. J. Sidebottom, *J. Chem. Soc., Chem. Commun.*, 753 (1978).
2305. A. I. Meyers, P. D. Edwards, W. F. Rieker, and T. R. Bailey, *J. Am. Chem. Soc.*, **106,** 3270 (1984).
2306. J. C. L. Armande and U. K. Pandit, *Tetrahedron Lett.*, 897 (1977).
2307. T. M. Williams and H. S. Mosher, *Tetrahedron Lett.*, **26,** 6269 (1985).
2308. D. Seebach, W. Lubosch, and D. Enders, *Chem. Ber.*, **109,** 1309 (1976).
2309. D. Enders and D. Seebach, *Angew. Chem.*, **12,** 1104 (1973); *Int. Ed. Engl.*, **12,** 1014 (1973).
2310. D. Hoppe, *Angew. Chem.*, **96,** 930 (1984); *Int. Ed. Engl.*, **23,** 932 (1984).
2311. W. A. Thaler, *J. Am. Chem. Soc.*, **88,** 4278 (1966).
2312. T. A. Hase, L. Lahtinen, and A. Klemola, *Acta Chem. Scand., Ser. B,* **31,** 501 (1977).

INDEX

Under "Synthesis", end products are indexed. Under "Synthons", intact moieties appearing in the synthon *and* in the end product are indexed. Thus the following entry will be indexed under "Synthesis, of γ-hydroxy acids" and "Synthons, for carboxylic acids":

$$R\text{—}CHO + {}^{-}CH_2CH_2COOH \longrightarrow R\text{—}CHOH\text{—}CH_2CH_2COOH$$

On the other hand, the following entry would be indexed under "Synthons, for γ-hydroxy acids":

$$R\text{—}X + {}^{-}CHOH\text{—}CH_2CH_2COOH \longrightarrow R\text{—}CHOH\text{—}CH_2CH_2COOH$$

The ubiquitous alkyl halides, aldehydes and ketones have not been indexed as electrophiles. For individual halogens, see also 'halo'.

INDEX